W9-AWN-777

Build Your Own

TRULOVE • *Build Your Own Wireless LAN* (with projects)

Crash Course

LOUIS • *Broadband Crash Course*
VACCA • *i-Mode Crash Course*
LOUIS • *M-Commerce Crash Course*
SHEPARD • *Telecom Convergence, 2/e*
SHEPARD • *Telecom Crash Course*
BEDELL • *Wireless Crash Course*
KIKTA/FISHER/COURTNEY • *Wireless Internet Crash Course*

Demystified

HARTE/LEVINE/KIKTA • *3G Wireless Demystified*
LaROCCA • *802.11 Demystified*
MULLER • *Bluetooth Demystified*
EVANS • *CEBus Demystified*
BAYER • *Computer Telephony Demystified*
HERSHEY • *Cryptography Demystified*
TAYLOR • *DVD Demystified*
BATES • *GPRS Demystified*
SYMES • *MPEG-4 Demystified*
CAMARILLO • *SIP Demystified*
SHEPARD • *SONET / SDH Demystified*
TOPIC • *Streaming Media Demystified*
SYMES • *Video Compression Demystified*
SHEPARD • *Videoconferencing Demystified*
BHOLA • *Wireless LANs Demystified*

Developer Guides

VACCA • *i-Mode Crash Course*
GUTHERY • *Mobile Application Development with SMS*
RICHARD • *Service and Device Discovery: Protocols and Programming*

Professional Telecom

BATES • *Broadband Telecom Handbook, 2/e*
COLLINS • *Carrier Grade Voice over IP*
CHERNOCK • *Data Broadcasting*
HARTE • *Delivering xDSL*
HELD • *Deploying Optical Networking Components*
MINOLI • *Ethernet-Based Metro Area Networks*
BENNER • *Fibre Channel for SANs*
BATES • *GPRS*
SULKIN • *Implementing the IP-PBX*
LEE • *Lee's Essentials of Wireless*

BATES • *Optical Switching and Networking Handbook*
WETTEROTH • *OSI Reference Model for Telecommunications*
RUSSELL • *Signaling System #7, 4/e*
MINOLI • *SONET-Based Metro Area Networks*
NAGAR • *Telecom Service Rollouts*
LOUIS • *Telecommunications Internetworking*
RUSSELL • *Telecommunications Protocols, 2/e*
MINOLI • *Voice over MPLS*
KARIM/SARRAF • *W-CDMA and cdma2000 for 3G Mobile Networks*
BATES • *Wireless Broadband Handbook*
FAIGEN • *Wireless Data for the Enterprise*

Reference

MULLER • *Desktop Encyclopedia of Telecommunications, 3/e*
BOTTO • *Encyclopedia of Wireless Telecommunications*
CLAYTON • *McGraw-Hill Illustrated Telecom Dictionary, 3/e*
RADCOM • *Telecom Protocol Finder*
PECAR • *Telecommunications Factbook, 2/e*
RUSSELL • *Telecommunications Pocket Reference*
KOBB • *Wireless Spectrum Finder*
SMITH • *Wireless Telecom FAQs*

Security

NICHOLS • *Wireless Security*

Telecom Engineering

SMITH/GERVELIS • *Cellular System Design and Optimization*
ROHDE/WHITAKER • *Communications Receivers, 3/e*
SAYRE • *Complete Wireless Design*
OSA • *Fiber Optics Handbook*
LEE • *Mobile Cellular Telecommunications, 2/e*
BATES • *Optimizing Voice in ATM/IP Mobile Networks*
RODDY • *Satellite Communications, 3/e*
SIMON • *Spread Spectrum Communications Handbook*
SNYDER • *Wireless Telecommunications Networking with ANSI-41, 2/e*

BICSI

Network Design Basics for Cabling Professionals
Networking Technologies for Cabling Professionals
Residential Network Cabling
Telecommunications Cabling Installation

Network
Manager's
Handbook

Nathan J. Muller

McGraw-Hill

New York   Chicago   San Francisco   Lisbon   London
Madrid   Mexico City   Milan   New Delhi   San Juan
Seoul   Singapore   Sydney   Toronto

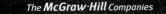

The McGraw-Hill Companies

Library of Congress Cataloging-in-Publication Data

Muller, Nathan J.
Network manager's handbook / Nathan J. Muller.
    p.   cm.
Includes index.
ISBN 0-07-140567-4
1. Computer networks—Management—Handbooks, manuals, etc.   I. Title.

TK5105.5.M843   2002
004.6—dc21   2002032553

1 2 3 4 5 6 7 8 9 0   DOC/DOC   0 8 7 6 5 4 3 2

ISBN 0-07-140567-4

The sponsoring editor for this book was Marjorie Spencer, the editing supervisor was Stephen M. Smith, and the production supervisor was Sherri Souffrance. It was set in Century Schoolbook following the MHT design by McGraw-Hill Professional's Hightstown, N.J., composition unit.

Printed and bound by R R Donnelley.

This book is printed on recycled, acid-free paper containing a minimum of 50% recycled, de-inked fiber.

To my wife of 33 years: Linda…you're a keeper!

Contents

Part 3   Keeping the Network Healthy

Chapter 14. Maintenance and Support Planning   381

Chapter 15. Network Monitoring and Testing   407

Preface

There is no question that corporate networks have become the lifelines of today's information-intensive businesses. With ever more information being entrusted to corporate networks, often for global distribution via the public Internet or private intranets, keeping these lifelines free of congestion and disruption has become an ongoing challenge—and for good reason. The inability to keep data moving across the network quickly and efficiently can result in huge financial penalties:

- A Wall Street brokerage house can lose as much as $100,000 in commissions per minute when buy/sell instructions from customers are disrupted during peak hours.

- An insurance company can lose its *Fortune* 500 accounts if it cannot live up to specified levels of network uptime to process the claims of its clients' employees.

- A large financial services firm can lose $200 million in transactions if its network were to experience an outage for only 1 hour.

If these ramifications of network failure are not dramatic enough, consider that if a major outage were to occur on any of the backbone networks in the Federal Reserve Bank system for only an hour, the movement and processing of as much as $5 trillion in monetary transactions could be seriously delayed and have a ripple effect throughout the global economy.

On a smaller, but no less important, scale, businesses have similar concerns about their networks. Distributed computing environments—distinguished by desktop processing and resource sharing via LANs and global interconnectivity via WANs—have corporate managers clambering for resources that will keep their networks up and running, especially in the face of such security threats as viruses, hackers, disgruntled employees, and, more recently, terrorists. After all, in not providing adequate levels of network security, reliability, and uptime, companies risk not only financial losses, but diminished employee productivity, slower responses to competitive pressures, erosion of customer goodwill, and, if problems persist, loss of market share and investor confidence.

There is more to distributed computing and networking than merely connecting various products and hoping that they will work together. This book describes the planning, design, procurement, maintenance, security, and management requirements for today's increasingly sophisticated systems and networks. Often these infrastructural requirements have more to do with the successful implementation of communications systems and networks than the technologies themselves. But the best technologies will be useless if implemented in an environment characterized by poor planning, mistakes in procurement, the inability to implement comprehensive management, or not having the right people with the right skills. Additionally, the organization will not realize anticipated returns on its technology investments if the right decisions are not made in such areas as lease versus purchase, availability of vendor maintenance and support, and service restoration options.

This book is a practical guide that puts these and other important infrastructural issues into proper perspective to greatly increase the chances for success in building and operating advanced communications systems and networks. As such, it is suited to both experienced and entry-level professionals, as well as college and university students who are preparing for careers in the field of telecommunications or information technology (IT) management.

The primary reader is assumed to be a telecom or IT professional at a company that is seeking to build, upgrade, replace, or expand its network or integrate it with the Internet. This book is also of value to those who are already in the process of acquiring equipment and services in support of their organization's business objectives. Many potential readers will be technically oriented and have the responsibility for recommending, planning, or implementing various pieces of the network, or integrated legacy systems with local- and wide-area networks. Interconnect vendors and service providers will find this book useful as a tutorial for new hires and salespeople, and as an economical means of educating potential customers on various network issues that their products or services address.

The information contained in this book, especially as it relates to specific vendors and products, is believed to have been accurate at the time it was written and is, of course, subject to change with continued advancements in technology and shifts in market forces. The mention of specific products, services, and vendors is intended for illustration purposes only and does not constitute an endorsement of any kind, expressed or implied, by the author or publisher.

Nathan J. Muller

Acronyms

AAL	ATM Adaptation Layer
ABATS	Automated Bit Access Test System
ABM	Accunet Bandwidth Manager (AT&T)
ABR	Available Bit Rate
ABS	Adelphia Business Solutions
ac	alternating current
AC	Access Control
AC	Address Copied
ACD	Automated Call Distributor
ACTA	Administrative Council for Terminal Attachments
ADCR	Alternate Destination Call Routing (AT&T)
ADM	Add-Drop Multiplexer
ADN	Advanced Digital Network (Pacific Bell)
ADPCM	Adaptive Differential Pulse Code Modulation
AES	Advanced Encryption Standard
AI	Artificial Intelligence
AM	Amplitude Modulation
AMI	Alternate Mark Inversion
ANR	Automatic Network Routing (IBM Corp.)
ANSI	American National Standards Institute
AP	Access Point
APC	Access Protection Capability (AT&T)
API	Application Programming Interface
APPC	Advanced Program-to-Program Communications (IBM Corp.)
APPN	Advanced Peer-to-Peer Networking
ARB	Adaptive Rate Based (IBM Corp.)

ARP	Address Resolution Protocol
ARS	Action Request System (Remedy Systems Inc.)
AS	Autonomous System
ASCII	American Standard Code for Information Interchange
ASIC	Application-Specific Integrated Circuit
ASN.1	Abstract Syntax Notation 1
ASP	Application Service Provider
ASTN	Alternate Signaling Transport Network (AT&T)
AT&T	American Telephone & Telegraph
ATIS	Alliance for Telecommunications Industry Solutions
ATM	Asynchronous Transfer Mode
ATM	Automated Teller Machine
AWG	American Wire Gauge
B8ZS	Binary Eight Zero Substitution
BBS	Bulletin Board System
BCP	Business Continuity Plan
BECN	Backward Explicit Congestion Notification
Bellcore	Bell Communications Research
BER	Bit Error Rate
BERT	Bit Error Rate Tester
BGP	Border Gateway Protocol
BIOS	Basic Input-Output System
BMC	Block Multiplexer Channel
BMS-E	Bandwidth Management Service—Extended (AT&T)
BOC	Bell Operating Company
BootP	Boot Protocol
BPDU	Bridge Protocol Data Unit
BPS	Bits per Second
BPV	Bipolar Violation
BQB	Bluetooth Qualification Body
BQTF	Bluetooth Qualification Test Facility
BRI	Basic Rate Interface (ISDN)
BSC	Binary Synchronous Communications
CAD	Computer-Aided Design
CAM	Computer-Aided Manufacturing
CAN	Campus Area Network
CAP	Competitive Access Provider

CASE	Computer-Aided Software Engineering
CATV	Cable Television
CBQ	Class-Based Queuing
CBR	Case-Based Reasoning
CBR	Constant Bit Rate
CBT	Computer-Based Training
CC	Coordination Center
CCC	Clear Channel Capability
CCITT	Consultative Committee for International Telegraphy and Telephony
CCR	Customer-Controlled Reconfiguration
CD-ROM	Compact Disk–Read Only Memory
CDN	Content Delivery Network
CDP	Content Delivery Provider
CDPD	Cellular Digital Packet Data
CEO	Chief Executive Officer
CERT	Computer Emergency Response Team
CHAP	Challenge Handshake Authentication Protocol
CI	Component Interface
CIO	Chief Information Officer
CIR	Committed Information Rate
CIS	Center for Internet Security
CLEC	Competitive Local Exchange Carrier
CLEI	Common Language Equipment Identifier
CMOS	Complementary Metal Oxide Semiconductor
CO	Central Office
CPE	Customer Premises Equipment
CPU	Central Processing Unit
CRC	Cyclic Redundancy Check
CSA	Certified Security Administrator
CSE	Certified Security Engineer
CSMA/CD	Carrier-Sense Multiple Access with Collision Detection
CSU	Channel Service Unit
CTI	Computer-Telephony Integration
CTO	Chief Technology Officer
D/E	Debt/Equity (ratio)
DA	Destination Address

DACS	Digital Access and Cross-Connect System (AT&T)
DAP	Demand Access Protocol
DASD	Direct Access Storage Device (IBM Corp.)
dB	Decibel
DBMS	Database Management System
DBU	Dial Backup Unit
dc	direct current
DCE	Data Communications Equipment
DCE	Distributed Computing Environment
DCF	Discounted Cash Flow
DCS	Digital Cross-Connect System
DDS	Digital Data Service
DDS/SC	Digital Data Service with Secondary Channel
DEC	Digital Equipment Corp.
DECT	Digital Enhanced Cordless Telecommunication
DES	Data Encryption Standard
DFSMS	Data Facility Storage Management Subsystem (IBM Corp.)
DID	Direct Inward Dial
DIF	Digital Interface Frame
DiffServ	Differentiated Services (IETF)
DIP	Debtor in Possession
DLC	Digital Loop Carrier
DLSw	Data Link Switching (IBM Corp.)
DM	Distributed Management
DME	Distributed Management Environment
DMI	Desktop Management Interface
DMTF	Desktop Management Task Force
DNS	Domain Name System
DoD	Department of Defense (U.S.)
DOD	Direct Outward Dial
DOS	Disk Operating System
DOV	Data over Voice
DS0	Digital Signal—Level Zero (64 kbps)
DS1	Digital Signal—Level One (1.544 Mbps)
DS1C	Digital Signal—Level One C (3.152 Mbps)
DS2	Digital Signal—Level Two (6.312 Mbps)
DS3	Digital Signal—Level Three (44.736 Mbps)

DSL	Digital Subscriber Line
DSLAM	Digital Subscriber Line Access Multiplexer
DSSS	Direct-sequence spread-spectrum
DSU	Data Service Unit
DTE	Data Terminal Equipment
DTMF	Dual Tone Multifrequency
DWDM	Dense Wavelength Division Multiplexing
DXI	Data Exchange Interface
E-mail	Electronic Mail
ECSA	Exchange Carriers Standards Association
ED	Ending Delimiter
EDI	Electronic Data Interchange
EDRO	Enhanced Diversity Routing Option (AT&T)
EDS	Electronic Data Systems
EEOC	Equal Employment Opportunity Commission
EFT	Electronic Funds Transfer
EGP	External Gateway Protocol
EIA	Electronic Industries Association
EISA	Extended Industry Standard Architecture
EMI	Electromechanical Interference
EMS	Element Management System
EOT	end-of-transmission
EP	Extension Point
EPS	Encapsulated PostScript
ESF	Extended Super Frame
FASB	Financial Accounting Standards Board
FASTAR	Fast Automatic Restoral (AT&T)
FAT	File Allocation Table
FC	Fibre Channel
FC	Frame Control
FCC	Federal Communications Commission
FCIA	Fibre Channel Industry Association
FCS	Frame Check Sequence
FDD	Frequency Division Duplexing
FDDI	Fiber Distributed Data Interface
FECN	Forward Explicit Congestion Notification
FEP	Front-End Processor

FFDT	FDDI Full Duplex Technology
FHSS	Frequency-Hopping Spread-Spectrum
FIFO	First-In, First-Out
FM	Frequency Modulation
FOD	Fax on Demand
4GL	Fourth-Generation Language
FRAD	Frame Relay Access Device
FRF	Frame Relay Forum
FS	Frame Status
FT1	Fractional T1
FTP	File Transfer Protocol
GAAP	Generally Accepted Accounting Procedures
GUI	Graphical User Interface
HASP	Houston Automatic Spooling Program
HDLC	High-Level Data Link Control
HEC	Header Error Check
HFC	Hybrid Fiber Coax
HP	Hewlett-Packard Co.
HPR	High-Performance Routing (IBM Corp.)
HSM	Hierarchical Storage Management
HTML	Hypertext Markup Language
HTR	High-Speed Token Ring
HTTP	Hypertext Transfer Protocol
HTTPS	Hypertext Transfer Protocol Secure Sockets
HVAC	Heating, ventilation, and air-conditioning
I/O	Input/Output
IAD	Integrated Access Device
ICMP	Internet Control Message Protocol
ICP	Integrated Communications Provider
ICS	Intelligent Calling System
ID	Identification
IDP	Internetwork Datagram Protocol (Xerox Corp.)
IEEE	Institute of Electrical and Electronic Engineers
IETF	Internet Engineering Task Force
IF	Intermediate Frequency
IGP	Interior Gateway Protocol
ILEC	Incumbent Local Exchange Carrier

IMA	Inverse Multiplexing over ATM
IMS/VS	Information Management System/Virtual Storage (IBM Corp.)
IN	Intelligent Network
INMS	Integrated Network Management System
IOC	Interoffice Channel
IP	Internet Protocol
IPX	Internet Packet Exchange
IR	Infrared
IRQ	Interrupt Request
IRR	Internal Rate of Return
IS	Information System
ISA	Industry Standard Architecture
iSCSI	Internet-Protocol Small Computer System Interface
ISDN	Integrated Services Digital Network
ISDN PRI	Integrated Services Digital Network Primary Rate Interface
ISO	International Organization for Standardization
ISP	Internet Service Provider
IT	Information Technology
ITR	Intelligent Text Retrieval
ITU-TSS	International Telecommunication Union–Telecommunications Standardization Sector (formerly, CCITT)
JPEG	Joint Photographic Experts Group
k or K (kilo)	One thousand (e.g., kbps)
KB	Kilobyte
LAN	Local Area Network
LANE	Local Area Network Emulation
LANRES	LAN Resource Extension and Services (IBM Corp.)
LAT	Local Area Transport (Digital Equipment Corp.)
LAVC	Local Area VAX Cluster (Digital Equipment Corp.)
LCD	Liquid Crystal Display
LCN	Logical Channel Number
LCN	Logically Connected Node
LD	Long Distance
LDAP	Lightweight Directory Access Protocol
LEC	Local Exchange Carrier
LED	Light-Emitting Diode
LEO	Low Earth Orbit

LIB	Label Information Base
LLC	Logical Link Control
LMDS	Local Multipoint Distribution Service
LSB	Label Switched Path
LSI	Large-Scale Integration
LSP	Label Switched Path
LSR	Label Switch Router
LU	Logical Unit (IBM Corp.)
M (mega)	One million (e.g., Mbps)
MAC	Media Access Control
MAC	Moves, Adds, Changes
MAN	Metropolitan Area Network
MAU	Multiple Access Unit
MB	Megabyte
MCA	Micro Channel Architecture (IBM Corp.)
MDS	Multipoint Distribution Service
MES	Master Earth Station
MI	Management Interface
MIB	Management Information Base
MIC	Management Integration Consortium
MIF	Management Information Format
MIPS	Millions of Instructions per Second
MMDS	Multichannel Multipoint Distribution Service
MOS	Mean Opinion Score
MPLS	Multiprotocol Label Switching
ms	millisecond
MTBF	Mean Time between Failures
MTSO	Mobile Telephone Serving Office
MTTR	Mean Time to Repair
MTTR	Mean Time to Response
MVPRP	Multivendor Problem Resolution Process
MVS	Multiple Virtual System
NAP	Network Access Point
NAS	Network Attached Storage
NAT	Network Address Translation
NAU	Network Addressable Unit (IBM Corp.)
NAUN	Nearest Active Upstream Neighbor

NCP	Network Control Point
NCP	Network Control Program (IBM Corp.)
NEBS	New Equipment Building Specifications
NetBIOS	Network Basic Input/Output System
NFS	Network File Server
NFS	Network File System
NIC	Network Interface Card
NIPC	National Infrastructure Protection Center
NIST	National Institute of Standards and Technology
NIU	Network Interface Unit
NLM	NetWare Loadable Module (Novell Inc.)
NM	Network Manager
NMS	NetWare Management System (Novell Inc.)
NMS	Network Management Station
NNI	Network-to-Network Interface
NNM	Network Node Manager (Hewlett-Packard Co.)
NOC	Network Operations Center
NOCC	Network Operations Control Center
NOS	Network Operating System
NPC	Network Protection Capability (AT&T)
NPV	Net Present Value
NSA	National Security Agency
NSOC	Network Security Operations Center
NT	Network Terminations
OAM	Operations, Administration, Management
OC	Optical Carrier
OC-3	Optical Carrier Signal—Level Three (155 Mbps)
OCR	Optical Character Recognition
ODBC	Open Data Base Connectivity (Microsoft Corp.)
OEM	Original Equipment Manufacturer
OID	Object Identification (or Identifier)
OLE	Object Linking and Embedding
OLTP	Online Transaction Processing
OMA	Object Management Architecture
OMF	Object Management Framework
OMG	Object Management Group
OOP	Object-Oriented Programming

ORB	Object Request Broker
OS	Operating System
OS/2	Operating System/2 (IBM Corp.)
OSF	Open Software Foundation
OSI	Open Systems Interconnection
OSPF	Open Shortest Path First
OTDR	Optical Time Domain Reflectometry
PA	Preamble
PAD	Packet Assembler-Disassembler
PAP	Password Authentication Protocol
PAR	Peak-to-Average Ratio
PBX	Private Branch Exchange
PC	Personal Computer
PCB	Printed Circuit Board
PCM	Pulse Code Modulation
PCX	ZSoft PC PaintBrush Bitmap
PDA	Personal Digital Assistant
PDS	Premises Distribution System (AT&T)
PDU	Payload Data Unit
PEM	Privacy Enhanced Mail
PGP	Pretty Good Privacy
PIM	Protocol Independent Multicast
PnP	Plug and Play
POP	Point of Presence
POS	Point of Sale
POTS	Plain Old Telephone Service
PPP	Point-to-Point Protocol
PRI	Primary Rate Interface (ISDN)
PSN	Packet Switched Network
PSTN	Public Switched Telephone Network
PTT	Post Telephone & Telegraph
PU	Physical Unit (IBM Corp.)
PVC	Permanent Virtual Circuit
QC	Quality Control
QoS	Quality of Service
R&D	Research and Development
RADIUS	Remote Access Dial-In User Service

RAID	Redundant Arrays of Inexpensive Disks
RAM	Random Access Memory
RBES	Rule-Based Expert Systems
RBOC	Regional Bell Operating Company
RDBMS	Relational Database Management System
RF	Radio Frequency
RFC	Request for Comment
RFI	Radio Frequency Interference
RFI	Request for Information
RFP	Request for Proposal
RFQ	Request for Quotation
RIP	Routing Information Protocol
RISC	Reduced Instruction Set Computing
RJE	Remote Job Entry
RMON	Remote Monitoring
RMON MIB	Remote Monitoring Management Information Base
ROA	Return on Assets
ROI	Return on Investment
ROM	Read Only Memory
ROR	Rate of Return
RPC	Remote Procedure Call
RSVP	Resource Reservation Protocol
RTNR	Real-Time Network Routing (AT&T)
RTP	Rapid Transfer Protocol (IBM Corp.)
RX	Receive
SA	Source Address
SAN	Storage Area Network
SAR	Segmentation and Reassembly
SCSI	Small Computer Systems Interface
SD	Starting Delimiter
SDH	Synchronous Digital Hierarchy
SDLC	Synchronous Data Link Control (IBM Corp.)
SDM	Subrate Data Multiplexing
SDN	Software Defined Network (AT&T)
SEC	Securities and Exchange Commission
SFD	Start Frame Delimiter
SFT	System Fault Tolerance

SIG	Special Interest Group
SIIA	Software & Information Industry Association (formerly, Software Publishers Association)
SLA	Service Level Agreement
SLIP	Serial Line Internet Protocol
SMBIOS	System Management Basic Input/Output System
SMDR	Station Message Detail Recording
SMDR-P	Station Message Detail Recording to Premises
SMDS	Switched Multimegabit Data Services
SMR	Specialized Mobile Radio
SMS	Systems Management Server (Microsoft)
SMT	Station Management
SNA	Systems Network Architecture
SNIA	Storage Networking Industry Association
SNMP	Simple Network Management Protocol
SoIP	Storage over Internet Protocol
SONET	Synchronous Optical Network
SPA	Software Publishers Association
SPX	Synchronous Packet Exchange (Novell Inc.)
SQL	Structured Query Language
SS7	Signal System 7
SSCP	System Services Control Point (IBM Corp.)
SSCP/PU	System Services Control Point/Physical Unit (IBM Corp.)
SSL	Secure Sockets Layer
SSP	Storage Service Provider
STDM	Statistical Time Division Multiplexing
STP	Shielded Twisted Pair
STP	Signal Transfer Point
STP	Spanning Tree Protocol
STS	Synchronous Transport Signal
STX	start-of-transmission
SVC	Switched Virtual Circuit
SWC	Serving Wire Center
T1	Transmission service at the DS1 rate of 1.544 Mbps
T3	Transmission service at the DS3 rate of 44.736 Mbps
TCP	Transmission Control Protocol
TCP/IP	Transmission Control Protocol/Internet Protocol

TDD	Time Division Duplexing
TDM	Time Division Multiplexer
TDMA	Time Division Multiple Access
TDR	Time Domain Reflectometry (or Reflectometer)
TE	Terminal Equipment
TIA	Telecommunications Industry Association
TIF	Tag Image-file Format
TIMS	Transmission Impairment Measurement Sets
TOS	Type of Service
TPDDI	Twisted-Pair Distributed Data Interface
TQM	Total Quality Management
TSANet	Technical Support Alliance Network
TSR	Terminal Stay Resident
TTS	Transaction Tracking Service
TX	Transmit
U/L	Upper/Lower
UBR	Unspecified Bit Rate
UDP	User Datagram Protocol
UDP/IP	User Datagram Protocol/Internet Protocol
UML	Unified Modeling Language
UNE	Unbundled Network Element
UNI	User-to-Network Interface
UPS	Uninterruptible Power Supply
URL	Uniform Resource Locator
USB	Universal Serial Bus
UTP	Unshielded Twisted-Pair
VAN	Value-Added Network
VAR	Value-Added Reseller
VBR	Variable Bit Rate
VBR-nrt	Variable Bit Rate—non-real-time
VBR-rt	Variable Bit Rate—real time
VC	Virtual Circuit
VCR	Video Cassette Recorder
VF	Voice Frequency
VG	Voice Grade
VLSI	Very Large Scale Integration
VMS	Virtual Machine System (Digital Equipment Corp.)

VOFDM	Vector Orthogonal Frequency Division Multiplexing
VoIP	Voice over Internet Protocol
VP	Virtual Path
VPN	Virtual Private Network
VSAT	Very Small Aperture Terminal
VT	Virtual Terminal
VT	Virtual Tributary
VTAM	Virtual Telecommunications Access Method (IBM Corp.)
WAN	Wide Area Network
WBT	Web-Based Training
WDM	Wavelength Division Multiplexing
WEP	Wired Equivalent Privacy
WFQ	Weighted Fair Queuing
WLAN	Wireless Local Area Network
WWW	World Wide Web
XNS	Xerox Network System (Xerox Corp.)

Assembling the Infrastructure

Role of the Communications Department

1.1 Introduction

Communications networks have become indispensable instruments for business survival, growth and competitiveness—in good economic times and in bad economic times. Whatever the general economic climate, there is no denying that communications networks speed information flow, improve the quality and timeliness of decision making, permit internal operations to be streamlined, enhance customer service, and, through these efficiencies, reduce operating costs. It is the job of a properly staffed, equipped, and managed communications department to keep corporate networks—voice and data—secure, available, and reliable so these and other benefits can be realized.

There are several compelling reasons for organizations to invest resources in establishing or expanding a communications department:

- An internal communications department gives the company maximum control over its network resources in terms of mixing and matching best-of-breed equipment and services to meet constantly changing business needs and market dynamics.

- An internal communications department offers the best response time to trouble calls, which report performance problems. In fact, with technicians on staff to continually monitor network performance, many problems can be identified and fixed before end users have a chance to call them in.

- With an internal communications department, there are more opportunities to exert leverage, since the company can pick and choose equipment and services from among several competing vendors and service providers, negotiate favorable contract terms and conditions, choose the most advantageous Service Level Agreement (SLA) for guaranteed performance, and take optimal advantage of bundled services for discounted pricing.

■ The communications department can fine-tune the network on a daily basis, deploy advanced technologies to improve network performance, and draw on relationships with vendors and carriers to obtain knowledge, expertise, and advice in new areas.

The failure of businesses to invest in information and communications systems can have some serious consequences. Employee productivity can suffer, which breeds job dissatisfaction. This, in turn, may cause missed deadlines and poor-quality work, which can affect the performance of others in the work group. At the departmental level, this lack of investment could result in not getting essential information fast enough, which can prolong the decision-making process in competitive situations and cause missed market opportunities. At the corporate level, any breakdown in the decision-making process could result in loss of customer loyalty, diminished investor confidence, and upset relationships with strategic partners.

Because the business world understands the advantages of technology, organizations of all types and sizes in all categories of business and industry continue to invest billions of dollars annually to build, maintain, upgrade, and expand their information systems and networks. Typically, it is the responsibility of the communications department to propose, fine-tune, and implement such plans, in consultation with other departments and executive management.

Putting together a communications department that can do all this is no small undertaking. As a network manager, you must have a diversity of expertise to assess needs, draw up plans, formulate budgets, hire and work with consultants, write RFPs and analyze vendor proposals, evaluate products and vendors, negotiate contracts, install equipment and cabling, troubleshoot problems, perform routine moves and changes, interface with carriers and suppliers, analyze pricing, assist end users with training and documentation, and respond to trouble calls.

A self-sufficient communications department should be divided into specialized areas of expertise: network planning and design, network management, help desk (technical support), administration, security, and operations management (see Fig. 1.1). This range of expertise, properly managed, enables the communications department to become a strategic asset that improves the performance of the entire organization, regardless of the location of specific offices, work groups, telecommuters, or mobile professionals.

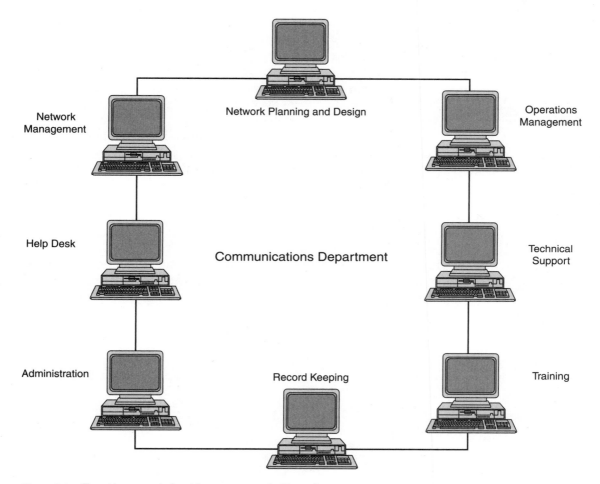

Figure 1.1   Functions carried out by a communications department.

1.2   Network Planning and Design

Network planning and design are important functions, especially for dynamic organizations that are continually adding new sites to the backbone or expanding the geographical reach of their networks to better compete in the global economy. These functions also are important for companies that are early adapters of new technologies.

1.2.1   Responsibilities

Your planning and design staff should use specialized modeling tools to accurately simulate a planned network's performance under a variety of conditions. For data networks, these tools allow designers and planners to view the network topology and test possible network design scenarios by simulating traffic flow and protocol stacks. Traffic is simulated using techniques such as recorded traces of actual traffic, statistical packet generators, or scripts describing typical network behavior. Tools can also measure performance of the model under selected traffic loads and its results are displayed in graphical or tabular format.

Such modeling tools can test a planned network's response to congestion and node failure, or perform "what-if" analyses to test the effect of adding or subtracting lines or equipment from the initial configuration. They can also analyze bandwidth allocation on the network, in terms of increments that will yield the best performance at the lowest cost, given the traffic load at any given time. Using traffic monitors attached to various network segments, they can filter and analyze different types of packets for their impact on network performance.

Network planners and designers also take into consideration the types of protocols to support, the number of nodes on the network (current and planned), and the performance parameters that ensure an acceptable quality of service for each application. Attention must also be given to the potential causes of congestion and the mechanisms that can be employed to prevent it from occurring on the Wide Area Network (WAN). Failure to address these issues can cause applications to run too slowly, frustrating users and diminishing their productivity. Therefore, network planners must decide on the best means of extending mission-critical applications across the enterprise, including remote branch offices, telecommuters, and mobile professionals. The choice for transport will usually boil down to the Internet Protocol (IP), frame relay, or Asynchronous Transfer Mode (ATM), depending on the mix of applications and their performance requirements in terms of such metrics as cost, reliability, availability, and latency.

With the plummeting price of bandwidth, there is the temptation to simply add more of it to improve application response time over these networks. But under this approach, certain applications would simply have more bandwidth to hog, leaving mission-critical and real-time applications gasping. Without careful bandwidth management, routine Web traffic, for example, can make it impossible to implement Voice over Internet Protocol (VoIP) with any degree of efficacy. Getting ahead of the performance curve requires a more practical solution, which entails the use of bandwidth management tools that can enable the network to effectively support many more users and applications than it could otherwise.

With such tools, network planners and designers give each application a set amount of bandwidth in accordance with their priority, which is determined by each application's performance requirements. Not only does this save on the cost of bandwidth, however cheap, it also eliminates the need to buy more equipment, manage more hardware "boxes," and add more skilled personnel. Containing operating costs is an important consideration, especially during times of slow economic growth, because it affects competitiveness, investor confidence, and access to capital markets.

Large enterprises usually want to take responsibility for the Quality of Service (QoS) function and require only basic connectivity from a carrier. But with so many devices to manage, administrators can easily get bogged down performing manual QoS configurations to fully optimize the enterprise network. The use of policy-based bandwidth management solutions makes this task less tedious and error prone.

Some of these policy-based tools are available as software solutions installed on existing routers located on the edges of the network, while others are implemented in hardware, requiring the purchase of dedicated devices that are also deployed at the network edges. Either way, a rulesbased interface allows network administrators to set up dedicated classes of service for the various applications, which instruct the edge devices on how to handle the different traffic types. Depending on the specific solution, they can configure the rules separately for each device or download them to a group of devices in one operation.

In smaller firms that are more resource-constrained, you may not have staff that can engage in network planning and design. These companies might be better off subscribing to the managed services of an Integrated Communications Provider (ICP) that can support IP, frame relay, and ATM to handle the growing number of diverse business applications that are extending to distributed locations. The ICP not only installs and manages the access equipment for the customer but the access links as well, ensuring the optimal performance of all applications. Partnering with the right ICP can free smaller firms to focus on core business issues. The ICP's management and life-cycle services ensure that the firm never has to worry about the details of network infrastructure.

With an ICP, small and midsize companies can get managed IP, frame relay, and ATM services from the same carrier with levels of performance to suit each application's performance requirements, even for voice calls. All of these protocols can even be supported out of the same Integrated Access Device (IAD) connected to T1, NxT1, T3, or OC-3 access facilities. The consolidation of multiple traffic types and the assignment of QoS to each application result in the optimal blend of bandwidth efficiency, response-time performance, and cost savings—with one carrier, one point of contact, and one monthly invoice.

A true ICP can also split the traffic by protocol to separate networks. This may be advantageous when your company wants 10 Mbps connectivity to the Internet, for example. In this case, the ICP would take the company's IP traffic and pass it through its multiservice network as an ATM

Permanent Virtual Circuit (PVC) configured for a Constant Bit Rate (CBR) class of service for smooth, consistent traffic flow to the nearest Network Access Point (NAP) on the Internet backbone.

Finally, ICPs give smaller companies the opportunity to take advantage of the most appropriate technology, or easily migrate between technologies as needs change, without having to deal with multiple service providers, equipment vendors, and integrators.

The availability of policy-based network management tools from a variety of vendors has made it easier for large enterprises to implement QoS policies with enough granularity to ensure that all applications are served in an appropriate manner without the company having to constantly shell out for more bandwidth in a futile effort to stay ahead of the performance curve.

Smaller companies that appreciate the value of bandwidth management, but do not have the resources to manage it themselves, can opt for the managed services of an ICP, which configures, installs, and maintains the customer premises equipment as well as the access and transport links for the optimal performance of all applications. Performance should be backed up with a Service Level Agreement (SLA), management reports, and complete life-cycle services.

The SLA is a performance guarantee for a given service. In the case of an Internet connection, the SLA specifies such things as network availability, round-trip delay, and the data transfer rate over the network. If the service provider fails to meet the performance obligations described in the SLA during any given calendar month, the customer's account will be credited. Usually SLAs only apply to business-class services. In addition to Internet connections, business-class services include private lines, frame relay, ATM, Virtual Private Networks (VPNs), and metropolitan-area Ethernet. There are also SLAs for managed application, security, and storage services. Thus, business connections and services are typically more reliable, but also cost more.

Among the other responsibilities of your planning and design staff is site engineering, which is the preparation of new sites to ensure that all environmental requirements are met before the installation of communications equipment. These requirements may include the installation of heating, ventilation, and air-conditioning (HVAC) systems that are specifically designed to support communications equipment. Other requirements include power, power distribution, and emergency backup facilities to ensure a continuous, reliable, uninterruptible energy source. You will sometimes have to knock down walls and reinforce floors to accommodate communications equipment cabinets. Terrestrial microwave and satellite systems, for example, involve roof mounts for antennas, the installation of which may require permission from building owners and compliance with local building codes. These systems, particularly multihop microwave, might require the procurement of right-of-way arrangements to achieve a clear line of sight between relay stations.

Many companies that do not have the resources to set up their own sites for network equipment, Web servers, and data storage are pursuing the alternative of collocation in which a service provider furnishes and manages the controlled environment. There are a number of reasons for considering collocation. Cost savings come from not having to invest capital in network infrastructure, recruiting and keeping qualified technical staff, and acquiring network management platforms and tools. Collocation also offers access to complementary services, such as direct connection to a metro fiber ring and access to a national data backbone capable of handling IP, frame relay, and ATM traffic, as well as voice services through a local central office switch to the Public Switched Telephone Network (PSTN).

With collocation sites available nationwide from the same provider, it is easier and faster for companies to set up Points of Presence (POPs) with less capital and fewer human resources. On-site technical support, offered by the collocation provider, allows immediate response to trouble conditions, resulting in less service downtime and less inconvenience to end users. Another key advantage of collocation is that it provides geographical diversity to protect IT assets against the effects of earthquakes, fires, floods, weather, brown-outs, and rolling blackouts, as well as terrorist attacks and workplace violence.

Your planning and design staff also interface with local and interexchange carriers for service and facilities provisioning, acceptance testing, and ongoing management. They must assess the impact of additional users and new applications to the enterprise network, and take steps to accommodate them so that peak performance can be maintained. Achieving peak performance involves the continual evaluation of offerings from equipment vendors and service providers, staying abreast of advancements in technologies and standards, and keeping informed of new tools that facilitate network planning and design.

1.2.2   Alternatives to in-house staff

There are several alternatives to having in-house planners and designers, but they all involve giving up control to outside firms that may or may not act in the best interests of the company. Many hardware vendors and carriers, for example, routinely offer planning and design assistance. However, the advice they offer is generally biased in favor of their own products and services, which may not be the most efficient or cost-effective solutions.

Another alternative to in-house planning and design staff is the systems integrator. Although systems integrators profess a more objective approach to network planning and design, many have strategic relationships with vendors and carriers that tend to undermine their objectivity. Before choosing a systems integrator, ask the firm to disclose these relationships and agree to thoroughly justify its selection of products and services from its partner vendors and carriers in terms of performance and value. On the other hand, if you need to set up network nodes quickly, a systems integrator that has prepackaged solutions immediately available may be the wisest choice.

You can also obtain planning and design assistance from traditional engineering consulting firms. Such firms tend to specialize in a particular technology such as T-carrier, microwave, optical fiber, or satellite. These firms are also among the most knowledgeable and objective, but are usually the most expensive. Unlike equipment vendors and carriers—which often provide free planning and design assistance and recover these costs in the sale of equipment and services—independent consulting firms rely strictly on their experience and expertise for their fees. Large systems integrators, on the other hand, can leverage their relationships with vendors and service providers to receive discounts, which allows them to moderate their fees to customers.

Of course, with every new technology, consulting firms spring up to address a whole new range of corporate needs. There are now consulting firms that specialize in such areas as the Internet, frame relay, ATM, and Integrated Services Digital Network (ISDN). Many of the consultants come from the vendor and carrier communities and have a tendency to recommend the products and services of their previous employers. Before choosing such firms for planning and design assistance, ask them to disclose the backgrounds of the individual consultants and the products and services they recommended in the last five to ten projects. If there is bias in product and service selection, this information will reveal it.

1.3 Network Management

Once the network is installed, upgraded, or expanded, performance is monitored by your network management staff at one or more specially equipped workstations located in the network control center or in remote locations. Network management staff typically engage in the following activities:

- Fault detection and isolation
- Maintenance tracking

- Performance measurement
- Configuration management
- Applications management
- Security enforcement
- Inventory/accounting

Responsibility for one or more of these activities may be parceled out to other specialists. For example, you may assign maintenance tracking to the help desk operator and configuration management and inventory/accounting to a Local Area Network (LAN) administrator. Each function may entail the use of a specialized management application purchased from a different vendor. Regardless of who has daily responsibility for a particular activity, and how many different applications are involved, in most cases you can integrate all of them under a single enterprise-level network management platform such as Hewlett-Packard's OpenView, IBM's NetView, or SunSoft's Solstice SunNet Manager. One network manager may have enterprisewide responsibility, while others might be assigned regional or domain responsibility for carrying out daily operations. Problems that cannot be solved at these levels can be escalated to the enterprise level for resolution, where engineering-level assistance may reside.

Smaller companies, of course, may not be able to afford a dedicated staff for managing voice and data services, much less have an in-house engineer. In such cases, you may outsource management of the telephone system and data to the equipment vendor or to a third-party management firm for a monthly fee. Such firms monitor their customers' equipment and communications lines remotely from a Network Operations Center (NOC) on a 24×7 basis. For problems that cannot be resolved from the NOC, a technician is dispatched to the customer location with the appropriate tools and replacement components. Services that fall outside the scope of the management contract are billed on a time and materials basis.

1.3.1   Fault detection

With fault detection and isolation capabilities, your company's network management staff can find out whether problems are caused by equipment failures, line outages, or both. Today's network management systems can detect problems by continuously monitoring system and line performance and comparing it to baseline performance. A baseline is a set of performance characteristics that indicates proper operation of a system or communications line. If performance falls outside of the baseline—in terms of latency, for example, or Bit Error Rate (BER)—these are indicators of service deterioration that can slow application response time and frustrate users and affect their productivity. With a monitoring capability, you can quickly identify problems and take steps to return the system or network to baseline performance.

These and other abnormal events raise alarms at the management console, which displays a graphical map of the network. The console operator can obtain increasingly detailed information about the problem by drilling down from a node (e.g., switch or router) on the network map to a particular system, card, and LED indicator. The operator can disable the faulty component while diagnostic tests are being performed. A large part of this process can be automated, so that when problems are discovered, isolated, and fixed, the results are recorded to a log without operator intervention.

The kinds of events that can be automatically detected and reported differ from one management platform to another, but the following types of events typically are detected and reported regardless of the particular platform:

- *Threshold events.* A predefined performance threshold has been exceeded.

- *Network topology events.* An object or interface has been added or deleted from the network.

- *Error events.* An inconsistent or unexpected behavior has occurred.

- *Status event.* An object or interface has changed, or an object or interface has started or stopped responding to echo requests—a process that indicates whether a remote node is incapable of being reached.

- *Node configuration event.* A node's configuration has changed.

- *Application alert event.* A management application has generated an alarm or alert.

- *All events.* All of the above events and other events are listed in one dialogue box.

These and other types of events are displayed in a color-coded event window according to their level of severity. By obtaining an indication of a problem's severity, the operator can distinguish between the failure of a backbone router, for example, and less important problems such as congestion notifications. If congestion is building up in the network, getting the backbone router back into proper operation will likely solve that problem as well.

1.3.2   Maintenance tracking

Maintenance tracking is related to fault detection and isolation. It is accomplished through a database that accumulates trouble ticket information. A *trouble ticket* notes the date and time a problem occurred, the specific devices and facilities involved, and the vendor from which it was purchased or leased. It includes the name of the network management system operator who responded to the alarm and summarizes short-term actions taken to resolve the problem. Maintenance tracking also involves scheduling preventive maintenance activities and keeping maintenance records.

As a network manager, you can use a comprehensive trouble ticket database for long-term planning and decision support. You can also call up reports on all outstanding trouble tickets, such as those involving particular segments of the network, those recorded or resolved within a given period, those involving a specific type of device or vendor, or those that have not been resolved within a specific time frame and had to be escalated.

This information can be used in a variety of ways. By categorizing trouble tickets by network segment, for example, you can find out which segments seem to be experiencing the most problems. The problems can then be categorized to determine the most likely cause of a persistent problem so that an appropriate solution can be applied. If the problem cannot be isolated to a specific segment but appears on multiple segments that share a common equipment type, use that information to help the vendor isolate a hardware or software problem and fix it with a replacement component or a software patch.

The trouble ticket database can also be adopted to track the performance of in-house technicians. If certain types of problems are resolved quickly by certain technicians and not by others, based on the problem and its severity you can call on the most appropriate individual. At the same time, if there is a technician who seems to lag behind all the others in resolving certain problems, it may indicate the need for additional training, perhaps at the vendor's training facility where the technician can earn a certification.

1.3.3 Performance measurement

Another function of the network management staff is performance measurement, which has two aspects: response time and network availability. *Response time* refers to how long computer users must wait for the network to deliver requested information from the server or host. The network management system displays and records response time information, and generates response time statistics for a particular server, terminal, line, network segment, or the network as a whole. *Network availability* is a measure of actual network uptime, either as a whole, or by devices or segments. Such information may be compiled into statistical reports that summarize such measures as total hours available over time, average hours available within a specified time, and Mean Time Between Failure (MTBF).

With response time and availability statistics compiled and formatted by a network management system utility, you will have objective information at your disposal. This information can help you establish current trends in network usage, make comparisons with baseline performance, predict future trends, and plan the assignment of resources for specific present and future locations and support new applications. Also, if the company has made SLAs with internal departments or with carriers, this kind of information can be used to verify that performance guarantees are being delivered in accordance with the agreement.

1.3.4   Configuration management

Network management systems also provide the means to configure lines and equipment at remote locations. If a WAN link becomes too noisy to handle data reliably, for example, the system will automatically reroute traffic to another line or send it through the public network. When the quality of the failed line improves, the system will reinstate the original configuration. The network management systems of some T1 multiplexers, for example, are even capable of rerouting high-speed data, but leaving unaffected voice traffic and low-speed data where it is. This helps prevent the disruption of routine communications, especially if not enough bandwidth is available during a reroute situation.

Configuration management also refers to the capability of altering circuit routing and bandwidth availability to accommodate applications that change depending on the time of day. Voice traffic, for example, tends to diminish after normal business hours, while data traffic may change from transaction-based to wideband applications that include inventory updates and large printing tasks. And when it comes to data, configuration management capabilities allow the interface definition of a circuit to be changed so that the same circuit can alternatively support both asynchronous and synchronous data applications. It also includes having the capability to determine appropriate data rates in accordance with response time objectives, or to conserve bandwidth during periods of high demand.

Configuration management not only applies to the voice and data circuits of a network but to the equipment as well. For example, in the WAN environment, router features may be enabled or disabled. The features and transmission speeds of software-controlled modems may be changed. If a nodal multiplexer fails, the management system can call its redundant components into action, or invoke an alternative configuration. And when nodes are added to the network, the management system can come up with the best routing plan for the traffic that the nodes can handle.

Configuration management can be applied to servers and workstations too. With the growing population of these systems deployed across widely dispersed geographical locations—each potentially using different combinations of operating systems, applications, databases, and network protocols—

at your disposal. This information can help you establish current trends in network usage, make comparisons with baseline performance, predict future trends, and plan the assignment of resources for specific present and future locations and support new applications. Also, if the company has made SLAs with internal departments or with carriers, this kind of information can be used to verify that performance guarantees are being delivered in accordance with the agreement.

1.3.4   Configuration management

Network management systems also provide the means to configure lines and equipment at remote locations. If a WAN link becomes too noisy to handle data reliably, for example, the system will automatically reroute traffic to another line or send it through the public network. When the quality of the failed line improves, the system will reinstate the original configuration. The network management systems of some T1 multiplexers, for example, are even capable of rerouting high-speed data, but leaving unaffected voice traffic and low-speed data where it is. This helps prevent the disruption of routine communications, especially if not enough bandwidth is available during a reroute situation.

Configuration management also refers to the capability of altering circuit routing and bandwidth availability to accommodate applications that change depending on the time of day. Voice traffic, for example, tends to diminish after normal business hours, while data traffic may change from transaction-based to wideband applications that include inventory updates and large printing tasks. And when it comes to data, configuration management capabilities allow the interface definition of a circuit to be changed so that the same circuit can alternatively support both asynchronous and synchronous data applications. It also includes having the capability to determine appropriate data rates in accordance with response time objectives, or to conserve bandwidth during periods of high demand.

Configuration management not only applies to the voice and data circuits of a network but to the equipment as well. For example, in the WAN environment, router features may be enabled or disabled. The features and transmission speeds of software-controlled modems may be changed. If a nodal multiplexer fails, the management system can call its redundant components into action, or invoke an alternative configuration. And when nodes are added to the network, the management system can come up with the best routing plan for the traffic that the nodes can handle.

Configuration management can be applied to servers and workstations too. With the growing population of these systems deployed across widely dispersed geographical locations—each potentially using different combinations of operating systems, applications, databases, and network protocols—

devices and facilities involved, and the vendor from which it was purchased or leased. It includes the name of the network management system operator who responded to the alarm and summarizes short-term actions taken to resolve the problem. Maintenance tracking also involves scheduling preventive maintenance activities and keeping maintenance records.

As a network manager, you can use a comprehensive trouble ticket database for long-term planning and decision support. You can also call up reports on all outstanding trouble tickets, such as those involving particular segments of the network, those recorded or resolved within a given period, those involving a specific type of device or vendor, or those that have not been resolved within a specific time frame and had to be escalated.

This information can be used in a variety of ways. By categorizing trouble tickets by network segment, for example, you can find out which segments seem to be experiencing the most problems. The problems can then be categorized to determine the most likely cause of a persistent problem so that an appropriate solution can be applied. If the problem cannot be isolated to a specific segment but appears on multiple segments that share a common equipment type, use that information to help the vendor isolate a hardware or software problem and fix it with a replacement component or a software patch.

The trouble ticket database can also be adopted to track the performance of in-house technicians. If certain types of problems are resolved quickly by certain technicians and not by others, based on the problem and its severity you can call on the most appropriate individual. At the same time, if there is a technician who seems to lag behind all the others in resolving certain problems, it may indicate the need for additional training, perhaps at the vendor's training facility where the technician can earn a certification.

1.3.3   Performance measurement

Another function of the network management staff is performance measurement, which has two aspects: response time and network availability. *Response time* refers to how long computer users must wait for the network to deliver requested information from the server or host. The network management system displays and records response time information, and generates response time statistics for a particular server, terminal, line, network segment, or the network as a whole. *Network availability* is a measure of actual network uptime, either as a whole, or by devices or segments. Such information may be compiled into statistical reports that summarize such measures as total hours available over time, average hours available within a specified time, and Mean Time Between Failure (MTBF).

With response time and availability statistics compiled and formatted by a network management system utility, you will have objective information

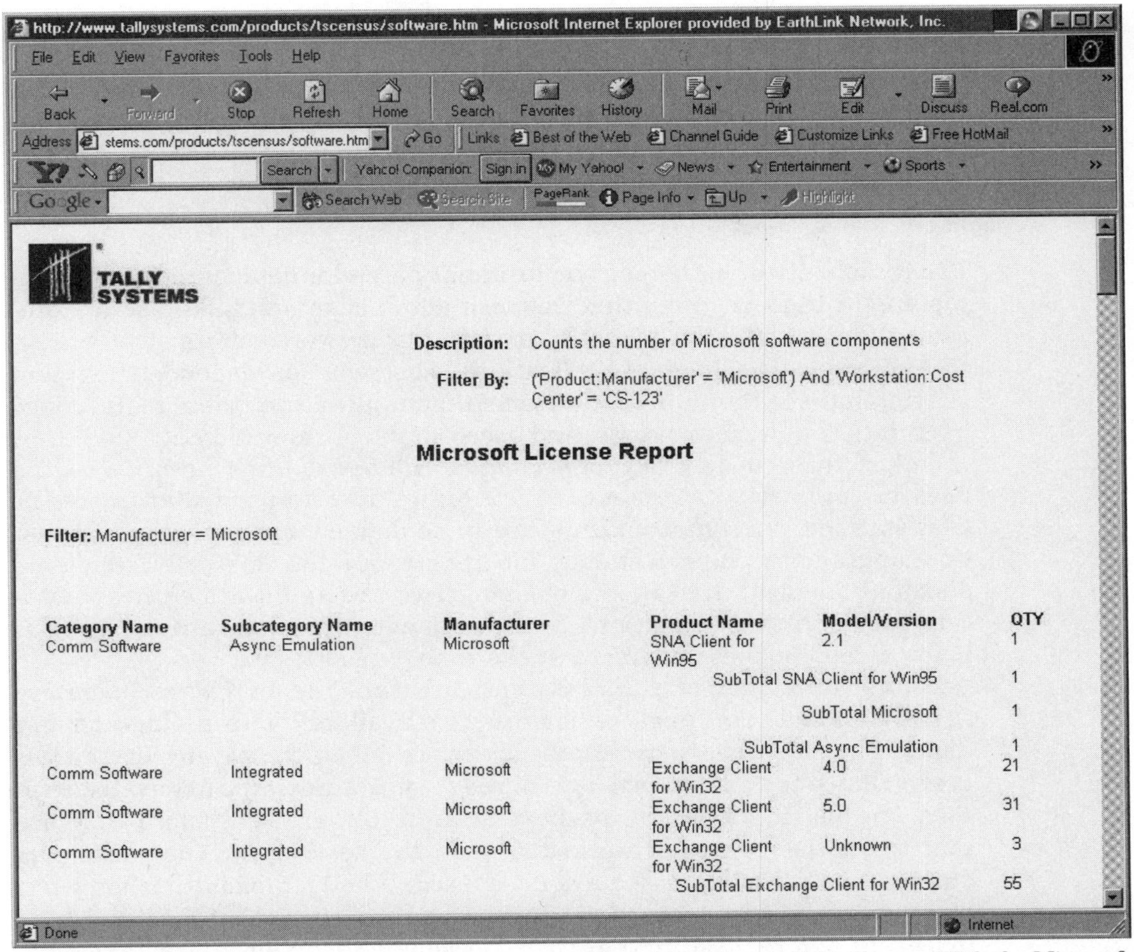

Figure 1.2 Software license compliance summary report from Tally Systems' TS. Census. With the Microsoft filter applied, the report shows which Microsoft software components are installed on the network.

in the most cost-effective manner. Determining vulnerability is the critical first step in protecting the enterprise network, particularly from threats originating from the Internet. You can often make an initial vulnerability with the same tools that hackers use to discover and exploit breach points. Depending on the type of network, protocols, operating systems, and applications a company has, you will be able to select appropriate tools for testing the known vulnerabilities associated with these elements.

After submitting the enterprise network to a battery of tests and identifying the potential breach points, you can devise an effective solution, such as implementing a firewall or upgrading a router with firewall capabili-

software has become more complex and difficult to install, maintain, and meter. The ability to configure the systems and perform all these tasks over a network from a central administration point can leverage investments in software, enforce vendor license agreements, qualify the organization for discounts on network licenses, and greatly reduce network support costs.

1.3.5   Applications management

The complexity of managing the distribution and implementation of desktop applications requires that network administrators make use of automated file distribution tools. By assisting a network administrator with tasks such as packaging applications, checking for dependencies, and offering links to event and fault management platforms, these tools reduce installation time, lower costs, and speed problem resolution.

One of these tools is a *programmable file distribution agent,* which is used to automate the process of distributing files to particular groups or workstations. A *file distribution job* can be defined as a software installation, upgrade, or patch; a startup file update; or a file deletion. Using a file distribution agent, these types of changes can be applied to each workstation or group of workstations in an automated fashion during business hours or nonbusiness hours, or at the time of next logon.

Before automated file distributions are run, the hardware inventory agent is usually run to check for resource availability, including memory and hard disk space. Distributions are made only if the required resources are available to run the software. To reduce network traffic associated with software distributions, the packages can be compressed before they are sent to another server or workstation. At the destination, the package is automatically decompressed when accessed. When a file distribution job is about to run, target users receive a message indicating that files are about to be sent and requesting that they choose to either continue or cancel the job, or postpone it to a more convenient time.

Maintaining a software inventory allows the network administrator to quickly determine what operating systems and applications are installed on various servers and clients. In addition to knowing what software components are installed and how many are in use on the network (see Fig. 1.2), the network administrator can track application usage to ensure compliance with vendor license agreements.

1.3.6   Security

Another responsibility of the communications department is to assess the network's potential vulnerability to unauthorized access and exposure to viruses, and devise solutions that can be worked into the network design

ties, which enforces security. A firewall regulates access to and from the Internet, holding back unidentifiable content, and implementing countermeasures to thwart suspected break-in attempts (see Fig. 1.3). These functions are carried out by rule sets that are loaded into the firewall and fine-tuned as new threats are identified.

Companies that cannot afford the expertise required to properly configure and manage a firewall and respond to various crises on a 24 × 7 basis can outsource network security to a service provider that takes full responsibility for configuring and fine-tuning the firewall and for ongoing management. The service provider furnishes the customer with management reports that track the performance of all monitored resources. To maintain the effectiveness of the firewall on an ongoing basis, the service provider employs certified security professionals who stay informed of the latest threats in order to respond quickly with new rule sets that implement appropriate countermeasures.

Network management systems are probably the most important assets that must be protected because they provide a view into the entire corporate network and can be used to gain entry into virtually every information system. Fortunately, network management systems have evolved to address this concern. Workstations employed for network management may be password protected to minimize disruption to the network through database tampering. Various levels of access may be used to prevent accidental damage. A senior technician, for example, may have a password that allows him or her to make changes to the various databases, whereas a less experienced technician's password allows only a review of the databases

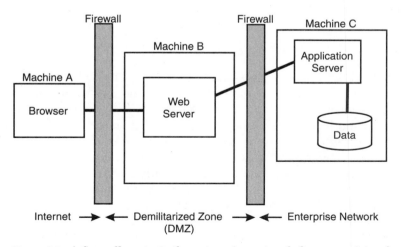

Figure 1.3 A firewall protects the enterprise network from a variety of attacks, safeguarding mission-critical resources.

without allowing him or her to make any changes. The network management system will also log successful and unsuccessful attempts to access any network resource, providing the first step in tracking down abuse.

With an increasingly decentralized and mobile workforce, organizations are coming to rely on remote access arrangements that enable telecommuters, traveling executives, salespeople, and branch office personnel to dial into the corporate network with an 800 number or a set of regional or local access numbers. In such cases, appropriate security measures must be taken to prevent unauthorized access to corporate resources from the remote access server. Security methods that can be employed include authentication, access restriction via password or directory profile, time restrictions that drop the connection, limiting the number of consecutive connection attempts and/or the number of times connections can be established on an hourly or daily basis, and callback systems to ensure that connections are being established from only authorized phone numbers.

1.3.7   Inventory and accounting

The network management system also allows staff to keep an inventory of the network, including the number and types of lines that are serving various locations, and what capabilities exist for alternative routing. Even the cards in the equipment cabinets at remote locations can be accounted for, as well as the spares at each location. Some systems track the purchases and depreciation of components to facilitate corporate accounting. All of this information may be displayed graphically at a color terminal and archived on disk or output to a printer.

1.4   Help Desk

Today's distributed computing environment, characterized by desktop processing and resource sharing via LANs and global interconnectivity via WANs, requires tools and expertise that will satisfy the growing requirement for end-user assistance. The reason is as simple as it is compelling: Ignoring requests for help can result in lost corporate productivity, slowed responses to competitive pressures, and eventual loss of market share.

One way to efficiently and economically service the needs of a growing population of computer and communications users is to set up a dedicated organization that is equipped to handle a wide variety of problems. This group is often called the *help desk*.

Briefly, the help desk acts a central clearinghouse for support issues, and is manned by a technical staff that fields support problems and attempts to solve them in-house before calling in contractors, carriers, or in-house specialists. The help desk operator logs every trouble call and, if possible, attempts to isolate the cause of the caller's problem. Experienced help desk operators can answer 90 percent of all calls without having to pass them to another authority. Problem solving is often done with the aid of a knowledge base containing all the problems and solutions encountered in the past. A high-speed search engine provides easy access to this information. If the problem cannot be solved over the phone, the operator dispatches an appropriate technician and monitors progress to a satisfactory conclusion before closing out the transaction.

The specific responsibilities of help desk operators include responding to user inquiries, isolating problems, preparing and administering trouble tickets, dispatching technicians, monitoring trouble report resolution and escalation, verifying problem resolution with users, and maintaining the trouble log. If the organization maintains a knowledge base of problems and solutions, then this too must be kept current to speed up troubleshooting, especially among less experienced help desk operators.

Aside from handling trouble calls from users, help desks can provide such services as order and delivery tracking, asset and inventory tracking, preventive maintenance, and vendor performance monitoring. The help desks themselves may be manageable via a separate management tool and tied into the enterprise network management platform. Depending on the degree of integration, there can be data sharing between the help desk and the enterprise management system. For example, if the help desk takes the responsibility for routine moves, adds, and changes then the network management system can make use of the updated information to populate various network configuration screens on the console.

1.5   Administration

The major administrative functions of the communications department typically include routine equipment moves, adds, and changes; software distribution and license management; and networkwide backup and data recovery. The administrative function is often integral to the help desk. Large companies, however, often have a separate LAN administrator because these responsibilities could easily overwhelm a help desk, which is better used to render assistance to problems and dispatch technicians when problems are isolated to a system or the network.

1.5.1   Moves, adds, and changes

Many of the resources of the LAN administrator go into fulfilling routine requests for moves, adds, and changes. The specific tasks associated with this function include the following:

- Processing orders for moves, adds, and changes
- Assigning due dates
- Providing information required by technicians
- Monitoring service requests, scheduling, and completions
- Maintaining the equipment and spares database
- Creating equipment orders
- Maintaining equipment orders and receiving logs
- Preparing monthly summary reports of moves, adds, and changes

Fortunately, there is a good selection of LAN utilities available that automates much of this activity, including service request processing, scheduling, and work order status monitoring.

1.5.2   Software distribution and license management

In addition to moves, adds, and changes, the administrative process includes keeping corporate PCs updated with the latest versions of operating systems and applications software. These functions can be implemented from a central workstation using tools that make use of the Windows Graphical User Interface (GUI) and Object Linking and Embedding (OLE) techniques like "drag-and-drop" to simplify and automate these operations.

On the administrator's workstation, the network appears as one or more logical trees showing the various nodes, which represent a network server directory and/or individual PC hard drives. The administrator can arrange the tree in a number of ways to facilitate administration. For example, the logical tree can match the organizational hierarchy or be arranged according to various categories, such as by users of UNIX, DOS, and Windows.

Once the logical tree is created, files can be copied to PCs or network directories by simply dragging them from one side of the administrative tool's application window to the desired place in the logical tree. With a single action, one or more files can be copied to a single PC, specific groups of PCs, or to all PCs on the network. The drag-and-drop technique also can be used to make changes to DOS autoexec.bat and config.sys files as well as Windows .ini files and Program Manager program groups. From the administrator's workstation, commands can be issued to restart Windows, reboot the PC, and perform other actions, such as to check a node's available memory and disk space. Sequences of actions can be scripted to simplify future file distribution. Multiple scripts can be combined into batch files that automate software installation and file distribution over the entire network.

Another responsibility of the LAN administrator is software license management. Allowing unlicensed copies of software to proliferate over a corporate network can trigger copyright infringement lawsuits and result in heavy financial penalties. There are software metering tools available that permit the LAN administrator to control the distribution, use, and access to software in accordance with the vendor's license agreement. There are also tools that detect the presence of unlicensed software anywhere on the network so that it can be tracked down and removed. The person(s) responsible for such violations can then be warned.

1.5.3   Network backup and data recovery

An important responsibility of the LAN administrator is network backup and data recovery. Protecting mission-critical data stored on LANs requires backup procedures that are well defined and rigorously enforced. Proper procedure includes backing up data in a designated rotation, using appropriate media, and testing the storage methods periodically to ensure that data can be easily and quickly restored.

Network backup is not a simple matter for most businesses. One reason is that it is difficult to find a backup system capable of supporting different network operating systems. And if midrange systems and mainframes are added to the equation, no vendor's solution is adequate for meeting the backup needs of today's information-intensive companies. The client-server environment has special backup needs: Back up too often and network throughput suffers; back up too infrequently and some data can become lost or corrupted between backups.

Whether the need is for online or offline storage—or a combination of the two—cost-performance, scalability, reliability, and investment protection are among the key factors influencing the choice of solution. Of course, there are many types of products to choose from that greatly ease the job of the LAN administrator. Some products automate the entire job of network backup and data retrieval according to administrator-defined schedules, and are capable of operating across different types of storage media and of being configured for hierarchical storage management.

An effective network backup strategy might be to use all three media types—hard disk, magnetic tape, and optical disk—in a hierarchical arrangement that is based on how readily data needs to be available (see Fig. 1.4). For example, data that is used frequently can be stored on a server's hard drives, whereas data that is used only occasionally might go to a tape library, and data that has not been used in several months could be archived to an optical disk.

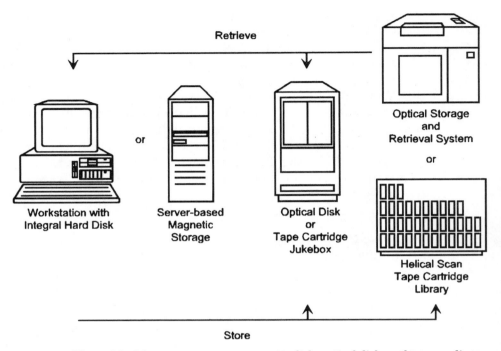

Figure 1.4   Hierarchical data storage spans magnetic disk, optical disk, and tape media.

Distributing storage resources via a Storage Area Network (SAN) can protect data from loss owing to a local outage or disaster. Essentially, a SAN is a specialized network that enables fast, reliable access among servers and external or independent storage resources, regardless of physical location. In a SAN, a storage device is not the exclusive property of any one server. Rather, storage devices are shared among all networked servers as peer resources. Just as a LAN can be used to connect clients to servers, a SAN can be used to connect servers to storage, servers to each other, and storage to storage for load balancing and protection. Fibre Channel or Gigabit Ethernet links can provide high-speed transfers of data between systems distributed within a building, on a campus, or throughout a metropolitan area. For longer distances, ATM and IP technologies can be used to transport data over the WAN.

1.6   Record Keeping

The record-keeping function of the communications department involves maintaining a central library of all operations logs and reports. Among the variables that need to be tracked are the following:

- Switch equipment repair and replacement
- Station equipment repair and replacement
- Software alarms, bugs, patches, and versions
- Trunk- and line-related transmission problems
- Wiring and jack problems
- Vendor response and performance
- User-related requests and problems

The following information can be extrapolated from these variables:

- Levels of parts replacement and repair
- Frequency of software-related tasks
- Grade of service from telephone companies and interexchange carriers
- Frequency of wire and/or cable-related tasks
- Frequency of move, add, and change activity
- Frequency of database modifications
- Vendor reliability
- User training requirements

Proper system and network management requires accurate record keeping. Daily, weekly, and monthly logs should be maintained so that periodic analyses can be performed. By periodically reviewing these logs and reports it is often possible to track such things as vendor execution and the propensity of certain brands of equipment to fail, or which types of circuits from certain carriers degrade in performance. This kind of information may be useful in choosing vendors and carriers or negotiating with them for better contract terms and conditions.

1.7   Training

The training function of the communications department includes providing on-the-spot user training via the help line; planning, organizing, and conducting formal training classes (often with assistance from the human resources or training department); and following up to ascertain the effectiveness of training sessions and making modifications as necessary.

A potential problem is that many communications professionals lack the qualifications or background to be instructors. When these people are allowed to go unprepared into classroom training settings, little or no learning is the likely result. If your communications department cannot find or cannot afford experienced trainers, you must develop them from within. Many times, any shortcomings can be remedied by "train the trainer" sessions in which communications staff provides professional trainers with the information necessary to construct courses and deliver them to end users via classroom or online sessions. Sometimes needs assessment surveys can justify training costs that can be picked up by other departments. And if there is a need to train users on specific systems or applications, many times vendors can be persuaded to provide training on site, or supply training materials in the form of videocassettes, CD-ROMs, or Web pages.

1.8 Technical Support

The technician is a key member of the communications department. The technician's responsibility is to provide routine and remedial service for systems and networks through quick, efficient, and cost-effective problem identification, isolation, diagnosis, and restoral. The technician also performs upgrades of various types of hardware and installs new hardware and connections. Of course, there may be more than one technician—perhaps dozens, depending on the company's size, number of locations, and geographical distribution. In cases where it is not possible to have an on-site technician, third-party maintenance firms, carriers, and vendors might have to be relied upon.

Typically, technicians work closely with help desk personnel, LAN administrators, and network managers who provide them with work orders and much of the documentation they need to begin the troubleshooting process. This information may consist of the hardware or software configuration of a workstation or server, a floor plan showing wiring and jack locations, or the maintenance history of a system. With this information, the technician can arrive at the location equipped with the right tools, test instruments, software patches, or reference materials.

A standard tool of the technician is the protocol analyzer, which is used for monitoring and troubleshooting both wireless and wireline networks. The analyzer can be connected to a network segment to view every packet and produce summary information on various types of packets, such as undersized packets, and events, such as packet collisions. Alternatively, the technician may install monitoring agents on the devices throughout the network. Either way, the tool captures packets according to predefined criteria and stores the information for further analysis. For instance, the analyzer can be set to examine only NetWare packets to track down the source of an intermittent problem with opening files at a remote server. According to an established threshold, it will generate an alarm if there is more than one file-open error per 100 file opens. Packets that match the filter criteria are captured, while those that do not match it are discarded. Once captured, the packets can be played back to pinpoint where and why an error occurred so that an appropriate solution can be applied.

Web-based network management and trend-reporting tools are available that enhance the protocol analyzer's usefulness, allowing the network manager to generate scheduled and on-demand reports that can be accessed with a browser from any location. These reports provide insight into traffic trends, such as the Top 10 applications and the worst response time by user. Such reports deliver information that is essential to minimizing network downtime, and they provide trend-reporting data to assist in future network planning.

1.9   Operations Management

A large communications department may have an operations manager who is responsible for ensuring system integrity and an optimum level of service for all users. This involves overseeing all operations procedures and department resources to ensure maximum system and network availability and reliability. The typical job responsibilities of an operations manager include the following:

- Managing department personnel and supervising workloads

- Determining budgets for maintenance, equipment, and personnel

- Evaluating system performance to identify areas of vulnerability and potential problem areas

- Developing and periodically testing disaster recovery plans

- Evaluating vendor performance to determine if internal standards and contractual obligations are being met

- Determining equipment and transmission facility needs and implementing system reconfigurations to accommodate corporate expansion and emerging applications

- Establishing required service levels and response times in consultation with department heads and top management

- Managing relationships with hardware vendors, carriers, consultants, systems integrators, maintenance firms, and contract programmers

- Assisting top management with contract negotiations and leasing arrangements with equipment vendors and service providers

In addition to monitoring the performance of maintenance and support services, the operations manager is responsible for providing reports to top management that address cost-benefit analysis and account for allocated resources. Such information is typically used to validate the current approach to maintenance and support, make changes that will bring about additional efficiencies and economies, and/or expand the nature and scope of maintenance and support activities.

Toward these ends, it is necessary for the operations manager to provide top management with annual and long-range plans, which generally involve the following activities:

- Maintaining records that provide concise information about the current status of information systems and networks

- Auditing the performance and progress of various vendors and service providers on major projects

- Gathering information about the current and future equipment and applications requirements of corporate divisions, departments, branch offices, work groups, telecommuters, and mobile professionals

- Assisting department heads with advice on the alternatives for supporting their near- and long-term business objectives

- Keeping personnel records up to date, including all pertinent information about technical and management skill levels, knowledge areas, continuing education, and incentive plans

- Keeping track of developments in information systems and networking technologies and their potential impact on the organization's competitive position

1.10 Importance of Staff Continuity

Staff continuity is a key factor in maintaining information systems and communication networks. A technician, for example, who stays with the same employer has time to become familiar with the various idiosyncrasies of the network, such as the propensity of certain equipment or links to fail. In staying with the same company, a technician has the time to fully understand the relationship of the network toward achieving the company's business objectives. This means understanding the performance requirements of mission-critical applications, as well as the current objectives and future expansion plans of each department.

The continuity of technical staff is vital to improving the response time to network problems. With accumulated skills and experience, a technician does not necessarily have to start from square one to track down the source of a problem. Knowing exactly where to look for the cause of a problem greatly improves network availability, since less time will be spent testing the various lines and hardware components of the network.

Staff continuity is also important for establishing and sustaining good relationships with critical external constituents such as vendors, carriers, consultants, and systems integrators. A technician who has been with the company for many years will also have had time to build a trusting personal relationship with his or her counterparts from these organizations. These relationships can go a long way toward improving their responsiveness when problems do arise and fostering a climate of cooperation, rather than finger-pointing, when problems span multiple areas of responsibility.

In the effort to keep qualified technical professionals in your organization, try several incentives. For example, after a specified time with the company, staff members may qualify for full or partial reimbursement for college tuition, including textbooks and lab supplies. Provide a notebook computer with a dial-in capability for remote diagnostics, which will permit flexible working hours and greater job autonomy.

An annual three-day leave to attend a career-related seminar or trade show with all expenses paid might prove to be another valuable incentive, as would subscriptions to various technical journals and book clubs.

Supporting technicians' continued training is important for several reasons. First, it keeps their skills up to date, which translates into direct and tangible benefits to the organization. Second, training improves their career prospects within the organization. This is especially true of programmers, who are typically evaluated in terms of depth and breadth of technical knowledge at the time of hire. However, if a programmer wants to advance to systems analyst, project leader, or manager, the employer will most likely put a higher value on overall industry knowledge as well as specific business management and interpersonal skills. In preparing suitable candidates to assume higher levels of responsibility, your company not only recovers its investment in training but also reaps the additional advantage of their corporate experience. Finally, training provides technicians with a sense of growth and fulfillment. These are the psychic rewards that contribute greatly to job satisfaction. With a high level of job satisfaction, there is less likelihood that a technician will be easily lured to another company, where such rewards may not be available even though the pay may be higher.

Another incentive would be an advisory role in formal corporate working groups, where the technician's visibility would be heightened within the organization and a fair amount of ego gratification would be provided as well. And if the individual has strong instructional skills as well as interpersonal communications skills, these could open up a role in corporate new-hire orientation, in-house training programs, and in cross-training colleagues in the communications department.

Any one or a combination of these incentives is more economical than continually replenishing qualified personnel. If these suggestions are not very appealing to a specific valuable employee, it is worth the effort to ascertain what the technician values most, so that an effective incentive program can be tailored to his or her career needs.

1.11   Role of the CIO

The organization's technical staff may be quite varied and include programmers, network managers, database administrators, consultants, analysts, maintenance and repair technicians, help desk operators, telephone and computer systems experts, and cable installers. Some of these individuals may report to the Information Systems (IS) manager or to the telecom manager, with both departments coming under the purview of a Chief Information Officer (CIO).

The primary responsibility of the CIO is to ensure the delivery of reliable and available information and communications services to the enterprise at reasonable cost. This often means that the CIO must act as a resource arbiter between the traditionally separate realms of IS (data) and telecom (voice).

Aside from budgeting, resource allocation, and IS and/or telecom oversight, the responsibilities of the CIO typically include one or more of the following:

- Re-engineering

- Modernization

- Total Quality Management (TQM)

- Enterprisewide systems integration

- Systems interoperability

- LAN interconnection

- Multimedia applications development

- Standards compliance

- Best practices

The position of CIO was originally conceived as a way for senior computer executives to attain a special status equal to that of the vice presidents of engineering, manufacturing, sales, marketing, and human resources. Early advocates of the CIO role claimed that the position should be the equivalent of the chief financial officer, which is second in power only to the chief executive officer. In practice, only about 10 percent of the Fortune 500 companies have CIOs that report directly to the CEO, and in reality there are very few success stories of CIOs revamping organizational structures and improving corporate profit margins through innovative uses of technology. There is also very high turnover among CIOs. On average, a CIO stays on the job only 15 months. There are two reasons for this state of affairs. One is that the position of CIO lacks definition: The position is hard to justify when many of the job responsibilities are already being carried out by IS and telecom managers. Second, even if the mission is clear, often people who are hired as CIOs are political types who are more focused on becoming the next CEO rather than tending to their primary job responsibilities. The frequency of CIO failure in recent years has been so great that the position suffers a credibility gap from which it may not recover anytime soon.

1.12   Conclusion

Today's operating environment often places corporations in a squeeze between high support costs and high levels of dependence on network equipment, facilities, and services. The communications department plays a pivotal role supporting the business objectives of the entire organization and doing so in the most cost-effective manner. The extent to which this can be achieved will be determined in large part by the quality of technical resources available in the local job market and the tools with which qualified personnel can be equipped. Even when most of the maintenance and support functions of a communications department can be outsourced to third-party firms, there will still be the need for internal resources to manage these relationships and to ensure that these firms continue to act in the best interests of the company.

In taking responsibility for managing their communications networks, however, companies can better meet their changing business needs by mixing and matching best-of-breed solutions from a variety of sources. An internal communications department offers the best response time to trouble calls, since technicians continually monitor the network and are able to resolve many problems before end users become aware of deteriorating network performance. A communications department also offers more opportunities for cost savings, since it can keep the network optimized and take advantage of advanced technologies and new services to obtain greater efficiencies and higher performance levels.

An internal communications department keeps accurate records that can result in better identification of recurring problems, easier determination of vendor performance, and the pinpointing of areas that can be improved with more user training.

Operating, maintaining, and managing a corporate network is an ongoing challenge. When properly staffed and equipped, the communications department becomes a strategic asset that improves the performance and responsiveness of the entire organization. Many of the themes mentioned in this chapter will be discussed in greater detail throughout the book.

2

The Procurement Process

2.1 Introduction

Investigating, evaluating, and ultimately investing in communications products and services have become a confusing and costly process for many organizations. Many of the challenges associated with the procurement of products and services are a direct result of deregulation, which has been going on in the United States for more than 30 years. The process started with competition in the provision of Customer Premises Equipment (CPE) and continues today with the provision of local and long-distance services, Internet access, and broadband communications. The premise behind deregulation is that users should not be compelled to buy or lease telecommunications equipment or services from monopolistic companies at inflated prices.

The impact of deregulation includes the following:

- It promotes competition, giving users more choices and lower prices.

- With more choices, users are not forced to rely on unresponsive vendors and carriers.

- Competition accelerates technology innovation as each vendor and carrier struggles to differentiate its offerings from those of its rivals.

Vendors and carriers that cannot compete based on their products' performance often find a lucrative niche for themselves by competing on price. Thus, customers can choose products and services based on performance or price, depending on their needs and priorities. Performing a trade-off analysis of price versus performance has become a standard procedure for deciding which products and vendors to choose.

With competition, the pace of innovation has greatly accelerated. Too often, network planners complete the evaluation and implementation of an optimal network only to discover that new products or services have become available, rendering the network obsolete before it is even fully implemented. The competitive climate has also made many vendors and service providers vulnerable to an acquisition or a merger—even dismemberment through bankruptcy—leaving their customers in doubt about future product support.

2.2   Managing Change

Managing change is probably the most difficult task for those planning, operating, and maintaining an enterprise network. Under ideal conditions, buying established products and services is difficult enough. But in today's deregulated, highly competitive marketplace, with so many variables to consider, the chance for error is much greater. Choosing the wrong vendor or service provider can result in severe penalties, including:

- Wasted capital investment
- Missed corporate objectives
- Underutilization of personnel
- Lost sales
- Damage to competitive position

Corporate management must develop a strategy for minimizing the risk inherent in technology procurements and convey this strategy to communications department staff to ensure that costly mistakes are avoided at every point in the decision-making process.

As applied to technology procurements in a highly competitive marketplace, risk management has several concurrent objectives:

- Address organizational needs
- Ensure network efficiency
- Save money
- Protect existing investments in technology
- Safeguard credibility in communications department staff

Before investing in any one vendor's implementation of a new communications technology, an organization's network design staff should carefully examine both the stability of the vendor and the nuances of its implementation of the technology. The following sections offer a framework that identifies all elements in the decision-making process and demonstrates how to minimize risk at every point along the way. This framework is intended to help communications department staff anticipate problems so that costly mistakes can be avoided.

2.3   Organizational Objectives

All too often, network managers, anxious to implement a particular communications technology, initiate action without going through careful planning, evaluation, and testing processes. In the relatively static, regulated environment of the past, there was virtually no need to research vendors, compare implementation techniques, or do extensive testing because the AT&T–Bell Operating Company–Western Electric monopoly virtually dictated what equipment could be used and how it would be connected to the network. In today's competitive communications environment, however, the failure to adequately plan for the procurement of new network technologies and services invites disaster.

Building, upgrading, or reconfiguring a voice or data network begins with the articulation of organizational objectives followed by a detailed action plan. A specific objective, for example, might be "to serve customers (or employees) faster and more efficiently by increasing network availability to 99.999 percent and keeping application response time to under 5 seconds." A broader objective might be to support the company's reorganization effort to permit faster response to changing market conditions.

Whatever the objective, the communication network plays a pivotal role in the organization's ability to achieve it. To facilitate planning, the broad organizational objectives from top management must be translated into formal statements that communications department staff can understand and work toward. Typically, this will also require the participation of staff from other departments. Participation from all parties encourages shared responsibility for implementation and ownership of a successful outcome. A set of clear objectives also increases the likelihood of top management support. The objective-formulation stage is fundamental to any project because it sets the tone and direction for the organization. Failure to take these corporate acceptance issues into account introduces a high degree of risk that can jeopardize the success of any major project.

2.4   Needs Assessment

You, as the network planner, can keep abreast of the organization's present and future networking needs by studying the business plan. By knowing where the company plans to be in the next three to five years, how it plans to get there, and how corporate resources will be deployed, you can determine how the network might be used to support the organization's overall business objectives.

Once broad objectives have been articulated, they can be used as the basis for developing a Request For Proposal (RFP) that will be issued to vendors and service providers. With input from other departments, you, as the network planner, can ascertain specific requirements. Soliciting input from other departments also gives them a stake in the outcome of the project, promoting interdepartmental cooperation. The RFP can be continually refined as new information becomes available. Eventually, you will have the necessary information required to identify project milestones, delegate tasks to staff members, and specify time frames for task completion.

2.5   Selecting Potential Vendors

After determining present and future corporate network requirements, you will issue a final RFP to the appropriate vendors or service providers. Potential vendors for the project will respond with information on products, pricing, and availability. The process of identifying vendor candidates is generally based on a variety of criteria, including visibility in the company's specific industry, market share, and experience in similar projects, as well as current performance levels if a relationship is already established with a particular vendor.

To eliminate some of the risks that might arise if a technology acquisition plan is too rigid and myopic, you can use the RFP as an invitation to vendors and service providers to suggest alternative methods of implementing the project. Many vendors and service providers have specialized technical expertise and can offer suggestions for improving the design of a network. Often, this advice can help an organization avoid costly mistakes and place the proposed network in a better position for future growth. Some vendors and service providers have been known to develop very detailed analyses, comparing their proposed alternative with the one described in the RFP and, in the process, showing how the user can save thousands of dollars in equipment costs and/or line charges per month.

Vendors and service providers constitute a valuable resource that is frequently overlooked in planning a network. In the highly competitive communications market, every vendor is looking for ways to outmaneuver and outperform its rivals. Recognizing this, an astute network planner can draw upon this competitive urge in the initial stages of a project to help ensure its successful outcome.

After responses to the RFP are evaluated, the strongest proposals will require further consideration. In addition to describing efficient, economical, and technically sound solutions, the proposals should demonstrate organizational viability. Without a thorough evaluation, the technology investment may be at risk since an unstable vendor may discontinue products, cut back on support staff and response time, go out of business, or be acquired by a larger firm with different sales and support priorities. Because any number of potential dangers threaten vendors and service providers in the communications market, potential customers should exercise appropriate diligence in the vendor evaluation process before committing to a large-scale procurement. The following sections discuss the most important vendor characteristics that you should consider when performing the evaluation step for major technology procurements, including network equipment and carrier services.

2.5.1 Financial and credit information

Financial statements and credit references must be evaluated and verified. Failure to pass verification should disqualify vendors and service providers. However, you should not base a purchasing decision on a favorable verification alone. A complete financial picture of a company is only a snapshot in time. Many negative factors can influence a vendor's financial position between vendor selection and product delivery. Only those companies that are prepared to weather financial storms deserve consideration for a long-term technology investment. Financial and credit history should only be used to disqualify candidates—never to choose them. Other factors must be evaluated before a selection can be made.

Is there ever a good reason to be more lenient with regard to the financial health of a firm when choosing a product or service? After all, during the 2001–2002 recession even blue-chip companies like Lucent Technologies experienced huge losses. And among large and small Competitive Local Exchange Carriers (CLECs), a number of them sought Chapter 11 bankruptcy protection until they could reorganize and emerge debt free with new financing arrangements. Others went out of existence for lack of follow-up financing. The same fate befell many broadband companies that offered Digital Subscriber Lines (DSL) for high-speed Internet access. The plights of these and other types of vendors and service providers were complicated by the reluctance of investors to sink more money into the telecommunications industry after they had been stung by the collapse of the so-called dot-com industry in 2000.

In difficult economic times, there is certainly more risk in going with financially shaky vendors and service providers. But if a financially strapped candidate has a proven record of performance and reliability and offers the most viable solution, steps can be taken to minimize the risk. The following factors merit investigation:

- *Has the firm laid off employees to trim overhead and improve the balance sheet?* Although this is common practice and suits the short-term outlook of investors, it places enormous stress on employees who must shoulder the additional work burden. This situation inevitably diminishes customer service and has a ripple effect that has an impact on the performance of other operating units, which further destabilizes the company.

- *Is the company demonstrating progress toward "positive" reorganization?* Find out, for example, if the company is getting rid of assets that are unrelated to its core business. During vibrant economic times, many companies tend to bulk up with assets that are only peripherally related to their core business in the belief that they must become a "single-source" provider. Often these assets are not a proper fit, poorly understood from a management perspective, or come with problems of their own that cannot be easily fixed. The result is that these assets become a financial drain, which is magnified during tough economic times.

- *Does the company have "debtor in possession" financing to carry it forward as it conducts talks with potential long-term investors?* When a business seeks to reorganize under a Chapter 11 proceeding, it is required to submit a plan of reorganization to the bankruptcy court. It can then receive Debtor in Possession (DIP) financing, secured by its receivables and other assets, and hold off payments to prebankruptcy creditors. DIP financing provides a source of working capital during reorganization to enable the company to continue to do business. It is a verifiable source of funding for suppliers, thus maintaining their cooperation throughout the reorganization. It also sustains a positive credit relationship with a financial institution, which may be used as a bridge to bank financing. If the troubled company cannot formulate such a plan, however, the business may be forced into a Chapter 7 liquidation. In order to minimize risk, it is imperative to check the status of the Chapter 11 filing.

Of course, IS and telecom managers will not be expected to play a role in the financial evaluation of a vendor or service provider. This would be the responsibility of the CIO or the financial officer of the company, either or both of whom would deal with their counterparts at the candidate firm. But it is still worthwhile for IS and telecom managers to become familiar with these issues, if only to understand where the proper responsibility for these matters resides.

2.5.2   Product development

To determine a candidate firm's commitment to and continued success with its product or service, the evaluation team must delve into its development effort. Answers to the following questions should reveal the strategy and commitment toward product development:

- *Is there a formal, budgeted program for product development, and is it adequately staffed?* This provides some assurance that promised features, enhancements, and upgrades will be available in the future.

- *Are there unique features, industry "firsts," copyrights, patents, licenses to other companies, or OEM/VAR arrangements?* Answers to these questions validate claims of industry leadership and innovation, and lend credibility to the candidate firm's position and reputation in the industry. If too much of the intellectual property belongs to other companies and it is essentially a reseller, it should be able to point to the added value it provides. Otherwise, it could mean that the firm relies on thin margins that could leave it too exposed to vacillations in market demand.

- *How many product revisions have there been?* If the number seems too high or too low compared to other vendors, this factor can help determine such things as product quality, customer satisfaction, vendor responsiveness to changing customer requirements, and openness to customization.

- *What is the number of completed versus canceled projects?* This question probes a little deeper into the internal workings of the company and it can help gauge a vendor's success in planning and implementing new products.

- *What were the dates of the product's announcement and its first delivery?* Comparing these dates may provide an indication of the vendor's R&D capabilities, manufacturing capacity, and quality control.

- *What will be the method for implementing enhancements?* Do the enhancements require factory modification, on-site replacement of components, or software loading from a remote location? The answers to these questions help determine the life-cycle cost of the product, and how well the vendor has thought through the product's upgrade path.

- *Is the vendor using state-of-the-art componentry and circuit integration technology, such as Large Scale Integration (LSI) or Very Large Scale Integration (VLSI)?* What about the use of Application Specific Integrated Circuits (ASICs) that offload specialized tasks like encryption and compression from a system's main processor? This information can be used to predict performance and reliability, especially for new products with no history in a live environment.

2.5.3   Integration and engineering

Today's enterprise networks usually consist of a hodgepodge of different "boxes" purchased from numerous manufacturers or leased from different service providers. Even when vendors claim support for particular industry standards, however, customers would be making a serious mistake if they assume that they can merely plug these boxes together and that they will run properly on an existing network. Additionally, it should not be assumed that the vendor or service provider would hold the hand of a company's network support staff through every trial and tribulation of network configuration and implementation. There are cases where problems are never solved—until another vendor or service provider comes along with the solution, but by then money has already been wasted on the original purchase.

The simplest communications devices require proper integration into an existing network by qualified technicians who are familiar with a network's current configuration and the applications it supports. Therefore, it is important to determine the true integration and engineering capabilities of the vendor or service provider.

In addition to verifying the vendor's experience with similar projects, the evaluation team can minimize risk in identifying a vendor's integration and engineering capabilities by getting answers to the following questions:

- What specific experience does the vendor have with regard to site preparation, installation, coordination with third parties, acceptance testing, and cutover?

- Will the vendor commit to all promises in writing?

- Will the vendor accept a "weasel" clause that gets the customer out of the contract, with no penalty, should the vendor fail to perform?

- Will the vendor accept third-party acceptance testing to determine the functional performance of installed products or systems?

- If the product or system is found to be faulty, will the vendor agree to a penalty structure that discounts the outstanding balance by a percentage amount for every day the cutover is delayed?

- If it is determined that the product or system cannot comply with the functional specifications of the RFP without adding hardware or software, will the vendor pay the cost to bring the product or system into compliance?

- Can the customer's network support staff talk to the vendor's operations staff directly to determine the feasibility of undertaking engineering and integration projects within specified time frames?

- Can the customer obtain and verify the credentials of installers, engineers, and site supervisors to determine their qualifications for implementing the project?

Another way to determine integration and engineering capabilities of a vendor or service provider is to ask about specific problems they have solved for customers that could not be solved before they came along. A worthy vendor or service provider should have dozens of stories about how they helped customers out of jams nobody else seemed to know how to fix. Then ask for the names of those customers for a reference check and verify that the problem was solved to their satisfaction.

2.5.4   Quality control

A good indicator of product reliability is the vendor's Quality Control (QC) program. Today, many of the product reliability problems that vendors experience are rooted in inadequate vendor commitment to product testing and system integration testing. A visit to the vendor's manufacturing plant should include a walk through the QC facility. Some of the important items to check include:

- An incoming inspection station that checks batch components and other raw materials delivered at the receiving dock. The inspection station should monitor material containers for signs of shipping damage and order discrepancies.

- Throughout the manufacturing facility, multiple inspection points at various stages of the assembly process, including automated testing stations and visual inspections by operators.

- The implementation and enforcement of electrostatic safeguards at workbenches and assembly lines to protect sensitive components that must be installed manually from becoming damaged.

- Consistency (or disparity) of answers from plant supervisors and operators when questioned about Printed Circuit Board (PCB) failure rates.

- Integrity testing of all boards at the finished product level. With high failure rates, statistical sampling methods at this juncture should be viewed as unacceptable.

- QC procedures that include automated administrative support that allows operators to instantly calculate and display differences between test specifications and actual measurements. This allows potential problems to be spotted and corrected before products go out the door. With manual systems, the chance of human error increases and potential problems may go unnoticed until products are already installed at customer locations.

- Clean and orderly work areas that are free of potential safety hazards in the production facilities. Although this point may seem minor, it provides a good indicator for determining whether the floor managers lean toward thoroughness or complacency.

In addition to observations conducted at the vendor's production site, an evaluation team can also investigate one of the final steps in product development—the beta test site. Beta sites are special customer sites that have agreed to test new products under actual working conditions. When considering the purchase of new products that have no performance history, potential customers can obtain location and key contact information for the vendor's beta sites. The evaluation team should also verify that the results of the beta site tests come from an actual customer site and not the vendor's own laboratory. In addition to asking for the results of these tests, a network administrator should also request a detailed summary of the benchmark tests that the vendor chose to use and a brief explanation of why they were chosen.

Quality control is just as applicable to software as it is to hardware. However, only in recent years have software companies invited QC people to become involved during the product's design stage. Among the large vendors, it is now the responsibility of QC staff to help review specifications and establish a clear understanding of how the product is put together. Armed with this knowledge, the QC people assume the role of actual customers to uncover every conceivable way the product can fail. For software QC testing, the customer should confirm that the vendor's rigorous testing procedure not only applies to new products but to follow-on enhancements as well. A vendor's attention to a uniform standard of quality ensures that the software products meet customer expectations regarding ease of use, as well as functionality.

Early involvement of QC people in the product development cycle enhances the vendor's ability to solve customer problems. The evaluation team can minimize the risks inherent in major software purchases by establishing a prerequisite for a formal QC program as an integral part of a vendor's product development cycle. In addition, the vendor's QC program should be appropriately staffed and budgeted. If the vendor has experienced layoffs, find out if the QC staff was also trimmed.

A vendor's success in providing service and support to its worldwide customer base comes from its commitment to quality. To gauge this commitment, the evaluation team should determine if the vendor has modeled its quality systems to conform to the globally recognized ISO 9000 standard for quality management and quality control. The ISO 9000 system, established in 1987, is made up of a series of standards and supplementary guidelines created by the International Organization for Standardization (ISO). The quality standards are generic in nature and can be applied across industry lines. ISO 9000 has been adopted by more than 100 countries.

ISO 9000 certification provides assurance to a vendor's global customer base that the processes involved in the design, development, manufacture, installation, service, and support of its products adhere to the most stringent and comprehensive quality standards. Services can also be certified as compliant with ISO 9000 standards. In this case, the certified company must conduct an annual satisfaction survey of all contract customers, recording and tracking the quality of after-sales service and support. The service provider's policies and procedures must be well documented and distributed and adhere to the same quality standards.

2.5.5 Interoperability testing

When considering any vendor for a major purchase, it is important to consider interoperability, especially when the products are based on recently issued standards. Typically, vendors will interpret the same standard in slightly different ways, with the result that their products will not work together on the same network. Then industry groups will have to reconcile the differences, even to the point of setting up testing labs where vendors can plug their equipment into a network to evaluate interoperability with the equipment of other vendors.

With the growing popularity of Storage Area Networks (SANs), for example, vendors have been building switches, interfaces, and other components that allow companies to build distributed storage environments to improve response time and protect mission-critical data from a local disaster. A key enabling technology for SANs is Fibre Channel, which provides up to 2 Mbps of throughput between storage resources. But different interpretations of the Fibre Channel standards by vendors have resulted in products that do not work together on the same network. As a result, interoperability problems still plague the SAN industry.

Some large vendors such as EMC and IBM as well as the Storage Networking Industry Association (SNIA) have invested in interoperability labs where other vendors could see how well their equipment works with components from other vendors. If a problem is detected, the product is returned to the vendor so firmware or code can be upgraded. If no problem is found with the product during the retest, the lab certifies it as having passed a set of tests that is commonly accepted among SAN vendors. This gives companies assurance that the products they buy from one vendor will work with those of another vendor that have passed the same tests.

Another new technology called Bluetooth provides "always on" wireless communication between portable devices and desktop machines and peripherals. Among the many things Bluetooth enables users to do is swap data and synchronize files without having to cable devices together. Bluetooth technology can also be used to make wireless data connections from handheld devices to conventional Local Area Networks (LANs) via an access point equipped with a Bluetooth radio transceiver that is wired to the LAN. And since the Bluetooth baseband protocol is a combination of circuit- and packet-switching, it is suitable for voice as well as data. For example, instead of fumbling with a cell phone while driving, the user can wear a lightweight headset to answer a call and engage in a conversation without even taking the phone out of a briefcase or purse.

To prevent interoperability problems between different vendors' equipment, the Bluetooth Special Interest Group (SIG) early on set up a qualification process to ensure that products comply with the Bluetooth specification. Upon passage of this qualification process, products can display the Bluetooth brand mark, signifying to consumers that they will interoperate as expected. Any product that displays the Bluetooth brand mark must be licensed to use the mark and only products that pass the qualification test can be issued a license.

This qualification process entails product testing by the manufacturer and by a Bluetooth Qualification Test Facility (BQTF), which issues test reports that are reviewed by a Bluetooth Qualification Body (BQB). All hardware or software modifications to a qualified product are documented and reviewed by the BQB that issued the qualification certificate for the product.

The qualification requirements are not the same for all products. Those that are specifically designed and marketed as development tools or demonstration kits are exempt from testing requirements, and qualification is possible for these products by filing a simple declaration of conformance. And products that integrate a Bluetooth component that has been prequalified may be exempt from repeating tests for which the component is prequalified.

In sum, the issue of interoperability should be thoroughly investigated before a major purchase. Preferably, the vendor should be able to provide proof of interoperability through certification from a credible source. This will ensure that the company does not get stuck with proprietary technology that forces it to depend on one vendor as the source of supply and, as a consequence of no competition, pay inflated prices. Interoperability is also important for obtaining the flexibility to mix and match best-of-breed solutions on the same network. And having multiple sources of supply will give the company bargaining leverage on such things as price, maintenance, and training.

2.5.6 Repair and return

Sometimes equipment problems can only be solved by a site visit from a vendor's technician or a representative of a third-party maintenance firm. In most cases, technicians dispatched to a customer location are not trained or equipped to perform board-level repairs. Even if a company employs its own technicians to perform maintenance, they are usually trained only to isolate faulty boards and swap them with spares. The faulty boards are then sent to the vendor's depot test-and-repair facility. To minimize exposure to risk, the evaluation team should make sure that the vendor has properly staffed and equipped facilities with which to fulfill the product warranty, or has made arrangements with a third-party maintenance firm to perform this essential function. Otherwise, the vendor's warranty is practically worthless because the vendor is not adequately prepared to service products after the sale.

If the vendor does offer appropriate facilities for repairs, the average turnaround time for defective part repairs should be determined. At the least, in-warranty items sent in for repair should be returned within 10 working days and emergency repair service should take only three days. Critical items like control-logic boards should carry same-day support services, which require the repair center to supply "loaners" until a faulty unit can be repaired. In addition, fault-suspect components and devices that test positive should be returned to the customer without charge.

The evaluation team should be skeptical of vendors who "throw in" repair and return services as sales gimmicks to get business or to close a sale. In general, if the vendor is not wholly committed to the quality of both the product and after-sale services, no amount of presale negotiation will make them follow through in a timely manner. Additionally, past and present customers should be asked for their opinions on how they rate the vendor in these areas.

If possible, the evaluation of a vendor should extend to its repair center before a major purchase is made, even to the point of an on-site visit. While at the repair center, find out whether items returned for repair go through the same quality control procedures as newly manufactured products. If so, there may be opportunities to purchase refurbished network equipment and components at substantial cost savings. Check into the warranty period of refurbished items and whether there is a right of return and full refund if they do not perform to their original specifications.

2.5.7 Customer service

A vendor's commitment to customer service should go far beyond just having a 24-hour hot line to technicians who can resolve problems over the phone. The customer service unit should be staffed with people who will "own" a customer's problem until it is resolved to everyone's satisfaction. The evaluation of any equipment vendor or service provider should include inquiries about the size of this operation and the qualifications of the customer support people, including the number of years they have been in the industry. The qualifications of customer support staff are especially important when an organization has remote locations that are staffed entirely with nontechnical professionals. It takes special interpersonal skills, as well as in-depth technical knowledge, to guide such people through diagnostic routines and restoration actions.

Most vendors that stress the quality of their customer support have quality control procedures in place that record measured response times for servicing calls, with follow-up surveys which assess the level of customer satisfaction. This level of post-sale support is a good indication of a vendor's goal to maintain a continuing relationship with its customers. Any reputable vendor or service provider will be happy to share this information with potential customers.

The evaluation team should also check into the availability of local support. The shrinking demand for some products has forced many companies to pull back local field service staff into larger regional centers. Many times, customers pay a premium price for local support, which is bundled into the product price. Customers who pay a premium price for a product because it includes local support with response times of one or two hours do not want to be at the mercy of vendors who later decide to centralize support operations and delay response times by three to four hours.

The evaluation team can help ensure adequate maintenance response times by specifying a system of response time credits and component downtime credits. With *response time credits,* the vendor discounts maintenance charges for every hour that maintenance personnel fail to arrive within the agreed time. With *component downtime credits,* the vendor discounts maintenance charges for every hour that equipment is out of service. These types of guarantees should be covered in a Service Level Agreement (SLA), which, as explained in Chap. 1, is essentially a contract that specifies the responsibilities of the vendor or service provider and the penalties for noncompliance that are due the customer.

2.5.8   Technical documentation

Until recently, technical documentation typically received scant attention from vendors. As products moved from the design stage to the production stage, rudimentary documentation was hastily thrown together in the hope of placating customers who were not really accustomed to expecting anything more. Although this is changing, many vendors still do not appreciate the customer's need for quality documentation. Too many vendors, even some of the largest companies with seemingly vast resources, still try to smooth things over by delivering rough production drawings, circuit schematics, hastily bound and barely readable photocopied documents, internal memoranda, and parts lists that are of little or no use to customers.

Today's complex computer and telecommunications technologies require that vendors view documentation as an integral part of the product, inseparable from the hardware or software. Without a comprehensive documentation package from the vendor, customers could be leaving themselves—and their networks—vulnerable to the whims of the vendor, especially if a customer's organization experiences frequent staff turnover levels. All this could prolong system or network downtime.

The evaluation team should also review the vendor's product documentation to validate the claims of salespeople. The documentation should provide comprehensive installation procedures, initialization and/or setup instructions, and a complete explanation of the product's features and administrative capabilities. In addition to appendixes that amplify aspects of the product's operation, manuals should include detailed indexes. A good documentation package also includes a troubleshooting guide that will help the network administrator determine the nature and scope of problems before calling the vendor's customer support people. In addition, the vendor should have a Web page where customers can go for product documentation and updates, find troubleshooting advice, and submit queries that are answered in a timely manner.

Products typically evolve over time as a result of enhancements and technological advancements. Unfortunately, many vendors do not match this product evolution with professional, up-to-date documentation. Therefore, the evaluation team should find out how the product documentation will be maintained and distributed to customers as the products change. To cope with the frequency of changes and the sheer volume of information, many vendors no longer issue paper documentation, but distribute this kind of information on CD-ROM or Web pages. If this is the case, find out how the information will be presented and organized and, if possible, review it to determine ease of navigation. The use of electronic media should speed up, not slow down, problem resolution.

2.5.9   Customer training

A reputable vendor will offer complete information about its products and technology; dedicated training staff and facilities are an excellent indication of the vendor's appreciation of long-term customer relationships—not just quick sales.

Unfortunately, formal classroom training at the vendor's facilities is not always sufficient for products that require custom configurations. The evaluation team should investigate the availability of on-site training. Additionally, the evaluation team should identify extra costs, if any, of additional training for new employees hired after the original training and product enhancements developed after the initial purchase. It is often worthwhile to find out if the vendor provides self-paced instruction, including print-based self-study, Web Based Training (WBT), and Computer Based Training (CBT). When purchasing products from large companies, there may be third-party training programs available; some may offer industry-recognized certifications.

While investigating training support, the evaluation team should ask about the experience and qualifications of the trainers. The vendor should not simply send technicians to provide training; generally, technicians lack experience with customer applications and rarely make good instructors unless they have been specifically trained for that responsibility.

The evaluation team should also request review copies of the training materials before making a major purchase. The materials should provide clear and comprehensive learning objectives supported by well-organized lesson structures and descriptions that can be used as reference material after the training sessions. If the vendor does not provide this kind of depth in its training package, the customer may be getting less out of the capital investment than anticipated at the time of purchase.

2.5.10   Primary line of business

For any significant investment in a product or service, the evaluation team should find out if it is a major or minor business activity of the vendor or service provider. If it is only a minor aspect of the total operation, or if it is viewed as a means to gain entry into more lucrative markets, customers may not get the attention they deserve when problems arise. In addition, if the product or service is not related to the core business, it might be a prime target for abandonment when a financial crisis strikes, and the vendor or service provider must sell it or close down the operation to improve its financial performance.

Stability in this area is important because as customer needs become more diverse and sophisticated, vendors and service providers must be able to respond appropriately. These responses can take various forms, including:

- Internal development
- Venture partnerships

- Acquisitions and strategic alliances

- Third-party applications development programs

Each method offers advantages and disadvantages for prospective customers. For example, doing business with a newly acquired or merged firm carries some risks. An acquisition brings with it internal upheaval, cultural shock, turf wars, and political maneuvering among management—all of which can result in a change of organizational priorities. During the period of turmoil, staff attention is focused inward, rather than outward to customers. This situation may last for several years, depending on the management skills of the firms involved. Simple economics drives most merger and acquisition activity and, quite often, the parent company does not fully understand the business or the technology of the acquired firm, or lacks the management skills to leverage its diverse assets into a cohesive whole. Because of these and other related reasons, 70 percent of all acquisitions in the telecommunications industry fail.

Although mergers and acquisitions may pose unforeseen problems for unwary buyers, understanding the reasons behind such arrangements can minimize risk. Often, the smaller firm is ripe for a takeover or buyout because it is in financial difficulty. In such cases, it is easy for the larger firm to exploit the smaller firm for whatever purpose is deemed necessary, such as immediate visibility in a new technology market, which may or may not be in the best interests of its customers. However, if two financially healthy industry leaders get together in a strategic partnership, the relationship has the potential of not only dominating the market but also of stemming the rising tide of competition and, ultimately, causing prices to rise or remain artificially high.

Aside from high prices, there are other risks to consider. First, it is not always certain that either party will base its long-term product development, marketing, and distribution strategies on a company that it does not control. Second, there will always be the temptation for each of the parties to strengthen its own products, rather than to devote resources to helping the company it is partnering with. Third, once each partner has achieved its hidden agenda, there is always the chance that the alliance will dissolve, to the detriment of customers who may have made long-term commitments based on the promises behind the alliance.

On the other hand, not every acquisition or strategic alliance heightens risk. From the buyer's perspective, the best guideline under these circumstances is to look behind the corporate scenes and try to project the impact of the new relationship between the two vendors on one's own current and future network, resources, and competitive position. During the probe of these types of vendor relationships, the investing company must seek to confirm that the vendor's corporate ventures and associations with other firms will be beneficial to its current and near-term needs.

2.5.11 Vendor references

Generally, vendors are very willing to supply a list of references. Two or three carefully selected references, however, may not be sufficient. The evaluation team should ask for five or six randomly selected customer sites for a more representative cross-section of opinions about the vendor. Even if the vendor has carefully prescreened the list of references, the evaluation team can still uncover some valuable information by asking those references for both their frank opinions and the names of other users. In addition, many user groups and cooperative purchasing organizations can help in the decision process.

When calling vendor references, the evaluation team should ask about the timeliness with which installation, integration, or customization was completed. A vendor's cooperativeness in solving elusive hardware or software operation problems and whether the performance of the product or service matched the buyer's expectations are also important considerations that only a current user can describe. Obviously, if any reference no longer uses the vendor's products, it is advisable to find out why.

Occasionally, an application of a product or service is unique and the manufacturer or service provider may try to persuade a potential customer to provide some up-front money to complete product development, customization, or redesign of an existing product to fit the application. This situation is rare among very large firms, but smaller companies with limited resources might ask for sizeable deposits to start work. An organization must be cautious about entering into any agreement that does not include a detailed description of the nature and scope of such activities, along with a precise list of performance milestones.

Before matters reach this stage, however, the evaluation team should ask the vendor about its previous experience with such arrangements and obtain appropriate references. If this is the first such transaction for the vendor, it is a good idea to check the local media and the national trade press for any adverse publicity about the firm, its officers, or its products. Be especially alert for published evaluations of the company's products or its marketing efforts written by industry consultants or financial experts. The history of the vendor's product development efforts, especially the cancellation rate of development projects, should provide some indication of the vendor's commitment to tailored applications.

2.5.12   Escrow protection

With high-end software products, the evaluation team should find out whether the program's source code can be put into escrow in the event that the vendor goes out of business or closes out the product. In such cases, the source code, which reveals details of the software's architecture, should be deliverable automatically from an escrow account or from a third party specializing in such services. These arrangements require the assistance of an attorney who is experienced in matters of software protection; because without an experienced attorney, these arrangements can be easily overturned in court. When making these arrangements, it is also important to include considerations for product updates that also update the source code in escrow.

For hardware products, the evaluation team should find out how much of the product's technology is proprietary and what provisions have been made to provide customers with continuing support if the vendor should go out of business or discontinue the product. The vendor should have an arrangement with a third-party maintenance firm for delivering on-site support and an arrangement with a third-party repair firm to fix, replace, and refurbish equipment.

2.6   Select the Vendor

During the selection process, the evaluation team must assign priorities to each evaluation criterion according to the specific needs of the communications department, in addition to the entire organization. For example, if the organization's communications department is plagued by high turnover, the evaluation team should look more critically at the vendor's technical documentation, training, and customer service. If the organization requires a high degree of network availability, the evaluation team should place more emphasis on the service provider's disaster recovery mechanisms, response time to trouble calls, and escalation procedures.

At the same time, the evaluation team should select the vendor on the basis of the rationale of its proposal, the quality of its references, its financial resources, its creditworthiness, and the strength of its strategic partnerships. To further minimize risk, it is a good idea to name a second vendor as an alternative candidate. Should the first vendor fail to perform, for whatever reason, the organization then has the option to go with the second vendor, which has already gone through the evaluation process. In naming a second vendor, the buyer may also be able to lock in that vendor's pricing, thus eliminating the need to perform the evaluation process from square one.

2.7   The Action Plan

Together, the operations manager, network planner, and the vendor's technical staff formulate an action plan that includes delivery installation, integration, training, acceptance testing, and cutover. At this time, a contingency plan should be developed. This plan should implement automatically if the selected vendor fails to perform according to the project's primary plan. The contingency plan might include the invocation of penalty clauses in accordance with the purchase agreement. In anticipation of noncritical problems in one or more areas of the project's development process, the action plan should be flexible enough to accommodate refinements in any of the following areas without having to fall back on the contingency plan:

- Personnel assignments
- Work scheduling
- Minor events beyond the control of vendor or customer

The action plan must also take into account the following factors:

- Budget constraints for the project
- Availability of personnel to work with the vendor
- Cumulative technical expertise available to support the plan
- Level of company commitment to achieving the business objective

Inadequate support in any of these areas can jeopardize the success of the entire project and trigger a destructive blame game.

The contingency plan should take into account any missed deadlines, poor vendor performance, or possible events beyond the vendor's control. The plan should also specify alternative courses of action, agreed-upon remedies, and penalties. In addition, the provisions of the contingency plan should be the result of customer-vendor negotiation and should be worded in the purchasing agreement in such a way as to preclude misinterpretation by the vendor.

2.8   Feedback

Upon completion of the project, vendor performance should be evaluated according to the results of network monitoring and performance tasks. The results are compared to the vendor's original network performance expectations. Additionally, the help desk or network administrator should implement an ongoing evaluation system through which users can register their levels of satisfaction or dissatisfaction while using the new systems or network. These ongoing user evaluations not only help to measure the success of the new installation but they also help to determine network durability or degradation as communications traffic increases.

The process of evaluation should continue throughout a project's life. The initial stage of network operation provides the best proving ground for testing the new technology implementation, applications, the vendor's competence, and the network staff's support potential. The evaluation should include feedback that requires all of the involved parties to compare various objectives with actual outcomes.

If the actual outcomes conflict with the stated objectives in the action plan or the functional specifications of the RFP, several corrective options are available, such as invoking the contingency plan or increasing organizational resources to compensate for the deficiency. If the outcome of implementing a new technology does not satisfy the project's major objectives, the vendor will have to reevaluate its original proposal and commit more resources. In extreme cases, the organization's original project objectives may have to be redefined.

The operations manager performs the final evaluation. If a systematic evaluation of the vendor and its products was performed, there should be no surprises. The experience and knowledge gained in implementing the product or service can be used as an aid in determining future networking needs and applying the vendor evaluation criteria to new projects.

2.9   Coping with Change

With today's emphasis on competing in the Internet-enabled global economy, organizations of all types and sizes must continue to look for ways to streamline business operations while cutting costs and improving productivity. In this relatively new competitive environment, the corporate network takes on added significance because it affects the company's ability to service its current customers and reach out to potential new customers. A company's effectiveness in these areas will help position it for growth in terms of profit and market share.

Communications managers must meet these challenges. They must establish a formal program for tracking new technologies, adapt a structured approach to evaluating the need for new products and services, implement pilot tests to objectively evaluate new technologies, and focus on high-impact areas that will reduce expenditures while strengthening business operations.

2.9.1   Tracking new technologies

Technology changes faster than any single person can keep up with, yet for competitive reasons it is becoming essential that corporations have in place the means to track these changes. Failure to keep track of technology and its potential applications and benefits can cost a business money in the near term and competitive position in the long term. There are a number of ways corporations can efficiently and economically track emerging computer and communications technologies.

For example, an "advanced technology" group can be formed with representatives from various departments. The group can meet informally on a scheduled basis, with individuals being assigned topics to track and summarize for the entire group via such mechanisms as lunch-time presentations, monthly meetings, and postings that can be shared through the group's Web site on the corporate intranet. Alternatively, a private newsgroup could be established on the corporate intranet, where various discussions can be carried out as new information becomes available.

When enough information is gathered, the technology group can proceed in a more formal way. The group can turn its attention to nailing down the implementation details of a new product or service, matching them to specific corporate applications, working up a preliminary business case analysis, and ascertaining the cost of missed opportunities if the product or service is not actually implemented. The technology group could also explore the feasibility of limited implementations via prototyping and pilot tests.

Another way of staying current on new technologies is to draw on local universities, research centers, or consultants for periodic updates on technology trends and implementation issues. The advantage of using such resources is that they provide an independent assessment of proposed projects, rather than relying on vendors and service providers exclusively, who obviously have a stake in the outcome.

Whatever method is used to track emerging technologies, the likelihood of success can be increased by adhering to the following guidelines:

- Make sure top management supports the effort financially and organizationally.

- Keep end users informed of emerging technologies and encourage them to provide insights into possible applications and usage issues. Also encourage them to offer advice on how to measure current productivity so that a baseline can be developed with which to compare productivity when a new technology or service is put into place.

- Continually look for fresh perspectives by rotating people into and out of the initial discussions of the technology group. This ensures that the group does not become focused on short-term goals or that practical business considerations are ignored.

- When appropriate, approach vendors and service providers for their input. Many times they can offer insights based on their experiences gleaned from a broad customer base. Such insights can be useful in shaping pilot tests and implementation plans.

- Before full-scale deployment of a new technology or service, initiate a pilot test that tracks costs and determines whether the technology will work as planned in the real-world environment. Use the pilot test to develop a business case that includes the benefits the technology is expected to yield as well as the costs of missed opportunities if the technology is not implemented within a reasonable time frame.

2.9.2   Pursuing new products and services

The timing for and level of commitment to a new technology or service should be thought out as carefully as any new business initiative. This means that planning should start with the systematic review of the following:

- Carrier technology deployment schedules, including trials

- Vendor equipment migration plans, prototype offerings, beta test sites, and product roll-out schedules

- Internal applications, their operating parameters, performance requirements, and existing terminal equipment, as well as geographical locations

- Competitor technology deployment plans, noting applications involved, projected economies and efficiencies, and vendors and/or carriers that are involved

- Progress of international standards bodies and regulatory trends

In attempting to determine how a particular product or service will affect the enterprise, some of the questions that must be asked include:

- Will deployment improve customer service and response to new customer needs?

- Will deployment facilitate corporate expansion and entry into new markets?

- Will deployment lower a barrier that allows the enterprise to compete more effectively?

- Will deployment permit the enterprise to offer new services or expedite the delivery of existing services?

- Will deployment enable the enterprise to generate new revenues, or at least produce significant cost savings?

- Will deployment prematurely make current systems and networks obsolete before they have been fully depreciated? Will deployment affect long-term service agreements? Or will the advantages of immediate deployment result in savings that override such concerns?

- Will implementation of a particular product or service by the competition adversely affect the enterprise? If so, how long an interval may safely elapse before the enterprise starts experiencing negative results? How will the negative results likely manifest themselves? What are the possible ways the competitor will exploit its newfound advantage, and within what time frame?

2.9.3   Selling soft-dollar benefits

Senior executives are generally resistant to long-term network projects that cannot be justified on the grounds of cost savings. Therefore, the onus is on communications managers or the Chief Information Officer (CIO) to hone their ability to demonstrate the soft-dollar benefits of major network expenditures, such as improved corporate image and productivity, or increased customer satisfaction and loyalty.

With some projects, it is difficult to do a classic return on investment, particularly in an uncertain economy. Although some projects that lack well-defined cost savings can often fall victim to senior management myopia, more often than not this rejection stems from the failure of managers to adequately make the case for the soft-dollar advantages of the project. In the absence of a solid cost-benefits analysis, it is imperative that managers present a compelling strategic case.

For example, a plan for expanding the corporate network to include a number of overseas locations would be incomplete without emphasizing the benefits of adhering to international standards like the Open Systems Interconnection (OSI) reference model.* Senior management can often be sold on the long-term benefits of embracing an open system strategy, especially if it can be shown that money is already being spent on individual private nets and that the ability to interconnect those nets would save money in the long run.

With open systems expected to play an increasingly important role in the computing environment for many years to come, positioning the corporation

---

*Although OSI has languished for many years in the United States, it has asumed the status of law in much of the European community. There, corporate communications users have adopted an aggressive attitute in getting vendors to climb aboard the OSI bandwagon.

accordingly constitutes a strategic business move. After all, the ability to move applications from platform to platform results in richer, easier-to-write applications. And when a hardware platform changes, the current investment in software will not be jeopardized.

Typically, top management will not focus exclusively on a project's Return on Investment (ROI) if a good job is done selling the strategic benefits of a technology. When pitching network projects, for example, three metrics may be used: customer support, productivity improvements, and direct and indirect savings. The *customer support* metric might be used to show how an upgraded network would give customers faster access to corporate services. The *productivity* metric can be used to show how an integrated electronic mail system, for example, would improve staff productivity. And, of course, *cost savings* are always factored into the equation in some way. But it is the responsibility of the CIO or communications managers to educate executives on other strategic considerations. The likelihood of winning projects without guaranteeing immediate savings depends to a large extent on the sales ability of communications management staff and their relationships with top management. Do not ignore the competitive business environment—projects tend to fare better if it can be shown that competitors are using the same technology with apparent success or that a critical advantage can be gained over the competition with the project's successful implementation.

2.9.4 Pilot testing

Because emerging technologies lack a performance record under real-world operating conditions, it is advisable to conduct a pilot test before committing corporate resources to full-scale deployment.

Not only can a pilot test demonstrate the viability of new technologies, it can demonstrate their value to internal users, department heads, and top management. In obtaining and acting on their input, they will have a stake in the successful outcome of the test and be more forthcoming with support when the time comes to make a decision about fully committing to the technology.

Communications managers will be involved in the details of designing and implementing the pilot test with the vendor or service provider. Many variables contribute to the success of the pilot test. For example, do not assume the pilot test is free. Although vendors will typically supply their equipment and installation services at little or no cost, additional costs may be incurred for the communication lines. There may be other costs associated with floor space, test equipment, and staff hours needed to ensure a thorough pilot test. Staff priorities may have to be adjusted for the duration of the test.

Conduct the pilot test under real-world conditions. Design a series of tests that exercises systems at levels that closely approximate the network under a heavy load. Use real-world applications. Push the product to its limits not only to find out what the limits are but also to see how the system responds under such stress.

Keep detailed documentation of test results. They may be valuable later. For example, these results may be incorporated into the RFP or purchase agreement as the minimum performance level expected from all products purchased from the selected vendor. The results can also be used as the basis for acceptance testing.

Do not hesitate to call in the vendor when problems arise. The vendor should be able to explain the cause of the problem and take immediate corrective action. Using the vendor's product at no charge should not obligate the company in any way or inhibit the company from being as demanding as a paying customer. Often, the vendor comes away from the experience with more knowledge about its product's performance. So the vendor can benefit from this arrangement as well.

Use the pilot test as an opportunity to create a detailed implementation plan. This means taking note of things like cabling, interfaces, test equipment, and floor space requirements. There may also be the need for additional power and equipment racks, as well as a reconfiguration of facilities leased from carriers. It is also advisable to take note of what training will be required for technicians, network managers, system operators, and end users and to determine who will implement the training.

Understand at the outset that some vendors use pilot tests as the means to open doors to sales. They expect buyers to be passive and uncomplaining, especially when free use of the product is involved. It is imperative that you and your staff resist such mind games. Do not balk at rejecting the system and return it if the promised performance advantages do not meet expectations.

Finally, do not use the pilot test in lieu of standard purchasing procedures. Treat the pilot test as only one stage in the normal technology acquisitions process. This process typically includes needs analysis and development of an RFP, as well as proposal evaluation, vendor selection criteria, and acceptance testing that meets the agreed-upon performance specifications.

The pilot test should not be used merely to validate preconceived notions; it should be used as a tool for objective product evaluation and to eliminate risk in the acquisitions process. These are important considerations, which, if ignored, can squander limited corporate resources on nonproductive projects and activities.

2.9.5   High-impact activities

In the highly competitive global business environment, the enterprise must pick and choose its opportunities wisely. This means focusing on areas that will produce the best total return on investment. The contribution of the communications manager in this area would be to keep the corporate systems and networks optimized for maximum efficiency and cost savings. Accordingly, there are a number of areas worth looking into.

- *Service rates.* Keep up to date on changing service rates, volume pricing plans, and new carrier services. This can shorten decision cycles and enable the company to act quickly when carriers announce their service deployment schedules and migration timetables.

- *Review disaster recovery plans.* The locations of spare bandwidth capacity, redundant systems, and spare components should be known in order to minimize network downtime. Disaster recovery plans should be tested periodically to identify potential problems so they can be corrected before a disaster actually occurs.

- *Long-term agreements.* Look for opportunities to renegotiate long-term agreements for more favorable terms from carriers and equipment vendors. This might be possible by switching from one type of service to another, such as moving certain applications from IP to frame relay, or getting rid of digital subscriber lines in favor of T1 lines. If the vendor or service provider has an opportunity to upscale a product or service, it will be more inclined to restructure an existing agreement. Structure contracts in the future so you will not be locked out of better deals that come along in the interim.

- *Evaluate technology acquisitions in terms of leasing, rather than purchasing.* This can free up needed capital and reap possible tax advantages for the company. With technologies becoming obsolete faster, leasing might be preferable to buying in some cases.

- *Use technology acquisitions to leverage additional staff training from vendors.* To be successful, this must be discussed with the vendor during contract negotiations.

- *Outsourcing as an option to in-house management.* Investigate the advantages of outsourcing the maintenance and management of the network to a qualified firm that specializes in such services. This can free up resources to allow the company to focus on its core business.

- *Look for opportunities to upgrade or enhance the installed equipment base, rather than opting for wholesale replacement.* This may be possible with a software upgrade to a higher version, or the replacement of a lower-speed processor to a high-speed processor, the addition of more memory, or the insertion of an entire plug-in module to give the system new capabilities.

2.10 Conclusion

When it comes to major technology procurements, steps must be taken to minimize risk. Such steps ensure network efficiency, save money, and protect existing investments in personnel and technology. In adapting to the changing environment brought about by continued deregulation, increased competition, and the rapid pace of equipment obsolescence, communications managers can have an impact on the quality of corporate decision making as never before. In the process, communications professionals become indispensable members of the management team.

To deal effectively with these challenges, network planners must resolve how they will translate new corporate demands into networking solutions that take advantage of emerging technologies and, in the process, obtain the best price-performance ratio possible. Success may hinge on the extent to which communications managers are willing to round out their current skills with business knowledge and training on the emerging technologies that will support the enterprise well into the future. There must also be a willingness to help staff expand their perspective beyond particular technical disciplines so that they can become better equipped to assist in making sound judgments about technologies, applications, products, and vendors. The potential results are too compelling to ignore: The enterprise will be ideally positioned for growth and expansion in the ever-expanding, hyper-competitive global economy.

Chapter 3

Writing the Request for Proposal

3.1 Introduction

The Request for Proposal (RFP) is the most important document in the large-scale procurement process. In fact, the RFP is actually a combination of things. Since it details the requirements for new systems or networks, it is a plan as well as a solicitation. Since it describes the buyer's business processes as well as the operational requirements and performance criteria that the vendors' proposed solutions must meet, it is a contract. Since it provides a structure within which the formal responses are presented by competing vendors, it is also an evaluation tool. Among the goals of the RFP is to provide the buyer with the following:

- A consistent set of vendor responses, which are narrow in scope for easy comparison.

- A formal statement of requirements from which contracts can be written, and against which vendor performance can be benchmarked.

- A mechanism within which vendors, fostered by the implied competition of a general solicitation for bids, describe their terms and conditions.

A properly written RFP and carefully managed evaluation process can accomplish all of these goals. However, the RFP should not be so rigid that it locks out vendor-recommended alternative solutions that may be more efficient and economical. At the same time, an overly broad RFP that invites vendors to propose whatever is the optimum solution every step of the way is essentially no RFP at all. Not only does such an open-ended approach produce responses that are difficult to compare but it also leaves too much room for generalities and obfuscation on the part of vendors and carriers. An overly interpretive RFP can also open the door to challenges from the losing bidders, which can tie up corporate resources and delay installation. To avoid this situation, it is best to know as much as possible about the business objectives and feasible solutions, and describe them clearly and concisely so all vendor responses will be sufficiently focused to compare and evaluate fairly.

The proposal evaluations and vendor selections of federal, state, and local governments are particularly susceptible to challenge. This is usually because of the amount of money at stake, especially for local vendors and carriers. Companies that have lost bids have used numerous tactics, including the following:

- Challenge the fairness of the bid process
- Lobby officials and staff members for a reevaluation of the decision
- Publicly question the validity of the contract award
- Attack the reputation of the bid winner and its management team
- Disparage individual proposal-evaluation staff
- Submit new unsolicited bids, even after a thorough and completely audited bid process

It is typical in the telecommunications industry for vendors and carriers to come back and demand another chance. Depending on the dollar amount of the contract, losing bidders may even try to overturn the decision in court. The possibility of these outcomes can be greatly diminished with an RFP that is written clearly and concisely, especially with regard to the bid evaluation process.

3.2   Needs Assessment

Developing an RFP can be quite an involved process that encompasses all aspects of business operations. Depending on the nature and scope of the project, it can be a team effort that may include systems analysts, application developers, networking specialists, business analysts, technical writers, and a team leader. The team leader is usually a senior manager who has comprehensive knowledge of the organization and its existing systems and networks. One of the responsibilities of the team leader is to oversee the collection and integration of information from various sources and review the final draft of the RFP before it is reproduced and issued to vendors for bid. The team leader may also be the contact person who can answer questions from vendors and to whom vendors submit their proposals.

The RFP may have to take into account the different needs of other departments and work groups, as well as the enterprise as a whole. Consequently, a thorough needs assessment should be performed to obtain a complete understanding of all requirements. This can be done effectively by inviting representatives of other departments and work groups into the RFP development and vendor evaluation processes. This not only facilitates information gathering and decision making thereby helping to ensure that all needs are addressed but it also fosters a sense of ownership for the eventual solution and minimizes complaints later on.

There is no standard format for developing an RFP; in fact, each tends to take on a life of its own as input is gathered from a variety of sources, and as organizational needs change from one purchase to another. The RFP should be as thorough as possible, with everything expected of the vendor fully spelled out. If multiple solutions are possible, or if it is not known exactly what solutions are most appropriate for a given situation, vendors should be provided with as much raw data (traffic studies, baseline performance statistics, topology diagrams, etc.) as possible, so they can make appropriate recommendations. The RFP should also describe the anticipated growth of the company, the present network and attached systems, and communications facilities. It should also provide the reasons for their replacement (e.g., obsolescence, cost, technology, new applications) or expansion.

It is also advisable to keep the long-range plan firmly in mind. An RFP should go beyond soliciting solutions that meet only immediate needs. A solution that solves a company's problems effectively today can become a nightmare tomorrow. This is especially true of companies that are considering future expansion to international locations. It can be difficult to adjust systems to different standards, regulations, installation schedules, and maintenance response times. Differences in currency, language, culture, and work ethics add to these difficulties. Even if plans for international expansion are still embryonic, vendors with a global presence may be able to provide advice that can smooth the way when the time comes to extend network reach.

3.3   General Information

The RFP is usually broken down into sections, with Section I describing the intent of the RFP and establishing the ground rules for vendor participation.

3.3.1   Purpose

Section I of the RFP usually begins with a statement of purpose, such as:

- The XYZ Company requests proposals to provide products and services necessary to install, implement, and maintain a corporatewide data communications network, which will replace the existing network. This request will provide interested vendors with appropriate information with which to prepare and submit proposals for consideration by XYZ.

- It is the intent of XYZ to select the best proposal based on an evaluation of responses and other considerations described in this RFP.

- XYZ reserves the right to reject any and all proposals received as the result of this RFP prior to the execution of a contract.

The second item concerning "other considerations" is very important because it puts vendors on notice that factors apart from the proposal will play a part in the selection process. Such considerations might include findings that reflect negatively on the vendor, such as a poor performance report from references. They might include a bad credit rating from financial institutions, quarterly financial statements that portend bankruptcy, pending lawsuits that might jeopardize future performance, patent litigation to sort out intellectual property issues, or an investigation by the Securities and Exchange Commission (SEC) on accounting practices that could shake investor confidence. This statement will also discourage vendors from challenging the purchase decision based solely on point-by-point comparisons with the proposals of their competitors. Such challenges can drag on for months, causing missed project deadlines and consuming valuable organizational resources. In fact, the higher the value of the contract, the more likely that the decision will be challenged by other vendors, making "other considerations" a key element to include as part of the vendor selection criteria.

3.3.2 Scope

The RFP should provide vendors with an idea of how the RFP is organized. This can be done simply by listing the various sections of the RFP, including appendixes, with their respective page numbers in a table of contents format.

3.3.3 Schedule of events

A schedule of events should be included in the first section of the RFP. This consists of the decision-making milestones of the purchase along with their dates, starting with the RFP itself:

- RFP issued
- Vendor meeting
- Proposal deadline

- Vendor presentations

- Evaluation procedure and criteria

- Contract award

- Letter of intent

Each of these milestones deserves a subhead of its own in Section I of the RFP to relate additional information to the vendors.

RFP contact.   This section should include the name, address, and telephone number of the person to whom all questions, correspondence, and proposals should be directed. To avoid confusion, only one person should be listed as the point of contact. This will usually be the team leader.

Vendor meeting.   This section should include the date, time, and location of the vendor meeting, if one is required. The purpose of this meeting is to provide interested vendors with an opportunity to ask questions arising from their review of the RFP. Sometimes the RFP may contain ambiguous terms or statements—or not enough information. This meeting may also benefit the buyer, since vendors may bring up points not previously considered, and which merit inclusion into the RFP. It is important to issue addenda so that the entire RFP, as well as the vendor's proposal, can be included as part of the final contract. In this paragraph, the meeting's ground rules are explained, including the procedure for amending the RFP and distributing the changes to attendees, should the need arise.

Proposal deadline.   This section should state the due date as well as the number of copies of the proposal that must be submitted. Provide a point of contact, including telephone number, for deliveries. Specify the procedure vendors must use for obtaining extensions, if any. State whether multiple proposals will be accepted from the same vendor, and with what stipulations. Describe the procedure vendors should use to update their proposals. Define what constitutes a complete proposal. For example, if proposals will not be considered without pricing information, this should be stated. On the other hand, when pricing information will be accepted in a separate package if it arrives by the proposal due date should be mentioned as well. Take this opportunity to warn vendors that price increases will not be allowed after the proposal is submitted.

Vendor presentations.   If desired, specify the time frame that will be allotted to vendors for presentations in support of their proposals. If time and resources permit, consider going to the vendors' corporate offices for the presentations, which will provide an opportunity to evaluate the management team and various operations. And if a vendor's proposal has included subcontractors, specify that they have a representative in attendance at the presentation. (After narrowing down the list of vendors to a single candidate, the contract should be signed on home turf so that maximum leverage may be applied to the final negotiations.)

Evaluation procedure and criteria.   The process that will be used to evaluate vendors should be spelled out completely. Will one vendor be chosen from among the submitted proposals, or will the two or three strongest proposals be selected for follow-up presentations by the vendors? Will proposals be evaluated in-house, or will outside consultants play a role? What evaluation criteria will be used? Evaluation criteria may include any or all of the following items, which should be listed in the RFP:

- Prior experience of the vendor in successfully completing undertakings similar in nature and scope
- Understanding of the technical requirements and the magnitude of the work to be accomplished, as evidenced by the proposal and subsequent meeting(s) with the vendor
- Arranging the demonstration of a similar system currently in use at a customer site
- The completeness of the proposal with regard to information requested in the RFP, its level of detail, and conformance to specifications
- The vendor's ability to respond with a viable alternative solution, if it cannot precisely address the specifications described in the RFP
- The vendor's willingness to accommodate changes during installation
- The experience, qualifications, and professionalism of the vendor's staff assigned to the project
- The vendor's ability to comply with the terms, conditions, and other provisions of the RFP
- The vendor's work plan for delivery, installation, and acceptance testing
- The total cost of fulfilling the requirements described in the proposal
- The willingness of the vendor to provide information relating to organizational structure and departmental or work group capabilities that affect customer service
- Willingness of the vendor to demonstrate at any time during the evaluation process that all aspects of the RFP's requirements can be met or exceeded

A statement should be included in this section that lets vendors know that the evaluation procedure is intended to screen out nonresponsive and incomplete proposals so that the evaluation committee can concentrate its efforts on proposals that are responsive and complete. Also state that the reasons for rejecting vendor proposals will be carefully documented, but not released. A written record of such decisions holds more weight in court than vague recollections, should a vendor wish to challenge the procurement decision.

Most vendors dislike clauses known as "reservations," but they are absolutely required in the evaluation procedure and criteria section of the RFP as a protective measure. Inform the vendors that the final selection may not necessarily adhere to the stated evaluation criteria. Among the many possible contingencies, consider reserving the right to:

- Reject any and all proposals received in response to the RFP
- Enter into a contract with a vendor other than the one whose proposal offered the lowest price
- Adjust any vendor's proposed pricing based on a determination that selecting a particular vendor will involve incurring additional or hidden costs
- Waive or change any formalities, irregularities, or inconsistencies in proposal format or delivery
- Consider a late modification of a proposal, if the proposal itself was submitted on time, and if the modification makes the terms of the proposal more favorable
- Negotiate any aspect of a proposal with any vendor, and negotiate with more than one vendor at a time
- Accept any counterproposal or addendum submitted, whether or not there are contract negotiations with other vendors already in progress
- Extend the time for submission of all proposals
- Select the next most responsive vendor, if negotiations with the vendor of choice fail to result in an agreement within a specified time frame

The importance of the last clause cannot be overstated. In case negotiations with the first-choice vendor fail, it is a good idea to designate second and third choices so that the RFP does not have to be issued again. In naming alternative vendors, their interest will be kept alive and their proposals—including pricing and scheduling—will stay in force. At the same time, the first-choice vendor will have more incentive to negotiate unresolved issues in a fair and timely manner if it knows that other vendors are standing by.

It is quite common for organizations to have two committees perform separate evaluations of the same proposal: financial and technical. In this case, state that separate preliminary reviews will be conducted of the vendor's financial and technical packages to ensure that all mandatory requirements have been met.

For each package, describe the review process that will be used to qualify the vendors. If proposals will be assigned points for various evaluation factors, for example, list those factors and the maximum number of points that can be scored on each. If possible, provide an example of the scoring.

The purpose of describing the evaluation procedure is to convey an image of fairness to vendors so that the most qualified among them will be encouraged to respond with a proposal. This, in turn, will help ensure that organizational needs will be met with the best solutions available at the most reasonable cost.

Contract award.   In the description of the contract award, state how all parties will be notified of the final selection. For example, will the announcement be public, or will vendors be notified privately via letter? In any case, state that there is no obligation to disclose to any vendor the results of the evaluation process, or the reason why particular vendors were or were not successful. Also state how second- and third-place vendors will be notified of their bid status. This way, all parties will know that an alternative plan will be invoked if contract negotiations go awry with the first-choice vendor.

Include a statement about how proprietary information from vendors should be handled. Obviously, proposals contain sensitive information that vendors do not want falling into the hands of competitors. State that any specifications, drawings, documentation, pricing, and any other information pertaining to the business of the vendor that is submitted as a result of the RFP will be treated as confidential. Vendors will usually have a copyright on their proposals along with a caveat that obligates the recipient not to disclose its contents to a third party without prior written authorization.

Proposals submitted to government agencies, however, are usually considered to be in the public domain. If the purchaser is a government agency, vendors should be reminded that if they submit sensitive information as part of their proposals and they want it protected from public disclosure, they should appropriately mark the relevant pages at the top and bottom.

Letter of intent.   To clarify the ground rules for awarding the contract, state the subsequent steps in the contract award process. If the purchase entails a long delivery cycle, it might be worthwhile to state that a letter of intent will be issued at some point before or during the contract negotiations. As its name implies, this document is used only to establish the intention to purchase products from a specific vendor—it carries with it no obligation to follow through with an actual purchase and, as such, may be canceled at any time.

Consider inserting into the RFP a clause that vendors will not be permitted to issue press releases or issue public statements of any kind about the project under bid without prior approval. This will guard against the possibility of any vendor engaging in a premature or self-serving publicity campaign regarding its involvement in the project. Competitive information may be inadvertently revealed if the vendor happens to say too much in its zeal to leverage the relationship to make new sales.

3.4   Contract Terms and Conditions

Section II of the RFP is usually focused on contract terms and conditions. With guidance from internal financial and legal officers, this section of the RFP should outline the terms and conditions of the contract which the successful vendor will be expected to enter into. The reason for including this information in the RFP is to notify the successful vendor of the kind of contract it is expected to sign. The goal is to minimize time spent in over-the-table haggling, which may jeopardize the time frame for project completion.

This section of the RFP should include a clear statement to vendors that for their proposal to qualify for further consideration they must include a specific response to these terms and conditions. They may either indicate complete and unconditional acceptance, or include in the proposal specific language to replace those provisions to which exception is taken. Any differences can be taken up during contract negotiations, if the vendor makes it to that step.

With contract terms and conditions spelled out in the RFP, ideally, subsequent contract negotiations will end up being a mere formality.

3.4.1   Liabilities

The contract describes the terms and conditions of the procurement. It should include a set of liability clauses that specifies who is responsible for what, and who pays whom for the failure of the other, and under what circumstances. Such information is fairly standard in contracts, but including it in the RFP notifies the vendor of what to expect. The following sample clauses may help clarify such matters:

- *Proposal acceptance.* The vendor agrees that the submitted proposal, including separately submitted product pricing and proposal addenda, constitutes a part of the final contract.

- *Financial terms.* Neither party will assign this agreement or its rights or obligations, or subcontract its performance to any person, firm, or corporation without the prior written consent of the other party. This consent will not be unreasonably withheld.

- *Proprietary rights.* The vendor warrants that the products furnished under this contract do not infringe upon or violate any patent, copyright, trade secret, or the proprietary rights of any third party. In the event of any claim by any party against XYZ Company, the vendor will defend the claim in XYZ Company's name, but at the vendor's own expense, and will indemnify XYZ against any loss, cost, expense, or liability arising out of the claim, whether or not the claim is successful. If any product furnished is likely to or does become the subject of a claim of patent or copyright infringement then, without negating or diminishing the vendor's obligation to satisfy the final award, the vendor may, at its discretion, obtain for XYZ Company the right to continue using the alleged infringing product or modify the product so that it becomes noninfringing. In the absence of these options, or if the use of the product by XYZ Company is prevented by permanent injunction, the vendor agrees to take back the product and furnish a replacement that most closely matches the performance of the infringing product at no cost increase to XYZ Company.

- *Consent to jurisdiction.* The contract will be deemed to be executed in the City of XYZ, State of XYZ, regardless of the location of the vendor, and will be governed by and be interpreted in accordance with the laws of the State of XYZ. With respect to any action between XYZ Company and the vendor in XYZ State Court, the vendor waives any right it might have to move the case to Federal Court or move for a change of venue to an XYZ State Court outside the City of XYZ. With respect to any action between XYZ Company and the vendor in Federal Court located in the City of XYZ, the vendor waives any right it might have to move for a change of venue to a United States Court outside the City of XYZ.

- *Hold harmless.* The vendor will hold harmless and defend XYZ Company and its agents and assigns from all claims, suits, or actions brought for or on account of any damage, injury, or death, loss, expense, civil rights or discrimination claims, inconvenience, or delay which may result from the performance of this contract.

- *Injury or damage.* The vendor will be liable for injury to persons employed by XYZ Company, persons designated by XYZ Company for training, or any other person(s) designated by XYZ Company for any purpose who are not the agents or employees of the vendor. The vendor will be liable for damage to the property of XYZ Company or any of its users prior to or subsequent to the delivery, installation, acceptance, and use of the equipment either at the vendor's site or at XYZ Company or its users' places of business. Liability results when such injury or damage is caused from the fault or negligence of the vendor.

- The vendor will not be liable for injury to persons or damage to property arising out of or caused by an equipment modification or an attachment, or for damage to modifications or attachments that may result from the normal operation and maintenance of the vendor's equipment by XYZ Company or its agents.

- Nothing in this contract will limit the vendor's direct liability, if any, to third parties and employees of XYZ Company for any remedy which may exist under law in the event a defect in the manufacture of the vendor's equipment causes injury to such persons or damage to such property.

- *Force Majuere.* Neither party will be held responsible for delays or failures in performance caused by acts of God, riots, acts of war or terrorism, earthquakes, or other natural disasters.

- *Litigation expenses.* The parties agree that in the event of litigation to enforce this contract, or its terms, provisions, and covenants; to terminate this contract; to collect damages for breach or default; or to enforce any warranty or representation described in this agreement, the prevailing party will be entitled to all costs and expenses, including reasonable attorney fees, associated with such litigation.

In addition to the liability clauses listed above, government agencies typically include the following two protective measures in their RFPs:

- *Nonappropriation.* If the Department of XYZ does not receive adequate funding during the next succeeding fiscal period and is unable to continue lease, rental, or purchase payments covered by this contract, the contract will automatically terminate, without penalty, at the end of the current fiscal period for which funds have been allocated. Such termination will not constitute default under any provision of this contract, but the Department of XYZ will be obligated to pay all charges incurred through the end of such fiscal period, up to and including the formal notice given to the vendor. The Department of XYZ will give the vendor written notice of such nonavailability of funds within thirty (30) days after it receives notice of such nonavailability.

- *Performance bond.* Upon execution of a contract for lease, rental, or purchase, the Department of XYZ will require the vendor to furnish and maintain, until the product or system has been accepted, a performance bond in an amount equivalent to ten (10) percent of the purchase price.

3.4.2   Mechanical clauses

There are a number of mechanical clauses that should be included in the RFP's contract terms and conditions section. The following mechanical clauses clarify the relationship of the contract's format and individual clauses to the whole of the contract so that neither party can use it out of context to support a claim against the other.

- *Headings not controlling.* The headings and table of contents used in this contract are for reference purposes only and will not be deemed a part of this contract.

- *Severability.* If any term or condition of this contract or its application to any person(s) or circumstances is held invalid, this invalidity will not affect other terms, conditions, or applications, which will remain in effect without the invalid term, condition, or application. Only to this extent may the terms and conditions of this contract be declared severable.

- *Waiver.* Waiver of any breach of any term or condition of this contract will not be deemed a waiver of any prior or subsequent breach. No term or condition of this contract will be held to be waived, modified, or deleted except as mutually agreed in writing.

- *Authority.* Each party has full power and authority to enter into and perform this contract. The representative(s) signing this contract on behalf of each party has been properly authorized and empowered to enter into this contract. Each party further acknowledges that it has read this agreement, understands it, and agrees to be bound by it.

- *Compliance.* The vendor agrees, during the performance of work under this contract, to comply with all provisions of the laws and Constitution of the State of XYZ, and that any provision of this contract that conflicts with them is void. The parties also agree that any action or suit involving the terms and conditions of this contract must be brought in the courts of the State of XYZ or the United States District Court for the State of XYZ.

3.4.3   Technical specifications

The contract terms and conditions section of the RFP should include provisions that address system and/or network specifications. The following clauses are provided as examples only. As such, they are weighted to favor the buyer. If vendors would like to negotiate terms more favorable to them, they may do so by proposing alternative language in their proposals. Be careful not to word these clauses too restrictively; the object is not to keep reputable vendors from issuing a proposal, but only to ensure adequate protection under a variety of adverse circumstances that may arise in the future.

- *Vendor warranty.* The vendor warrants that the proposed equipment and any software, when installed, will be in good working order and will conform to the specifications described in the RFP, the vendor's official published specifications, the contract specifications, and the vendor's proposal. In lieu of this warranty of fitness, the procurement can be canceled within ninety (90) days of installation. XYZ Company shall pay a reasonable lease charge for the time the products were used.

- *System configuration.* The equipment and any software components to be supplied under this RFP and contract, for purposes of delivery and performance, will be grouped together in one or more configurations, as defined in this RFP (cite the appropriate section and/or addendum of the RFP). Any such configurations will be deemed incomplete and undelivered if any component in that configuration has not been delivered, or if delivered, is not operable.

- *System performance.* Vendor will certify in writing the date the equipment will be installed and ready for use. The performance period will commence on the first day following acceptance testing, at which time the vendor will relinquish operational control and responsibility.
  - No payments will be made to the vendor until all systems have been in satisfactory operation for at least thirty (30) days after installation.
  - If successful completion of the performance period is not attained within ninety (90) days of the installation date, the option to terminate the contract without penalty or to continue the performance tests will be exercised. The option to terminate the contract will remain in effect until such time as a successful completion of the performance period is attained. The vendor will be liable for all outbound preparation and shipping costs for contracted items returned under this clause.

- *Access to diagnostic information.* During the life of the equipment the vendor will provide access to diagnostic procedures and the information derived from them.

- *Equipment interfacing.* The vendor acknowledges the right of the buyer to connect equipment manufactured or supplied by others, which is compatible with the vendor's system. Such equipment includes, but is not limited to, peripheral equipment, terminal devices, computers, and communications equipment. The vendor will supply interface specifications and supervise the connection of equipment, if called upon.

- *Field service.* Vendor will warrant that, in any case where equipment is installed or modified on the premises of XYZ Company, which is contracted for under this agreement, vendor will make such installation at charges in effect at the time of the request by XYZ Company.

3.4.4   Project support

Additional protection may be built into the RFP with project support clauses inserted into the contract terms and conditions section of the RFP:

- *Staff quality.* The vendor will exercise due care to choose and manage its personnel so that only suitably qualified, disciplined, and responsible representatives will be operating at XYZ Company and user locations.

- *Training.* The vendor will provide appropriate training to XYZ Company on the operation, maintenance, and management of the vendor's products, as described in the proposal and its attachments and appendixes.

- *Documentation.* The vendor will provide XYZ Company with three (3) sets of documentation required to effectively operate the system as described in the vendor's proposal. This documentation may be in paper form or electronic media such as CD-ROM. Vendor represents that these documents are the only ones necessary for the operation of the system. The vendor will include any other materials and program descriptions it considers helpful to XYZ Company. XYZ Company may reproduce all documentation and materials provided by the vendor, provided that such reproduction is made solely for the internal use of XYZ Company and that no charge is made to anyone for such reproductions.

- *Emergency response.* XYZ Company will be provided with access to an answering service or operator for the purpose of requesting vendor assistance during times of emergency. A vendor representative must have a response time of one hour or less during nonwork hours, weekends, and holidays until full acceptance of the installed system by XYZ Company.

3.4.5   Costs and charges

The contract terms and conditions section of the RFP should include a set of provisions that clarifies costs and charges so that all parties understand their financial obligations under the agreement:

- *Term of agreement.* The terms, provisions, representations, and warranties contained in the contract will survive the delivery of the equipment; payment of any lease, rental, or purchase price; and transfer of title.

- *Payment procedure.* All payments otherwise due under this contract will not be payable until thirty (30) days after receipt of invoice from the vendor.

- *Transfer of title.* Before any payment is made, the vendor will provide a statement warranting that all equipment and materials, including those of its subcontractors, are free of mechanical liens or encumbrances.

- *Failure to perform.* In the event that the vendor fails to perform any substantial obligation under this agreement and the failure has not been satisfactorily remedied within thirty (30) days after written notice is provided to the vendor, XYZ Company may withhold all amounts due and payable to the vendor, without penalty, until such failure to perform is remedied or finally adjudicated.

- *Default.* XYZ Company may, with thirty (30) days prior written notice of default to the vendor, terminate the whole or any part of this contract in any one of the following circumstances:
  - If the vendor fails to perform the services within the time specified in the contract or within the time specified under subsequent extensions.
  - If the vendor fails to perform any of the other provisions of this contract, or fails to make satisfactory progress in the performance of this contract in accordance with its terms. Or the vendor does not remedy such failure within the thirty (30) days—or as mutually agreed in writing—after receipt of notice from XYZ Company specifying the failure.
  - If this contract is terminated pursuant to the provisions above, XYZ Company's sole obligation will be to: (a) continue any installment contracted payments due for products previously delivered and accepted, (b) purchase for title, as agreed, any products previously delivered and accepted for payment with principal outstanding, or (c) XYZ Company may, in addition, procure from the vendor goods specifically acquired by the vendor for the performance of such part of this contract as has been terminated.

- *Taxes.* XYZ Company will not be responsible for any taxes coming due as a result of this agreement, whether federal, state, or local. The contractor will anticipate such taxes and include them in the proposal. In the case of leased products, the lessor will be responsible for any personal property taxes and will adjust prices accordingly.

- *New equipment warranty.* The vendor warrants that all equipment and software, when installed, will be new and in good working order and will perform to the vendor's official published specifications and the contract specifications. Further, the vendor will make all necessary adjustments, repairs, and replacements without charge to maintain the equipment in this condition for a period of not less than one year after the standard of performance has been met and the product accepted by XYZ Company.

- *Prices and terms.* All prices, terms, warranties, and benefits granted by the vendor in this contract are comparable to or better than the equivalent terms offered by the vendor to any other public or private entity purchasing equipment of the same quality and quantity. If the vendor offers, during the term of this contract, greater benefits or more favorable terms to any other public or private entity, those benefits and terms will be made available to XYZ Company upon their effective date. Failure to do so will constitute a breach under this contract.

3.4.6   Reliability and warranty

The contract terms and conditions section of the RFP should include appropriate clauses concerning the product's reliability and warranty. The following items are offered as essential requirements:

- *Equipment reliability.* In all situations involving performance or non-performance of equipment or software furnished under this contract, the remedy available to XYZ Company will consist of either:
  - The adjustment or repair of the system or replacement of parts by the vendor or, at the vendor's option, replacement of the system or correction of programming errors.
  - If the vendor is unable to install the system or replacement system or otherwise restore it to good working order or make the software operate as required under this contract, XYZ Company will be entitled to recover actual damages as set forth in this contract. For any other claim concerning performance or nonperformance by the vendor pursuant to, or in any other way related to provisions of this contract, XYZ Company will be entitled to recover actual damages to the limits set forth in this section.

- *Acceptance testing.* In addition to operational performance testing by the vendor, XYZ Company reserves the right to perform additional testing, prior to acceptance, to ensure compliance with the requirements and specifications of this contract. All attachments may be inspected for compliance with the Federal Communications Commission (FCC) Part 68 technical and safety requirements. All wiring may be inspected for compliance with state and local electrical codes.

- *Building modifications.* The vendor will perform all work required to make the product or its several parts come together properly to fit the space allocated for its placement and to make provisions for the equipment to be received for work by other vendors. This work will include all cutting of floors, walls, and ceilings which may be necessary to install equipment and cabling, as well as the restoration of such surfaces to an approved condition.

- *Building repairs.* The vendor will take all the necessary precautions to protect the building areas adjacent to its work. The vendor will be responsible and liable for any building repairs required as a result of its work and caused by the negligence of its employees. Repairs of any kind that may be required will be made and charged to the vendor or, at XYZ Company's option, deducted from its final payment.

- *Cleanup.* As ordered by XYZ Company, and immediately upon completion of the work, the vendor will, at its own expense, clean up and remove all refuse and unused materials from the work site. Upon failure to do so within forty-eight (48) hours after written notification, the work may be done by others, the cost of which will be charged to the vendor or, at XYZ Company's option, deducted from its final payment.

- *Additional work.* Without invalidating this contract, XYZ Company may order extra work or make changes by altering, adding to, or deducting from the work and causing the contract sum to be adjusted accordingly. All such work will be executed under the conditions of the original contract by a change order. Under no circumstances will extra work or any change be made in the contract unless through a written change order to the vendor stating that XYZ Company has authorized the extra work or change. Any change order involving a ten (10) percent deviation from the total contract amount may require a new agreement. In the event the extra work or change involves materials and labor for which unit prices have not been established, pricing will be determined by mutual agreement.

- *Use of premises by vendor.* The vendor will confine all apparatus, storage of materials, and operation of this work to the limits specified by law, ordinances, or permits, and shall not unreasonably encumber the premises with materials. The vendor will comply with the laws, ordinances, permits, or instructions of the State regarding signs, advertisements, fires, smoking, and vehicular parking. The vendor will not load or permit any part of the structure to be loaded with weight that will endanger its safety.

- *Use of premises by owner.* XYZ Company and its users reserve the right to enter upon the premises, to use same, and to have work done by other vendors, or to use parts of the work of this vendor before the final completion of the work, it being understood that such use by XYZ Company or its users in no way relieves the vendor from full responsibility for the entire work until final completion of the contract. XYZ Company reserves the right to enter into other contracts in connection with this work.

- *Recovery from disaster.* In the event the system or any component of the system is rendered permanently inoperative as a result of a natural occurrence or disaster, the vendor will deliver a replacement within thirty (30) days from the date of XYZ Company's request. In such event, vendor agrees to waive any delivery schedule priorities and to make the replacement system available from the manufacturing facility currently producing such equipment, or from inventory. The price for replacement equipment will be the price payable under this contract. If the inoperability is due to the negligence or fault of the vendor or its subcontractors, replacement equipment will be delivered at no cost to XYZ Company.

3.4.7   Maintenance

The following provisions concerning maintenance should be incorporated into the RFP's contract terms and provisions:

- *Vendor responsibilities.* Vendor will provide maintenance, including associated travel, labor, and parts, either under a maintenance contract, or on a time-and-materials basis at the prices listed in the proposal. This provision does not apply to the repair of damage resulting from accident, transportation between XYZ Company sites, neglect, misuse, or causes other than ordinary use.

- *Maintenance personnel.* Hardware maintenance will be performed by qualified maintenance personnel totally familiar with all of the equipment installed by the vendor at XYZ Company and its user sites. Maintenance personnel will be given access to the equipment when necessary for the purposes of performing maintenance services under the terms of this agreement.

- *Term of maintenance services.* Maintenance services will be provided at the prices quoted in the vendor's cost proposal, and may be renewed annually for up to two (2) years at the original prices. XYZ Company may elect to terminate maintenance services at any time upon thirty (30) days prior written notice to the vendor.

- *Maintenance documentation.* For purchased equipment, the vendor will, upon request, provide to XYZ Company such current diagrams, schematics, manuals, and other documents necessary for the maintenance of the system by XYZ Company or its subcontractor(s). There will be no additional charge for these maintenance documents, except for reasonable administrative costs involved for reproduction.

- *Right to purchase spares.* Vendor guarantees the availability of long-term spare parts for all equipment acquired under this contract for a minimum period of six (6) years following the date vendor provides written notification to XYZ Company that the equipment is out of production, but in no case less than ten (10) years from the date of this contract. Such sales will be made at the prices then in effect, except that prices will not be increased per year by more than the National Consumer Price Index, calculated at a simple rate of increase for each year between the date of acceptance of the equipment purchased under this contract and any order for spare parts.

- *Replacement parts.* The vendor warrants that only new standard parts or parts equal in performance to new parts will be used in effecting repairs.

- *Request for maintenance.* XYZ Company will be provided with continuous access to an answering service or operator for the purpose of notifying the vendor of the need for immediate maintenance services. The vendor will have a response time of two (2) hours or less, and have the ability to restore service within three (3) hours of notification.

- *Remote diagnostics.* It is desirable, but not necessary, that remote diagnostics be performed from the vendor's site. If this type of monitoring is not available, the vendor must describe to what degree its local point of contact will provide diagnostic support to XYZ Company.

- *Maintenance and repair log.* The vendor will keep a maintenance and repair log for recording each incident of equipment malfunction, as well as the date, time, and duration of all maintenance and repair work performed on the equipment. Each unit of equipment worked on will be identified by type, model, and serial number. A description of the malfunction will be provided, as well as the remedial action taken to restore the unit of equipment to proper operation. This report will be signed by the vendor's representative and XYZ Company's representative, with one copy sent to and retained by XYZ Company. All response time and downtime credits to XYZ Company will be based on this jointly signed document. Failure to provide XYZ Company with a properly completed and signed document will render any claims by the vendor invalid.

- *Response time credits.* If the vendor's maintenance personnel fail to arrive at the site requiring such services within the designated response time, the vendor will grant a credit to XYZ Company. The amount of creditable hours will be accumulated for the month and adjusted to the nearest hour. Each hour in excess of the specified response time will be computed at the rate of 1/30th of the monthly full-service maintenance agreement charge.

- *Component downtime credits.* If the faulty component cannot perform due to a malfunction through no fault or negligence of XYZ Company for a period of eight (8) consecutive hours or more than sixteen (16) nonconsecutive hours during a twenty-four- (24)-hour period, XYZ Company will be granted a credit toward monthly maintenance (or rental, if leased). For each hour of downtime, credit will accrue in the amount of five (5) percent of the total monthly charges for all components due under the proposed contract. Downtime will commence from the time of initial notification of the vendor that maintenance is required. The credit for component downtime will be computed to the nearest half or whole hour.

- *Equipment replacement.* If any unit of equipment fails to perform, and the total number of inoperative hours exceeds twenty-seven (27) hours over a period of three (3) consecutive calendar months, the vendor will, at the option of XYZ Company, provide:
  - A backup unit of equipment at no additional cost.
  - On-site technical support at no additional cost.
  - Replacement of the malfunctioning unit of equipment with a functionally equivalent unit of equipment in good operating condition at no additional cost to XYZ Company. In this case, accrued response time credits and downtime credits will be transferred to this unit of equipment.

- *Preventive maintenance.* Preventive maintenance, if required, will be scheduled by XYZ Company and the vendor at a mutually agreeable time. In the event XYZ Company decides that equipment performance warrants an increase or decrease in frequency or hours, vendor will so increase or decrease such maintenance, provided such request is reasonable.

3.4.8 Product delivery

In the contract terms and conditions section of the RFP, the vendor's responsibilities related to product delivery are discussed:

- *Installation responsibility.* The vendor will be responsible for unpacking, uncrating, and installing the equipment, including making arrangements for all necessary cabling, connection with power, utility, and communications services, and in all respects making the equipment ready for operational use. Upon completion, vendor will notify XYZ Company that the equipment is ready for use.

- *Risk of loss prior to installation.* During the period that the equipment is in transit and until the equipment is installed and ready for use on XYZ Company and its users' premises and acceptance tests are successfully completed, the vendor and its insurers, if any, relieve XYZ Company of all risks of loss or damage to the equipment. After the equipment is installed, ready for use, and has been accepted, all risk of loss or damage will be borne by XYZ Company, except where the damage is attributable to vendor's negligence or to defects XYZ Company could not reasonably have discovered.

- *Liquidated damages.* If the vendor does not install all the equipment specified in the agreement, including the special features and accessories included on the same order with the equipment, the vendor will pay to XYZ Company liquidated damages for each item of equipment, whether or not installed. For each day's delay, beginning with the installation date but not for more than 180 days, the vendor will pay to XYZ Company 1/30th of the basic monthly rental and/or maintenance charges or 1/1000th of the purchase price of all equipment listed in the order, whichever is greater.
  - If XYZ Company operates any units of equipment during the time that liquidated damages become applicable, liquidated damages will not accrue against the equipment in use.
  - If the delay is more than forty-five (45) days, XYZ Company may terminate the agreement with the vendor and enter into an agreement with another vendor. In this event, the terminated vendor will be liable for liquidated damages until the substitute vendor's equipment is installed, or one hundred eighty (180) days from the original installation date, whichever occurs first.

3.4.9 Rights and options

Various rights and options clauses should be included in the contract terms and conditions section of the RFP to take into account various contingencies that may arise in the future and that may have adverse consequences:

- *Equipment upgrades.* XYZ Company may at any time, upon demand, require the vendor to substitute upgraded equipment for any component purchased under the provisions of this contract, including spares and replacement components, with XYZ Company paying the base price of the original item as well as the difference between the price of the equipment installed under this contract and the price in effect for the upgraded equipment.

- *Equipment changes and attachments.* XYZ Company will have the right to make changes and attachments to the equipment and any software, provided that such changes or attachments do not lessen the performance or value of the equipment or prohibit the proper maintenance from being performed.

- *Software ownership.* The vendor agrees that any software and accompanying literature developed specifically to implement this agreement will be the sole property of XYZ Company. The vendor further agrees that all such material constitutes a trade secret, and must use its best efforts in the selection and assignment of personnel to work on the development of such software to prevent unauthorized dissemination or disclosure of information related to its development.

- *Rights to new ideas.* The parties acknowledge that the performance of this contract may result in the development of new proprietary concepts, methods, techniques, processes, adaptations, and ideas. XYZ Company will have unhindered right to use such processes and ideas for its own internal purposes. The vendor will have unrestricted right to use such processes and ideas for commercial purposes, including the right to obtain patents and/or copyrights.

3.4.10   Relocation

Sometimes it may become necessary to move equipment or whole systems from the original site to another site. The contract terms and conditions section of the RFP should include provisions for relocating purchased equipment without voiding vendor warranties or the terms of the agreement:

- In the event the equipment being maintained under the terms and conditions of this contract is moved to another location belonging to XYZ Company, the terms and conditions of this contract will continue to apply.

- Except in emergencies, XYZ Company will provide the vendor with at least thirty (30) days notice to move the equipment.

- Maintenance charges will be suspended on the date that the dismantling of the equipment in preparation for shipment is completed. Maintenance charges will be reinstated on the day the vendor completes equipment reassembly. XYZ Company will be charged for disassembly and reassembly at the vendor's then prevailing price for such services.

- Shipment to the new location will be by such means as normally used by the vendor, by padded van or air freight, or any means specifically requested by XYZ Company. XYZ Company may ship the equipment via its own transportation or by commercial carrier or, at its option, provide the vendor with authorization to ship by commercial carrier on a prepaid basis, in which case XYZ Company will be invoiced for transportation, rigging, drayage, and insurance costs.

3.5   Proposal Specifications

Part III of the RFP describes the format that vendors must follow in their responses. The purpose of mandating a particular format is to facilitate the review process. The objective is to minimize time spent in figuring out if the vendor has supplied the information requested in the RFP. The reviewer should be able to extract the relevant information quickly and make appropriate comparisons among all the vendor proposals. Vendors should be encouraged to include additional information that they may consider appropriate or helpful in evaluating their proposals. The following language is offered as the introduction to the proposal specifications part of the RFP:

- All documents submitted in response to XYZ Company's Request for Proposal must be clearly identified by title, volume, and/or document number with the pages numbered consecutively. Accessibility to the proper information is more likely to result in an accurate and complete assessment of the proposal during the evaluation process.

- All documents that comprise vendor proposals must be delivered to XYZ Company in sealed packages. Each package must be clearly labeled as follows:
  - Proposal for XYZ Company Data Communications Network
  - Vendor's name
  - Document name (Contractual proposal, Technical proposal, Financial proposal, Reference materials, etc.)
  - Date of submission

The vendor's meeting is the appropriate forum for requesting clarification of any elements of the RFP that remain unclear. Written requests for clarification submitted prior to the vendor's meeting will be appreciated. XYZ Company will treat such requests as confidential. Any delay in the schedule for receiving or evaluating proposals necessitated by a vendor's inquiry will be applied to all vendors.

3.5.1 Letter of transmittal

To ensure that there are no problems matching proposals with the proper vendors, specify the content of the cover letter that should accompany the proposal and each separate package that is considered a part of the proposal:

- Name and address of the vendor (or prime contractor)

- Name, title, and telephone number of the person authorized to commit the vendor to the contract

- Name, title, and telephone number of the person to be contacted regarding the content of the vendor's proposal, if different from above

- Name and address of any proposed subcontractors

- Time validity of the offer stated in the proposal (specify that the offer be valid for 90 or 180 days, or anything in between—whichever seems appropriate to the situation)

- A cover letter signed by an officer of the company

3.5.2   Proposal format and content

To facilitate the evaluation and comparison of proposals, plan a format and request that vendors adhere to it, possibly as a condition for acceptance. In writing an RFP for the first time, often help is needed with specifying the proposal format. Here are a few guidelines:

- *Executive summary.* This section will provide a summary of the proposal and should include a brief statement of the significant features of the proposal in its component parts. This section should include a statement of the vendor's capabilities and experience with projects of this nature and scope. Vendors may also include any additional information of a general nature that would aid the evaluation team in understanding the thrust of the proposal.

- *Contract terms and conditions.* Vendors must respond to the contract terms and conditions in Part II of this RFP, either by indicating verbatim acceptance or by including specific language for those provisions to which exception is taken. Failure to address the terms and conditions may result in the rejection of the proposal.

- *Project work plan.* The vendor must include with the proposal a detailed description of the work to be done to fulfill the requirements of this RFP, including the target dates for the completion of each task. The work plan must include, but should not necessarily be limited to the following items:
  - A statement of the vendor's understanding of the objective and scope of the requested work.
  - A detailed description of each major task associated with the project, including the total number of person days and elapsed time. This description will identify any anticipated decision points that will involve participation by XYZ Company.
  - A project organization chart that shows the involvement of XYZ Company and the vendor's staff.
  - A list of the vendor's staff available for the project and a statement of their qualifications, including relevant education, technical level, and similar past experience. Upon selection, the vendor's staff cannot be changed without notifying XYZ Company in writing.

- *Forms.* All forms included with this RFP must be completed and returned as part of the vendor's proposal. The forms are designed to aid the evaluation process and to demonstrate compliance with this RFP. Failure to complete all of the specified forms may result in rejection of the proposal.

- *Vendor qualifications.* The qualifications of vendors are addressed throughout this RFP. Responses to the contractual, technical, and financial parts of the RFP will be used to determine the vendor's capabilities to provide a data communications network to XYZ Company. In addition, the vendor must submit background statements to include:

- Financial statements for the last three (3) fiscal years.
- Three (3) references from financial institutions or creditors.
- A description of any litigation in which the vendor is currently involved.
- A list of subcontractors that the vendor intends to use for completing the project described in this RFP. This list will include the name of each subcontractor, as well as their addresses, phone numbers, and points of contact. Upon selection, the vendor may not change subcontractors without notifying XYZ Company in writing.
- Three (3) references from customers for whom the vendor has performed similar work. This list will include the name of each customer, as well as their addresses, phone numbers, and points of contact.

- *Technical proposal.* Vendors must respond to each of the system requirements. Failure to address each requirement may cause the proposal to be rejected from further evaluation. Since all evaluation team members are not technicians nor necessarily have technical backgrounds, it will be in the best interest of the vendor to keep descriptions in nontechnical language wherever possible.

- *Alternative proposals.* Alternative proposals may be submitted. Only those sections that are different from the original proposal need be submitted, provided all differences are clearly defined. Separate, sealed cost proposals clearly marked "Alternative Proposal" must also be submitted with each alternative proposal.

- *Reference materials.* Reference materials are those that are referred to in the proposal such as sales literature, technical manuals, or training manuals. Whatever materials are referenced in the proposal must be packaged separately and submitted as part of the proposal.

- *Financial proposals.* Vendors may submit separate pricing proposals that address one or more of the following options:
  - Lease price.
  - Straight purchase price.
  - Straight monthly long-term lease prices for five-year (60-month) and ten-year (120-month) periods.
  - In addition to the above, government agencies may want to consider the option of a tax-exempt installment purchase for five-year (60-month) and ten-year (120-month) periods.

Additional options or different time periods may be proposed at the vendor's discretion, provided the vendor responds to at least one of the four options described above.

3.6   Technical Requirements

Part IV of the RFP describes the general requirements of the system or network, including management, that the vendor must address throughout the proposal. The phrasing of the introductory paragraph may be simple. It should indicate that the company intends to purchase, for example, a digital PBX to replace existing analog equipment under lease from the ABC Leasing Company (see Appendix *X*) and that the new system will need to support intrabuilding wireless communication, voice messaging, local area networking, and include T1 interfaces that can be migrated to ISDN. Furthermore, the new system will provide greater configuration flexibility, scalability, and substantial cost savings over the system currently in use, and enable the organization to meet emerging applications needs and future growth.

In addition to a summary diagram of the current network, provide a separate diagram for each location showing the type and quantity of equipment in use. If growth is anticipated, provide separate diagrams showing the type and quantity of equipment that may be required. If growth is anticipated but the equipment requirements are unknown, supply enough data about current and projected traffic (data as well as voice), staffing levels, terminal stations, and type of transmissions (synchronous or asynchronous), as well as their breakdown by percentage so that the vendor has enough information to propose a solution.

3.6.1   General considerations

There may be some broad areas of concern that vendors may be expected to address in their proposals, such as:

- *Scope.* The successful vendor will be required to furnish, install, interface to telephone company equipment, test, maintain, and provide training for the system and individual hardware components.

- *Transmission speeds.* The vendor must be able to provide data transmission speeds of 8 to 512 kbps, which may differ from site to site, as well as 1.536 Mbps at some locations.

- *Transparency.* The data transmission capabilities of the PBX must be transparent to the user, with no alterations to data terminal equipment or networking software required to implement transmissions.

- *Accounting.* The vendor will convey appropriate technical information to assist a designated third-party software vendor to develop a customized call accounting system that will provide the call detail and summary reports listed in Appendix *Y*.

- *Data terminal equipment.* The data terminal equipment currently supported by the XYZ Company network is listed in Appendix *Z*.

- *Cabling/wiring.* The vendor must provide and install all cabling and station wiring for the new PBX system, including LAN connections. This requirement applies to all XYZ Company locations and all user workstations, from telephone interface to user terminals. Where feasible, the vendor may use existing user site cabling and/or wiring. In any case, the vendor must provide detailed diagrams of all cabling and wiring, and provide appropriate labeling at termination points.

- *Cabinets.* Vendors must supply equipment cabinets when installing the modem pool to support remote access. It is preferred that the modems use a card type that is usable for both standalone and rack-mount configurations, and that the card cage utilize a universal type backplane to accommodate any mix of modem types. Power distribution equipment must be included in the cabinets.

- *Security.* XYZ Company plans to implement a security system at each phone and/or terminal station under the following hardware and software constraints, which the proposed system must support:
  - For voice conversations, a personal password must be used to dial an external number. Long-distance call restrictions will be based on area code and/or dial "1."
  - For data transmissions through the PBX, a personal password must be used to access a host port on a dedicated or contention basis. Local and remote dial-in access to specific files, storage, and local area network resources will be granted and enforced in accordance with established company policy.
  - A terminal must be logged in to the message control program before any transaction will be passed to the host.
  - All transactions have the log-on code appended as a prefix before the data is passed from the message control program to the host.
  - The host processor uses the log-on code in building the key to access any and all on-line data files.
  - The message control program associates a hardware address with a specific user terminal identification. The security file relates terminal identification with valid log-on codes. If a user at any location attempts to log on using a code that is not valid for that location, the user will be denied access to data files after three unsuccessful log-on attempts.

- *Operating and maintenance procedures.* The vendor will be responsible for developing, for XYZ Company's approval, operating and maintenance procedures. These procedures will be prepared prior to cutover of the first site, and revised as necessary during system implementation.

- *Equipment labeling.* The vendor will label all racks, cabinets, equipment, and boards, as well as connectors and cross cabling. Such labeling must be in plain view.

The vendor will also include the following information about the proposed solution:

- Equipment requirements and costs by location
- Equipment configuration drawings by location
- Network configuration drawing(s) showing all equipment locations
- Space and power requirements for each location
- Environmental requirements for each location
- Description of technical documentation available for all equipment
- Complete description of circuit requirements for each location

3.6.2 Equipment specifications

The technical requirements section of the RFP provides an opportunity to request detailed information about vendor products. What follows is an example of a format that might be used to solicit vendor information about a centralized PBX management/control system.

- *Centralized management/control system.* XYZ Company requires a centralized network management control system to be installed at its present headquarters location. This system must be of sufficient capacity to support and control the entire XYZ Company network. Minimum components required include:
  - Central processing unit
  - Hard disk storage with tape backup
  - Network management terminal with graphics capability
  - On-line printer
  - Local and remote monitoring devices

The centralized network management/control system must support administrative activities at the operational and planning levels for:

- *Failure management* activities, which include problem determination and system restoration. Required operational level functions will include positive alarms (audible, visual, and printed) for network component failure or degradation. Alarm information will include nature of failure or degradation and location.

- *Performance management* activities, which will include usage and network availability parameters. All such data must be available for historical inquiry to aid in future planning and problem solving.

- *Configuration (change) management* activities, which combine data from failure management and performance management to support the long-range planning of the network's topology. In addition, configuration management features facilitate the scheduling and implementation of station moves, adds, and changes.

- *Inventory management* activities, which require that an inventory database be established that includes both active and spare parts. Inventory data combined with failure and performance data must provide the network manager with information to support critical network management decisions.

Vendors must provide the following information on their network management control system:

- *System characteristics,* which may include the following:
  - Number of processors
  - Processor type
  - Main memory capacity
  - Operating system
  - Storage capacity
  - Storage capacity expansion capability
  - Console display type

- *Technical control features and functions,* which may include the following:
  - Alarm conditions
  - Number of alarm levels
  - Alarm types
  - Monitoring
  - Remote monitoring devices
  - Type of monitoring signal

- *Network management features and functions,* which may include the following:
  - Data Base Management System (DBMS) supported
  - DBMS acquisition (bundled or separate)
  - Types of data recorded
  - Reports available (standard and customized)

- *Transmission specifications,* which may include the following:
  - Maximum transmit and receive rates
  - Transmission techniques supported
  - Interfaces supported
  - Maximum number of lines supported
  - Expansion increments

For each type of peripheral equipment, request appropriate information in similar detail from the vendors. If future migration to a new digital technology is a possibility, ask vendors to address the compatibility of their systems with the new technology or ask for information on how to upgrade their products with the new technology.

3.6.3   Appendixes

Include various appendixes that amplify key elements of the RFP. For example, include a summary diagram of the current PBX system or network, with supplementary diagrams showing specific details that vendors are expected to address, such as planned network locations.

Use a separate appendix to list equipment currently in use on the network. Include the quantity of equipment by model and manufacturer. Any forms or questionnaires also merit appendixes of their own, as do summary tables of voice and/or data traffic and any network modeling studies, including their assumptions. A glossary of acronyms used in the RFP may even be warranted.

In general, any information that will assist vendors in understanding and assessing organizational needs and developing a proposal that addresses those needs is appropriate for an appendix. These appendixes should be cited in the main body of the RFP.

3.7   RFP Alternatives

The purpose of the RFP is bid solicitation. Other types of documents are used when different forms of assistance are required. For example, the Request for Quotation (RFQ) may be used when planning the purchase of off-the-shelf commodity products such as PCs, printers, modems, and applications software. The RFQ is used when the most cost-effective solution is the overriding concern.

The Request for Information (RFI) may be used when seeking the latest information on a particular technology, service, or application, but there is no immediate need. Its purpose is merely to get briefed on the latest developments for long-term planning purposes, how vendors plan to exploit these developments in the future, or to get vendors' perspectives on the feasibility of using or integrating a particular technology in a current network. Compared to the RFP and RFQ, the RFI is a very informal document. Vendor responses tend to be brief, and they may or may not include information on product pricing and availability. Nevertheless, the RFI responses can be useful for planning purposes and for deciding which vendors might qualify for a future RFP.

3.8   Conclusion

Everyone has their own way of developing an RFP, but there are some things that can be done to aid in the development of a good RFP. Allow enough time for planning—not just the kind of planning required for daily operations, but strategic planning. Stay updated on the organization's business plan so that future requirements can be anticipated. Keep informed of new products and technologies. Read up on the latest merger and acquisition activity of current and potential vendors, as well as bankruptcy activity in the industry, and try to predict what impact this will have on the availability of system upgrades or network expansion. Learn to develop contingency plans that can be invoked virtually instantaneously if things do not go according to plan.

Take advantage of the creativity and problem-solving abilities of vendors and carriers. Instead of imposing specific solutions on vendors or carriers, use the RFP as the means to request possible solutions. Vendors and carriers continually complain that they are not given the opportunity to provide this kind of input. Invariably, they will come up with solutions not previously considered, if only because they have the benefit of being able to draw upon more expertise over a number of specialized fields. The proposed solution may even save money over the life of the contract. In sum, build into the RFP enough flexibility to allow input from vendors and carriers.

At the same time, have a technically knowledgeable team develop the RFP. Allow the team manager to attend and have authority over all meetings with vendors and carriers. And let that person play a pivotal role in evaluating vendor proposals. If there is no technical guru on staff, consider a qualified consultant for the role of team leader, especially if the RFP requires an in-depth knowledge of available products, technologies, and architectures. The competitive marketplace has become saturated with a seemingly endless variety of products, and the pace of innovation boggles the mind. A large capital purchase can be quite risky without some outside assistance. Qualified consultants can bring objectivity to the areas of needs assessment, vendor evaluation, and product selection. Beyond that, consultants provide extra staff and lend credibility to internal decision making.

With a multisite network, issue a single RFP for the entire network rather than issue separate RFPs for each site. Even though the time frame for completing the project may be as long as two or three years, volume discounts on equipment can be locked in by lumping all of the system or network requirements together under a single RFP.

Finally, package the RFP in a professional manner by organizing it simply and logically, thereby making it easy for vendors to follow and helping them to develop a timely response that addresses all of the important issues. Make every effort to eliminate typographical errors and ambiguous language. When reprinting the RFP for distribution, ensure that the pages are not spotty or streaked, reducing legibility. Make sure that pages are properly numbered, diagrams are properly labeled, and acronyms are spelled out. Do not let the binding interfere with the text. Display appropriate contact information in a prominent place at the beginning of the RFP. Attention to detail will facilitate the vendor selection process and help ensure a successful outcome for the organization.

Financial Planning

4.1 Introduction

As a network manager you not only play a key role in recommending, evaluating, and selecting the systems and networks that best satisfy corporate requirements but you are increasingly called upon to provide inputs relating to the most effective procurement and finance methods. This is understandable: With the cost of systems and networks consuming an ever-larger slice of the total operating budget, it is inevitable that the managers of these assets be given some sort of role in financial planning.

You bring to this decision-making process your knowledge of how technology can improve organizational performance and how hidden costs can affect the corporate budget. Without such inputs, it is difficult to prepare an accurate budget to support daily business operations and long-term organizational requirements. This accounts for the poor accuracy of cost projections on major system procurements and network upgrades: According to various industry estimates, about 60 percent of such projects come in significantly over budget.

Corporate accountants and budget analysts require the advice of network managers on several issues to produce an accurate financial analysis or budget proposal. For example, knowing the useful life of various systems and network components, as well as their propensities to fail and their frequencies of failure, may influence the decision about whether these systems and components should be purchased or leased. If there is a chance that the entire system or network might have to be scrapped in the near future and replaced with an emerging technology, this could affect the decision to lease or purchase, or to buy used equipment initially, with the goal of buying new equipment later. It could also affect the decision to expense or depreciate the purchase.

When it comes to enterprise applications, data warehouses, and decision support systems, the price of the software does not reflect the massive hidden cost of customization that is often required to integrate customer-facing applications with other enterprise applications such as supply chain, logistics, and resource planning. It also does not include the cost of staff required to provide ongoing support, which may include populating databases and running management reports to meet changing business requirements, implementing security features to extend or control access, and backing up the data to prevent loss. Every enterprise application requires a support infrastructure, the cost of which greatly exceeds the purchase price of the software. Your role as network manager is to uncover these hidden costs so that an appropriate finance method can be selected—outsource versus purchase—and a more accurate Information Technology (IT) budget can be developed.

At the same time, because network managers are involved with the financial planning process, you must learn to walk a fine line: Spending too much can raise the ire of top executives, who are increasingly concerned with cost containment. Spending too little can anger unit managers, who need the facilities and services to achieve the full productivity benefits of their applications, thereby meeting their performance targets. Spending on the latest technologies and services is a gamble because they may not work out; at the same time, neglecting new services, platforms, and tools may very well result in higher operating costs and longer application development cycles.

Further complicating financial planning is that a slow-growth economy and climate of downsizing virtually negates the possibility of a budget increase for many companies. For many network managers, gone are the days of straight-line budgeting, whereby a simple increase of 5 or 10 percent is tacked on to the previous year's budget. Not surprisingly, the expectations of top management and end users do not change: They still want greater access, faster response time, better network availability, higher reliability, and effective security—all without spending more money than in the previous year.

4.2 Asset versus Expense Management

The level of emphasis a company gives to the financial management of networks and information systems will vary among firms by the type of business and type of markets served, as well as by the types of services it provides and the demands of its customer base. A financial services or insurance firm, for example, generally will put a high value on its networks and information systems because they support the workflow processes that determine its ability to conduct business in a timely and efficient manner. For such companies, even a brief outage can have a severe and immediate impact on the bottom line.

In such environments, a critical issue that can complicate your job as a network manager is the tendency of the company to vacillate between the two approaches to managing technology investments: asset management and expense management.

Under *asset management,* the corporate network and information systems are viewed in strategic terms and there is a direct correlation between maintaining, enhancing, and expanding the network and growing the customer base. In using the network to continually improve customer service, reach new customers, and offer new services, competitive advantages accrue over those companies that have not yet awakened to such possibilities.

Under *expense management,* the network and everything attached to it are viewed as a necessary cost of doing business—something that can be cut back, or at least left as is, to save money. Such companies try to cope with competitive pressures by looking for ways to do more with less. Many times this results in a reduction of the communications budget, buying used equipment, forgoing hardware and software upgrades, and skimping on maintenance. While such actions may produce immediate savings, they can jeopardize the company's competitive position in the long run by limiting the organization's ability to exploit emerging opportunities.

Many well-intentioned companies that began with asset management are attempting to stay competitive in the global economy by downsizing operations. They now view their networks more in terms of expense management. Some companies continue to stick with asset management, believing that their networks can provide competitive advantages that can help them ride out market fluctuations at a time when everyone else seems to be cutting back. Either approach is continually subject to change at a moment's notice, but one thing is clear: Companies cannot afford to make costly purchasing and financing mistakes that constrain their ability to improve business operations in a timely manner.

4.3 The Planning Process

One way you as a network manager can improve the budget planning process is to consider the infrastructures you manage and perform a needs assessment before making any purchase or budget decisions. Needs assessment means reevaluating facilities and services, vendors and suppliers, tools and staff, and other key components of the infrastructure to determine how well they meet business needs. One important cost-saving measure might be to look for opportunities to replace only the hardware and software that have reached the end or are near the end of their useful lives, while upgrading other components to expand functionality and increase performance.

4.3.1   Needs assessment

Needs assessment entails quantifying the potential losses of nonimplementation, perhaps in competitive terms, as well as the benefits of implementation, relating them to the organization's overall business objectives. This information can be used to develop an appropriate network topology, expansion plan, or upgrade policy.

Many companies building or adding on to large information systems or networks can improve the cost-performance ratio of their operations by devoting more time to needs assessment. This can be made somewhat easier with such tools as project-management, security-analysis, and network-analysis software. These tools can generate summary and detail reports, often in graphic form, that can be used to validate network expansion, upgrades, configuration changes, and security requirements.

There are now some very advanced tools that can help managers ask "what-if" questions as they plan large-scale projects. These tools automate many of the tedious tasks involved with planning and design, including the estimation of network costs based on traffic patterns, usage, carrier-supplied rate information, and equipment depreciation. There are even tools that help managers sort through the advantages of buying or leasing equipment. The reports can be customized to show only the most relevant data.

And when it comes to planning for the future, such reports can pave the way for needed purchases. For example, based on current growth, the reports might reveal the need for additional bridges or routers at particular nodes within the next six months. The cost of this equipment can be included in the current budget cycle, rather than put off until the next one.

With network security becoming of paramount concern among companies of all types and sizes, there are software programs and services that can monitor ports to assess their level of vulnerability to hackers. This information can be used to determine an appropriate solution, which may include firewall appliances at telecommuter locations, software and memory upgrades to routers at branch offices, and dedicated firewalls for the enterprise network. Once the level of protection is determined, the choice may be to outsource security management to a qualified third party to spread the costs, or perform security management in-house for faster response to new security threats.

For large projects, it is advisable to use a software package that provides regular status reports on schedules, budgets, and personnel assignments, which a project team can review at weekly meetings. The program is updated daily. People who have been assigned to a project submit a time sheet each day with the number of hours they worked on a specific task. The information is keyed in to the program, which then generates new schedules and a variety of reports for project managers and senior executives.

4.3.2 Calculating true costs

A needs assessment from a network topology perspective can go a long way toward identifying the true costs of the network or upgrade. The local area network (LAN) topology should identify work group and departmental subnetworks and the connections between them; communities of interest and their local subnets as well as the number of attached hubs and workstations; and the locations of mission-critical databases, and whether they are located on minicomputers, mainframes, or file servers.

The wide area network (WAN) topology should identify all switching and feeder nodes; the speeds and locations of the lines and/or carrier services; the LAN interconnection equipment; any special transmission requirements that will improve performance and safeguard important data, such as compression and encryption; and network management systems, both primary and subordinate, located at domestic and international locations.

Other cost items for LANs and WANs may include provisions for disaster recovery, spare bandwidth to handle congestion, system modules that support specific protocols for various interoperability requirements and, of course, security mechanisms to prevent damage to or theft of corporate information.

The final part of this analysis includes backward tracing to determine how well the proposed network or upgrade meets specific requirements. From each workstation, subnet, work group, department, and the backbone network itself, the data traffic must be traced to the appropriate network elements to see if they meet the performance objectives. There is even software available that is designed to measure the user experience with network applications, including those on Web and e-commerce servers on the Internet. This type of software becomes more important as companies increasingly run enterprise applications over the public Internet, private intranet, or share extranet to empower employees, customers, suppliers, and partners.

The information from this analysis can be used to determine whether the network will meet the following enterprisewide requirements:

- Solve the major problems of information access, information flow, and information exchange that currently exist in the organization or between the organization and its various constituents

- Match the requirements likely to occur because of projected growth in the organization, including any special requirements of telecommuters and mobile professionals

- Anticipate the real projected cost of the system, based on available information from carriers, hardware vendors, systems integration firms, maintenance service providers, and in-house support staff

Meeting these requirements economically may entail further planning considerations, such as identifying what existing network elements and systems can be retained and used in the new network configuration. And, depending on the qualifications of in-house staff, there may be opportunities to save on the cost of systems integration, maintenance and repair, and user training.

Implementing client-server systems especially requires a careful cost-benefit analysis because of the potentially large investments in both new equipment and maintenance, and other less tangible tradeoffs that go into the decision to adapt an entirely new network architecture. In fact, the move from centralized information systems to the client-server architecture is most frequently cited as the reason for budget overruns. Additionally, support costs may also increase as more components are brought into the data center and office environments and such concerns as software maintenance and distribution become increasingly important. Security is another component of the client-server environment that increases costs, particularly when remote access is expanded to include telecommuters, mobile professionals, and, possibly, external constituents.

The accuracy of cost projections for a client-server network depends largely on the infrastructure requirements of the organization. Since most of the expense is incurred at the time of implementation, managers may have to distinguish between transient costs and long-term operational costs. The transient costs will be substantially higher in an organization that has not already equipped all potential users with the networks, interfaces, and desktop platforms required by the client-server model.

Companies may well find that their highest cost is preparing local sites for the client-server network. Even without large savings, the lower unit costs for client-server hardware components still come into play because the deployment of applications can be staged over a period of time as the IT budget allows.

Given the "do more with less" budgetary constraint that afflicts many organizations, a phased approach to implementation based on lower-cost systems and servers is more practical than adding mainframe capacity. At the same time, building client-server networks may require fairly substantial organizational adjustments. And many of those adjustments will involve additional expenses. This is counter-balanced by the fact that client-server applications are often faster and more economical to develop than applications for traditional host-centric systems, especially when object-oriented programming tools are used for the creation of reusable software modules.

Any cost-benefit analysis of new technology should be based on the intention to increase revenue, not simply to save costs. Companies should learn from their client bases what technological investments would solve client needs and return profits at the same time. For this and other reasons, many companies are turning to large systems integration firms to manage the risk of their projects. Expecting the systems integrator to act as a deep-pocket partner who absorbs the financial loss if the project is late or fails, however, is not realistic. Managing the risk together improves the successful outcome of complex projects. A firm is more likely to build quality systems by having clear goals, managing with those goals in mind, hiring the right people, creating a sense of teamwork, and devising a good plan.

The key to success is knowing how to put together and manage a cost estimate. The following guidelines can help prevent cost overruns:

- Prepare an initial cost estimate during the feasibility study, when user requirements are being determined.

- Assign the initial cost-estimating task to the network designers. Using different people to design the network and to provide cost estimates often leads to cost overruns on big projects and finger-pointing when things go wrong. Network designers who are involved in the project from cost estimation to design to implementation would be alert to poor user requirement reporting that could result in unreasonably low estimates.

- Delay finalizing the initial estimate until the end of a thorough study. Most projects should go through various evaluations to arrive at a final cost figure, but management and users typically remember only the first estimate and lock in on the low number.

The last point is especially important. Many times you will be under pressure from top management to provide a cost estimate, and you will just use a ballpark figure. A more realistic approach would be to get approval for a feasibility study, which will provide a more accurate indication of costs.

In any enterprise project, it is advisable to anticipate and minimize user changes. Without adequate controls, changes can become so numerous that the final network may look nothing like the one originally proposed—the one used as the basis for the cost estimate. Such a moving target makes it tough to keep costs under control. This situation is similar to the new home buyer who requests "small" changes throughout the construction process and then is shocked that the final price greatly exceeds the initial price of the home.

It is a good idea to monitor the progress of the proposed project. A formal monitoring process should include milestones, so managers can manage project costs. An independent auditor can even be used to keep tabs on the project's progress. This person could typically come from a separate department such as accounting or could simply be someone not immediately associated with the development of the project. Stringent, formal monitoring keeps network designers, integrators, vendors, and carriers on their toes as they strive to complete the project within the parameters of the budget.

4.3.3   Time value of money

In evaluating any technology acquisition, a key element is the time value of money. A dollar received today has more value than a dollar received a year from now. Also, a dollar that must be paid out today costs more than a dollar that can be held and paid out a year from now. The key factor is interest.

Based on a given interest rate, a capital investment that involves expenditures made over several years entails a cash flow that can be represented in terms of today's dollars. With this information, a comparison can be made of an investment that involves a large front-end payment and smaller ongoing payments with a lease that involves little or no front-end expenditures but higher monthly payments. The interest rate determines which method of payment is best: Low interest rates favor a purchase; high interest rates favor a lease.

There are tax advantages associated with purchasing and leasing. With leasing, the entire amount of the acquisition may be tax-deductible, depending on the type of lease. With purchasing, the acquisition can be depreciated over several years or depreciated in only one year under first-year expensing. Both methods of depreciation yield tax deductions to the company. The decision to lease or purchase must be considered within the context of the company's overall financial objectives. These decisions are usually made at the executive level.

4.3.4   Discounted cash flow analysis

There are two commonly used methods for analyzing technology investments that take into account the time value of money on cash flows: Net Present Value (NPV) and Internal Rate of Return (IRR). Both methods fall under the general heading of Discounted Cash Flow (DCF) analysis.

NPV assumes a specific interest rate and discounts future cash flows to their value in today's dollars based on that interest rate. IRR looks at the cash flows generated and determines what interest rate will yield an NPV of zero. That interest rate is the Rate of Return (ROR).

For analyzing technology investments, the NPV method is the most useful because it allows for a positive or negative result; that is, a purchase could have either a net cash outflow or a net cash inflow. Conversely, IRR looks at investments that yield a return or a net cash inflow—net cash outflow would yield a negative rate of return.

Routine acquisitions of data communications equipment generally do not yield net cash inflows. For example, in the case of some T1 multiplexers purchased to replace leased systems, there might be a net cash inflow. But more often than not, such purchases are forced by growth (e.g., the installed systems have reached capacity), and the purchase decision is based on which alternative will yield the smallest net cash outflow.

NPV analysis looks at the effects of an investment on a company's cash flow—how much cash the company will take in or pay out over the life of an investment in terms of today's dollars. The analysis is a two-step process: First the net cash flows must be computed, and then those cash flows must be discounted to their equivalent NPVs based on the interest rate applied.

4.3.5 Impacts on cash flow

There are three ways to acquire data communication systems and networks: rental, purchase, and installment purchase. Three main components affect cash flow in different ways: purchase cost, expensed items, and depreciation.

Purchase cost. This component is involved only in purchase and installment purchase transactions, and not in rentals. If a technology asset is purchased outright, the cash price of the asset is recorded as a direct cash outflow. In an installment purchase, the portion of the loan payments that represents principal is also recorded as direct cash outflow.

Expensed items. With this component, expenses are cash outflows that are applied as deductions against the company's gross profits. Expensed items are associated with almost all technology acquisitions, including rental charges, maintenance, and interest. While expenses are cash outflows, they also reduce the company's net income before taxes, and so reduce the company's income taxes. To determine the actual cash outflow of an expense, it must be adjusted (reduced) to reflect the tax effect (i.e., the reduction in income tax payable). The formula is:

$$\text{Expense amount} \times (1 - \text{income tax rate}) = \text{net cash outflow}$$

Expensed items are of two types: those that are fixed and those that are expected to increase. For example, the cost of equipment is fixed. But maintenance on that equipment tends to increase as the equipment gets older, and the charges may vary over time. Separating expensed items in this way helps determine the accuracy of the cash flow analysis based on the proportion of expensed items that are predictable compared with those that are not.

Depreciation.   Like other expensed items, depreciation reduces a company's net income before taxes and thus the amount of income tax payable. Unlike other expensed items, however, depreciation causes no cash outflow: The company reduces its tax liability, but does not actually pay any cash out for depreciation. Rather, depreciation is the recognition of the cash paid out to purchase the asset. Depreciation, therefore, is shown as cash inflow or savings on taxes, which is calculated as follows:

$$\text{Depreciation charge} \times \text{income tax rate} = \text{net cash inflow}$$

4.3.6 Loan amortization

Loan amortization is the process of computing what portion of the principal (i.e., the loaned amount) is canceled by each successive payment against the loan. In amortizing a loan, it can be determined how much interest has been paid on the outstanding balance of the loan during the year and how much has been paid as interest on the loan.

For illustration purposes, Table 4.1 shows a simple 5-year amortization table for a 12-port managed hub at a branch office.

TABLE 4.1   Simple 5-Year Equipment Amortization Table*

Year	Outstanding Balance at Beginning of Year	Interest Paid	Principal Paid	Outstanding Balance at End of Year
1	$1,200.00	$180.00	$ 177.98	$1,022.02
2	1,022.02	155.30	204.68	817.34
3	817.34	122.60	235.38	581.96
4	581.96	87.29	270.69	311.27
5	311.27	46.69	311.20	0
Totals:		589.88	1,199.93	

*In U.S. dollars.

4.4   Return on Investment

Return on investment (ROI) refers to the anticipated cost savings, productivity gains, or other benefits that will accrue to the organization as a result of implementing a new technology or service. The ROI is typically used to help cost-justify a capital investment. To help top management make confident and informed decisions, you, as a network manager, together with other department heads, should prepare an executive report that explains each option and its associated risks and benefits.

This report should address an organization's strategic business objectives, identifying potential targets for improvement and providing a high-level cost-benefit analysis. The objective of the report is to outline a preliminary plan for implementing the network, upgrade, or expansion plan within the existing work environment. The report should include a list of the departments that would benefit the most from the proposed plan. The criteria would be parameters such as traffic volume; geographic diversity; application requirements in terms of bandwidth, reliability, speed, connectivity, protocols, and delay; and customer-supplier linkages via such means as email, Electronic Data Interchange (EDI), extranets, document imaging, and Computer Aided Design/Computer Aided Manufacturing (CAD/CAM).

In addition, the report should address enterprisewide requirements in an expandable, modular fashion that protects existing investments while building for future needs, including the requirement for interconnectivity among operating groups and subsidiary companies, as well as trading partners.

To ensure the effective implementation of the plan, it is advisable to stress an environment that is both structured and flexible. Industry standards provide the structure, while flexibility results from building the solution on open platforms that can be tailored to specific user needs. In addition to maximizing integration potential, adhering to standards facilitates the incorporation of technological advances as they occur, regardless of their origin. So as new technologies emerge to better support business strategies, standards will permit the organization to take an early advantage without making the current investments obsolete. These new technologies may be refinements to existing network elements, such as routers, that can be accomplished with the addition of new modules, more memory, or upgrades to the operating system to add more functionality.

4.5   Procurement Alternatives

There are three methods of paying for equipment and systems: purchasing, leasing, or an installment loan. Each method has its advantages and disadvantages, depending on the financial objectives of the company.

4.5.1   Cash purchase

The attraction of paying cash is that it costs nothing extra. No interest is paid as with a loan, so the cost of the item from a financial perspective is just the purchase price. However, there are several reasons why a cash payment may not be a good idea, aside from the obvious reason that the company may have no cash to spare.

The first reason not to pay cash for a purchase is that the money to be taken from the reserves could probably be used to finance other urgent activities. The interest rate on a loan for those urgent activities may have higher finance charges if financing can be obtained at all. Sometimes a set of T1 multiplexers, for example, or other major purchases can be made based on a very low interest rate relative to prevailing commercial rates. To pay cash for equipment when it could be financed at 7 percent and then to pay 8.5 percent for money six months later is unsound financing. If financing is at an attractive rate, it is probably better to use that rate and finance the system, unless company cash reserves are so high that future borrowing will not be required for any reason.

Another reason not to pay cash is the issue of taxes and cash flow. If the company has accumulated a profit in cash and could use it to pay for a major network upgrade or expansion, for example, the cost cannot be deducted in a single year—it must be depreciated over a period of three or more years. The company will thus be placed in the position of owing tax on its profits, less first-year depreciation (and other operating expenses), and possibly not having enough cash to pay that tax.

A final reason for not using reserves to finance a large purchase concerns the issue of creditworthiness. Often a company with a short financial history can borrow for a collateralized purchase such as a network, but cannot borrow readily for such intangibles as ordinary operating expenses. Other times, an unexpected setback will affect the credit standing of the firm. If all or most of the firm's reserves have been depleted by a major purchase, it may be impossible to secure quick loans to meet new expenses, and the company may falter.

4.5.2   Installment loan

For the company that elects to purchase equipment but chooses not to pay cash, a loan financing arrangement will be required. These arrangements are often difficult to interpret and compare, so financial planners should review the cost of each alternative carefully.

If the company is unable or unwilling to pay cash for the system, the alternative is some form of deferred payment. The purpose of these payments is to stagger the effect of the purchase across a longer period. This reduces the cash flow in a given year. But as we all know, the interest is a substantial price to be paid. Whether the equipment is leased or installment purchased, interest will be paid. This may result in a conflict of accounting goals: Is cash flow or long-term cost, including interest, the overriding factor?

The question of financial priorities must be answered early in the equipment acquisition process. Providing that the interest premium for that longer term is not unreasonable, a purchase that is financed over a longer period will have a lower net cost per year, taking tax effects of depreciation into consideration.

If the company is in a critical position with cash flow, longer-term financing and minimum down payment will likely be the major priorities. The cash flow resulting from a Private Branch Exchange (PBX) purchase, for example, will be the annual payments for the system less the product of the company's marginal tax rate times the annual depreciation. If the system's payments are $5000 per month on a $200,000 note, and first-year depreciation is 20 percent, the company will pay $60,000 in the first year and have a tax deduction of $40,000. If the marginal rate is 30 percent, that deduction will be worth $12,000, so the net cash flow is negative at $48,000. Longer terms will reduce the payments but not proportionally owing to increased interest.

In cases where cash flow is not a major concern, the method of financing should minimize the interest payments to be made. Interest charges can be reduced by increasing the down payment, reducing the term of the loan, and by shopping for the best loan rates. Each of the ways in which a company can finance a system will affect its bottom line. Choosing the right financing method will depend on the company's financial priorities, interest rates, tax liability, and cash reserves. As a network manager, you may be called upon to provide budgetary pricing for new systems and incremental upgrades, but the means of financing will usually be determined at the executive level, in keeping with corporate goals.

4.5.3   Leasing

There are a number of financial and nonfinancial reasons for considering leasing over purchasing.

Financial incentives.   Leasing can improve a company's cash position, since costs are spread over a period of years. Leasing can free up capital for other uses, and even cost-justify technology acquisitions that would normally prove too expensive to purchase. Leasing also makes it possible to procure equipment on short notice that has not been planned or budgeted for in order to take advantage of new opportunities.

A large purchase will increase the debt relative to equity and have an adverse effect on the company's financial ratios in the eyes of investors, creditors, and potential customers. However, an operating lease can reduce balance sheet debt, since the lease or rental obligation is not reported as a liability. So at the least, an operating lease can represent an additional source of capital, while preserving credit lines.

Beyond that, leasing can help companies comply with the covenants in loan agreements that restrict the amount of new debt that can be incurred during the loan period. The purpose of such provisions is to make sure that the company does not jeopardize its ability to pay back the loan. In providing additional capacity for acquiring equipment without violating loan agreements or hurting debt-to-equity ratios, leasing allows companies to "have their cake, and eat it too."

With major improvements in technology becoming available every 12 to 18 months, leasing can prevent a company from becoming saddled with obsolete equipment. Leasing rather than purchasing can minimize potential losses incurred by replacing equipment that has not been fully depreciated. Furthermore, with rapid advancements in technology and consequent shortened product life cycles, it is becoming more difficult to sell used equipment. Leasing eliminates this problem too, since the leasing company owns the equipment and takes responsibility for getting rid of it through the gray market.

For organizations concerned with controlling staff size, leasing also minimizes the amount of time and resources spent in cost-justifying capital expenditures, evaluating new equipment and disposing of old equipment, negotiating trade-ins, comparing the capabilities of vendors, performing reference checks on vendors, and reviewing contractual options. There is also no need for additional administrative staff to keep track of configuration details, spare parts, service records, and equipment warranties. And since the leasing firm is usually responsible for installing and servicing the equipment, there is no need to spend money on skilled technicians or for outside consulting services. Lease agreements may be structured to include ongoing technical support and even a help hotline for end-users.

Another advantage to leasing is that it can minimize maintenance and repair costs. Because the lessor has a stake in keeping the equipment functioning properly, it usually offers on-site repair and the immediate replacement of defective components and subsystems. In extreme cases, the lessor may even swap out the entire system for a properly functioning unit. Although contracts vary, maintenance and repair services that are bundled into the lease can eliminate the hidden costs often associated with an outright purchase.

When purchasing equipment from multiple vendors, the customer often gets bogged down in processing, tracking, and reconciling multiple vendor purchase orders and invoices to obtain a complete system or network. Under a lease agreement, the leasing firm provides a single-source purchase order processing and invoicing. This cuts down on the customer's administration, personnel, and paperwork costs.

Finally, leasing usually allows more flexibility in customizing contract terms and conditions than normal purchasing arrangements. This is due to the fact that there are no set rates and contracts when leasing. Unlike many purchase agreements, each lease is negotiated on a case-by-case basis. The items that are typically negotiated in a lease are the equipment specifications, schedule for upgrades, maintenance and repair services, and training. Another negotiable item has to do with the end-of-lease options, which can include signing another lease for the same equipment, signing another lease for more advanced equipment, or buying the equipment. Many leasing firms will allow customers to end a lease ahead of schedule without penalty if the customer agrees to a new lease on upgraded equipment.

Nonfinancial incentives. There are also some very compelling nonfinancial reasons for considering leasing over purchasing. In some cases, leasing can make it easier to try new technologies, or the offerings of vendors that would not normally be considered. After all, leases always expire or can be canceled (a penalty usually applies), but few vendors are willing to take back purchased equipment. Leasing permits users to take full advantage of the most up-to-date products at the least risk and often at very attractive financial terms.

This kind of arrangement is particularly attractive for companies that use technology for competitive advantage because it means that they can continually upgrade by renegotiating the lease, often with little or no penalty for terminating the existing lease early. Similarly, if the company grows faster than anticipated, it can swap the leased equipment for new equipment and cover it with an addendum to the original lease.

It must be noted, however, that many computer and communications systems are now highly modular and scalar in design. Features and functions can be added, as can capacity, to keep pace with growth. Consequently, the fear of early obsolescence may not be as great as it once was. Nevertheless, leasing offers an inducement to try vendor implementations on a limited basis without committing to a particular platform or architecture, and with minimal disruption to mainstream business operations.

Companies that lease equipment can avoid a problem that invariably affects companies that purchase equipment: how to get rid of outdated equipment. Generally, no used equipment is worth more on a price-performance basis than new equipment, even if it is functionally identical. Also, as new equipment is introduced, it erodes the value of older equipment. These by-products of improved technology make it very difficult for users to unload older, purchased equipment.

With equipment coming off lease, the leasing company assumes the responsibility of finding a buyer. Typically, the leasing company is staffed with marketers who know how and where to sell used equipment. They know how to prospect for customers for whom state-of-the-art technology is more than they need, but a second-hand system might be a step up from the five-year-old hardware they are currently using.

There is also a convenience factor associated with leasing, since the lessee does not have to maintain detailed depreciation schedules for accounting and tax purposes, which would be required if the equipment was purchased outright. Budget planning is also made easier, since the lease involves fixed monthly payments. Pricing is locked in over the term of the lease, so the company knows in advance what its equipment costs will be over a particular planning period.

With leasing, there is also less of an overhead burden to contend with. For example, there is no need to stockpile equipment spares, sub-assemblies, repair parts, and cabling. It is the responsibility of the leasing firm to keep inventories up to date. Their technicians (usually third-party service firms) even make on-site visits to exchange smaller components and arrange for overnight shipping of larger components, when necessary.

Leasing can also shorten the delivery lead time on desired equipment. It may take six weeks or longer to obtain the equipment purchased from a manufacturer. In contrast, it may take from 1 to 10 working days to obtain the same equipment from a leasing firm. Often, the equipment is immediately available from the leasing firm's lease and/or rental pool. For customers who need equipment that is not readily available, some leasing firms will make a special procurement and have the equipment in a matter of two or three days, if the lease term is long enough to make the effort worthwhile.

Many leasing companies offer a master lease, giving the customer a prearranged credit limit. All of the equipment the customer wants goes on the master lease, and is automatically covered by its terms and conditions. In essence, the master lease works like a credit card.

4.5.4   Types of leases

Assuming that the decision has been made to lease rather than purchase equipment, it is important to know about the two types of leases available because they are treated differently for tax purposes. One type of lease is the operating lease, in which the leasing company retains ownership of the equipment. At the end of the lease, the lessee may purchase the equipment at its fair market value. The other kind of lease is the capital lease, in which the lessee can retain the equipment for a nominal fee, which can be as low as $1.

Operating lease. With the *operating lease* (also known as a *tax-oriented lease*), monthly payments are expensed, that is, subtracted from the company's pretax earnings. With a *capital lease* (also known as a *nontax-oriented lease*), the amount of the lease is counted as debt and must appear on the balance sheet. In other words, the capital lease is treated as just another form of purchase financing and, therefore, only the interest is tax deductible.

A true operating lease must meet the following criteria, issued by the Financial Accounting Standards Board (FASB), some of which effectively limit the maximum term of the lease:

- The term of the lease must not exceed 80 percent of the projected useful life of equipment. The equipment's "useful life" begins on the effective date of the lease agreement. The lease term includes any extensions or renewals at a preset fixed rental.

- The equipment's estimated residual value in constant dollars (with no consideration for inflation or deflation) at the expiration of the lease must equal a minimum of 20 percent of its value at the time the lease was signed.

- Neither the lessee nor any related party is allowed to buy the equipment at a price lower than fair market value at the time of purchase.

- The lessee and related party are also prohibited from paying, or guaranteeing, any part of the price of the leased equipment. The lease, therefore, must be 100 percent financed.

- The leased equipment must not fall into the category of "limited use" property; that is, equipment that would be useless to anyone except the lessee and related parties at the end of the lease.

With the operating lease, the rate of cash outflow is always balanced to a degree by the rate of tax recovery. With a purchase, the depreciation allowed in a given year may have no connection with the amount of money the buyer actually paid out in installment payments.

Capital lease. For an agreement to qualify as a capital lease, it must meet one of the following FASB criteria:

- The lessor transfers ownership to the lessee at the end of the lease term.

- The lease contains an option to buy the equipment at a price below the residual value.

- The lease term is equal to 75 percent or more of the economic life of the property. (This does not apply to used equipment leased at the end of its economic life.)

- The present value of the minimum lease rental payments is equal to or exceeds 90 percent of the equipment's fair market value.

From these criteria, it becomes quite clear that capital leases are not set up for tax purposes. Such leases are given the same treatment as installment loans; that is, only the interest portion of the fixed monthly payment can be deducted as a business expense. However, the lessee may take deductions for depreciation as if the transaction were an outright purchase. For this reason, the monthly payments are usually higher than they would be for a true operating lease. Depending on the amount of the lease rental payments and the financial objectives of the lessee, the cost of the equipment may be amortized faster through tax-deductible rentals than through depreciation and after-tax cash flow.

Although leasing can be used as an alternative source of financing that does not appear on the corporate balance sheet, the cost of a conventional lease arrangement generally exceeds that of outright purchase. Excluding the "time value" of money and equipment maintenance costs, the simple lease versus purchase break-even point can be determined by the formula:

$$N = P/L$$

where P is the purchase cost, L is the monthly lease cost, and N is the number of months needed to break even. Thus, if equipment costs $10,000, and the lease costs $250 per month, the break-even point is 40 months. This means that owning equipment is preferable if its use is expected to exceed 40 months.

As in any financial transaction, there may be hidden costs associated with the lease. If the lease rate seems very attractive relative to that offered by other leasing companies, a red flag should go up. Hidden charges may be embedded in the lease agreement, which would allow the leasing company to recapture lost dollars. These hidden charges can include shipping and installation costs, high-than-normal maintenance charges, consulting fees, or even a balloon payment at the end of the lease term.

The lessee may even be required to provide special insurance on the equipment. Some lessors even require the lessee to buy maintenance services from a third party to keep the equipment in proper working order over the life of the lease agreement. The lessor may also impose restrictions, such as where the equipment can be located, who can service it, and what environmental controls must be in place at the installation site.

4.6   Financial Management of Communications

The goal of financial management is to treat corporate communications just like any other segment of business. This requires the means to identify what is being spent for communications and where these expenses originate. After all, if communications costs cannot be identified, it becomes difficult, if not impossible, to understand which components can be optimized, what service costs can be reduced, and which components are totally out of control.

Most corporations do not adequately account for their communications costs. There are several reasons for this. One is that these costs are so far in the background that they are not relevant to most people. Another reason is that the responsibility for communications is distributed throughout the organization; consequently, there is no single point of accountability. To complicate matters, there seems to be no standard for determining what goes into the bucket called "communications costs." Some companies, for example, put the cost of lines and handsets in the telecommunications budget, while others do not. This means that the total cost for communications will differ greatly, even among companies of similar size and in similar lines of business.

Accounting for data communications costs can be even more of a problem. In the client-server environment, cost accounting can be pretty straightforward since discrete elements are involved: terminals, servers, and cabling. When the company starts to internetwork these elements over the wide area network, cost accounting gets very complicated.

The following questions illustrate the difficulty most companies have in accounting for all of their communications costs:

- How much of the hubs and routers are associated with the client-server nodes?

- When it comes to host resources, how far into the CPU do you go? Do you stop at the data center wall, or do you go into the communications controller?

- Do you assign some of the software overhead for managing the communication queries and routing?

- How much of the transaction processing software expenses should be included in the communications budget?

Even network managers who deal with communication services and equipment on a daily basis have difficulty accounting for the costs. With telecom, for example, it is easy to charge back the cost of outgoing calls and faxes to a department, work group, project, or individual using call accounting software, which processes the station message detail records collected by the PBX. But there is often confusion in how to spread the cost of the PBX, telephone sets, trunks and/or lines, and ongoing support costs across various categories of users. But data communications usage cannot be tracked and charged back in the same way as outgoing calls, which may be used for data as well as voice. It is even more difficult to account for data equipment costs. This is because a lot of the costs for data communications are buried in CPU costs. In other words, virtually any network element connected to the CPU is part of the CPU cost and should be charged out as such to end users. But if the user asks how much of the bill is for network usage and how much is for CPU hardware or software, it is often very difficult if not impossible to determine.

As noted, call accounting systems used with PBXs often provide a range of useful reports on line usage and call costs, which assist in allocating phone charges to individual departments, projects, and users. However, call accounting systems do not include the costs associated with data transmission, particularly when such services are billed per kilocharacter, as in email, or connection time as with fax transmissions, bulletin boards, and online information services. Nor do they take into consideration off-premises communications costs such as credit card charges accrued by traveling employees. At best, a call accounting system is useful for identifying only 30 to 40 percent of a company's total communications costs.

In order to manage the total cost of communications, a company affords its end users the opportunity to manage the quantity consumed. Considering the huge dollar amounts that may be invested in voice and data communications, an understanding of the costs and their aggressive management is critical to maintaining service quality and, consequently, corporate competitiveness.

4.7   Requirements for Strategic Operation

To get a firm handle on communications costs, a financial management structure for communications must be put in place. Although this is primarily the responsibility of the Chief Information Officer (CIO) or other executive with equivalent oversight responsibilities, you, as a network manager, will provide the key inputs that make such a structure work. A corporate communications organization must embrace three distinct missions to operate strategically.

First, there must be a *development mission,* which focuses on evaluating and implementing operationally sound communications networks and systems. This mission is already well understood, as demonstrated by the many extensive corporate networks that use high-capacity backbones, international gateways, and other advanced technologies to meet the growing demand for communications services.

Second, there must be a *service mission,* which is user-oriented and market-driven. This mission is about ensuring the delivery of adequate communications services, and includes planning service standards with users, developing service-level agreements with carriers, and providing user support in the form of help desks that troubleshoot problems and implement moves, adds, and changes.

Third, there must be a *management mission,* which centers on the use and conservation of a company's communications resources. This mission includes financial and information systems planning, specifically, such operational activities as communications, accounting, and budgeting.

4.8   Opportunity Assessment

As the communications industry becomes more (or less) competitive and more services become available, new cost-reduction opportunities are evolving. But if managers do not understand what is being spent and where it is being spent, it will be very difficult to assess whether these opportunities are important and whether they can be acted upon in a timely manner.

For example, having cost data at hand has allowed some companies to react quickly to the custom network agreements from AT&T, Sprint, and WorldCom, allowing them to achieve a reduction of their interexchange communications costs of as much as 25 percent. Considering that interexchange communications costs for many companies represent as much as 40 percent of their total communications operating costs, the savings and the competitive advantage to be gained with custom agreements can be quite substantial.

Negotiating a custom network agreement and then managing the cost as well as possible is essential for containing operating costs and translating the savings into competitive advantages. The fact is, unless the company is aware of its costs on an ongoing basis, it will not know whether it is paying too much from year to year. With all of the changes occurring in the communications industry, this is a very real possibility. Therefore, the financial management of communications consists of at least three functional areas: cost reporting, budgeting and control, and financial operations.

4.8.1   Cost reporting

Many times companies do not have an accurate picture of what they are spending for communications because the reporting structure is flawed. For example, management reporting includes capital expense reporting and operating expense reporting. However, all too often the capital and operating expenses for communications are distributed among several financial accounts and are not organized for effective use. Unless these expenses are discreetly identified and properly summarized, decision making could be hampered. For example, recording capital asset depreciation, communications equipment leasing, personnel occupancy, and salaries under noncommunications-related expense accounts usually results in underestimating the true size of the communications department budget.

More mundane factors also throw off communications budgets. Consider the simple telephone, for example. The mentality of looking at the telephone as a $39 instrument not worthy of any expertise—internal or external—is still common. What is often overlooked is the fact that a $39 telephone may actually cost several hundred dollars annually in ongoing maintenance charges in the form of moves, adds, and changes.

4.8.2 Budgeting and control

Budgeting and control involves capital and operating budget preparation, performance monitoring and forecasting, and project analysis. Financial management expertise in these areas is more essential in today's operating environment than in the past. The failure to properly calculate capital asset depreciation and taxes, for example, could result in a project that is not as cost-effective as it initially appeared. Economic conditions also must be factored into the cost equation. For example, a project that increases fixed communications costs may put the company at a disadvantage when current market conditions actually favor financial flexibility and variable costs.

4.8.3 Financial operations

The financial operations area includes service order processing, inventory management, and capital asset management. Since service orders involve operating and/or capital fund expenditures, they demand appropriate controls. If not controlled, service order processing can consume an inordinate portion of the communications budget.

Inventory management is necessary not only for tracking equipment and cabling but also to prevent excessive equipment purchases. It is not uncommon to find businesses spending more than 10 percent of their annual communications budget on unnecessary services, equipment, and cabling just because they do not keep accurate inventory records.

Accounting.   The accounting function includes maintaining a proper chart of accounts, performing bill reconciliation, and implementing a user chargeback system. A chart of accounts accumulates and reports financial information on various subclassifications of assets and expenses. To be effective, however, it must provide an adequate number of accounts and an adequate definition of accounts to manage the various components of the communications expense—and in a way that allows corporate management to react accordingly. Without a properly organized chart of accounts, it is often difficult to decide where to report different product and service charges. It will also be difficult to generate appropriate financial management reports. The CIO, for example, must have access to usable expense information with the right level of detail.

Bill reconciliation ensures that carriers and vendors are paid only for the services and products they deliver. The fact that there continues to be a lucrative business in helping companies recover money lost to carrier and vendor billing errors on a contingency fee basis demonstrates that bill reconciliation is either not being done, or that it is being done poorly. Thorough bill reconciliation can save corporations substantial amounts of money per year.

User chargeback allocates communications expenses to the appropriate business units using the services. However, there are potential roadblocks. Some companies get bogged down in the complexity of their chargeback systems in the quest to become accurate. Sometimes political issues sidetrack the chargeback system. Other times the problems are technical, as in the way voice and data are integrated. There must be a balance between the level of effort that goes into this process and what is going to be done with the information. Success hinges on the accumulation of only as much detail as can be reasonably managed.

Agreement management. Relationships with vendors and carriers are defined by contracts and agreements. Customized network services contracts require extensive financial analysis. These and other contracts must be reviewed periodically to ensure that defined performance standards, prices, payment schedules, and other provisions are still compatible with corporate objectives.

The potential pitfall of these and other types of agreements is that companies often abdicate their responsibilities for financial management to the carriers and vendors, who are probably not in a position nor should they be trusted to pick up that responsibility. Furthermore, when communications expenses lose their visibility they tend to get out of control: Either the company pays too much, or it gets too little. The challenge is to maintain the right level of control so that the company can make decisions that are in its best interests.

Depending on the dollar amount involved and whether corporate objectives are being achieved, corporate managers may want to consider alternative scenarios, renegotiate terms and conditions, or limit the length of service agreements to minimize exposure to financial risk.

Product management. Product management for communications focuses on such activities as product definition, unit cost analysis, and comparative product pricing analysis. It also requires that the company know how its communications costs compare with those of its competitors.

When a company defines products for delivery to users, it also develops a more accurate framework for comparative pricing. For example, in calculating loaded costs, such variables as vendor costs, depreciation, occupancy expenses, personnel salary costs, and other overhead expenses are taken into account. By comparing current loaded costs to equivalent alternatives, the company can more actively manage its communications requirements and associated expenses.

Financial management.   Today's asset management products and services do more than track hardware and software inventories and provide configuration information to aid in troubleshooting.—They can provide information for the financial management of technology assets. This includes accounting, acquisition, depreciation, and chargeback information. This information is entered into the asset management database, and the specific features of the database management system sort the data in the desired report format.

Several vendors offer asset management as either a product or service, which can be accessed through a systems or network management platform, or a more traditional enterprise accounting application. A range of data can be incorporated into such systems, including details about asset acquisition, depreciation, chargeback, and disposition. In the case of software, for example, acquisition information includes purchase date, cost, load date, bill date, last audit date, vendor, and vendor ID. Depreciation can be figured in several ways—for tax management, legal compliance, and internal accounting purposes. With this feature, an organization can maintain three sets of books concurrently to satisfy all of these reporting needs. With chargeback information, asset-related costs can be billed to appropriate departments, work groups, individuals, or even projects.

The accounting software or service also maintains tracking methodologies to account for obsolete assets that are given away, sold, or turned back to the leasing firm. This kind of information can help prevent organizations from continuing to pay for support on nonexistent equipment—a situation that is fairly common.

Some accounting products and services also support a feature called *obligation management,* which tracks maintenance, leases, and warranties. Maintenance tracking includes contract terms and conditions as well as program costs. The maintenance history of each asset is tracked, providing such information as date of last service, due date of next service, and the expiration date of the service contract.

Lease information includes the vendor, lease number, lease payments, terms and conditions, and buyout provisions. Warranty information is also tracked, providing the vendor, warranty provisions, and expiration date.

Such asset management is essential for controlling costs—most of which are "hidden costs" that surreptitiously drain telecom and IT budgets and divert scarce resources from technology acquisitions and service delivery.

4.8.4   How to start

Despite the seeming complexity of financial management, there is a way for you, as a network manager, to systematically go about contributing to this process, while also enhancing your value to the company.

The first step is to acquire an understanding of the company's business strategy, including products, markets, distribution channels, and critical success factors. These elements can help define the emphasis that should be applied to the primary communications missions: development, service, and management.

Next, identify and assess aggregate communications expenditures by reviewing the company's existing capital and operating expense reports, and summarizing the data. Compare these expenses to other major expense categories and to the budgets of other business units. This can help identify financial management requirements and view communications costs within the proper business perspective. For example, financial management usually needs more attention if expenditures are increasing or are budgeted to increase, or when expenses are recorded to a single account or to multiple accounts without being summarized separately for review.

Survey the current financial management infrastructure. This can be done by devising a matrix that identifies organizations, work groups, and information systems currently used to manage communications expenses. Identify what functions communications personnel or other personnel perform, or functions that are not performed at all. Identify how these functions are organized and supervised. Determine the kind of organizational infrastructure needed to provide the appropriate level of financial management emphasis. This determination can be made after the CIO develops a business functional model that includes a complete and prioritized description of all necessary financial management areas, processes, and activities. These can be compared with the existing organizational infrastructure to determine specific needs.

An information systems infrastructure should also be determined, which should include defining an optimal database to support communications financial management. The database architecture should identify the relationship between information systems. For example, bill reconciliation systems generally must interface with inventory records, and the architecture should describe how the financial database should be organized.

With these inputs, the CIO should be prepared to formulate a plan of action using the organizational and database support infrastructures described. The plan should identify the potential financial management alternatives, evaluate the relative costs and benefits, and recommend the actions to be taken.

4.9   Assessing Vendor Stability

In evaluating a vendor's proposal, the most important information may not be included. This is the financial data that can indicate the vendor's current financial health and future survivability. If the vendor has lost money in the last few quarters, it may indicate that it is about to institute cost-cutting measures that may affect customer service and long-term technical support. If it has lost money in the last few years, it may indicate that bankruptcy is a very real possibility. Most of the time, however, quarterly or yearly performance does not necessarily provide a true indication of a vendor's financial health.

Analyzing financial strength starts with an examination of the vendor's annual report, Form 10-K report, and quarterly 10-Q reports. The 10-K and 10-Q are required only of companies that are publicly traded on a stock exchange, so privately held companies will be unable to comply with a request for this document. Sometimes this information can be procured under a nondisclosure agreement. If not, then financial health information might be available from a market research firm. If there is no means available to check on a company's financial health, it is wise to drop the carrier or vendor from consideration.

In a time when a disappointing earnings number can cause the value of a company's stock to fall, more and more companies are emphasizing pro forma results and deemphasizing Generally Accepted Accounting Procedures (GAAP). Some companies even neglect to report GAAP results, preferring the pro forma method exclusively. Under pro forma, financial statements are adjusted to reflect a projected or recently completed transaction. For example, pro forma results are used to show the earnings that newly merged companies might have achieved had the merger occurred at the beginning of the reporting period. The term is often applied to income statements, balance sheets, and statements of cash flow. Pro forma quarterly results can sometimes be confusing because they may exclude information such as certain stock-based employee compensation costs, which would have a negative effect on results if reported as GAAP. When financial results are reported as both pro forma and GAAP, the latter provides a clearer picture of a company's financial health.

The larger a company, the more financially stable it is likely to be. This should not preclude a thorough financial analysis, however. Many of the nation's top firms have had serious financial problems in recent years. Poor management, questionable accounting and reporting practices, deregulation, global competition, sluggish economies, pull-back in consumer demand, and international currency fluctuations have all contributed in various degrees to the weakness of traditionally strong and stable companies. The important point is that size alone is no longer a reliable indicator of present or future viability.

Meaningful measures of size are sales, profits, cash flow, total assets, customer base, and number of service locations. The amount a company spends on research and development (R&D) can also be a measure of financial strength—especially future financial strength. Look for R&D expenditures to be between 7 and 12 percent of revenues. Anything less may mean that in the future the company will not be positioned to address the market with enhancements and new products.

With each company's financial statements, various performance ratios can be computed that will provide insight into potential problems. These ratios, however, offer wide latitude in interpretation.

4.9.1 Debt/Equity Ratio (Total Debt/Total Equity)

If the Debt/Equity (D/E) ratio is less than 1, the stockholders own most of the assets of the company; if the D/E is greater than 1, creditors hold more stake in the assets than the stockholders. Companies with high D/E ratios are in what is called a leveraged position and may have problems meeting payments on interest if profits fall. A very low D/E ratio may mean the company's management is too conservative, which may also have bad long-term effects. Other indications of a company's financial health include:

- *Current Ratio (Current Assets / Current Liabilities).* Current ratio is an indicator of short-term solvency. If small, the company could have serious problems paying its bills. If large, it could mean the company is not managing its assets to its best advantage.

- *Net Income as Percent of Sales (Profit Margin).* Generally, larger profit margins are better and indicate that the company is generating revenue. However, if a business is generating extremely large profit margins, it may be ill equipped to reposition itself to compete with newly attracted competitors. AT&T, IBM, and Microsoft, for example, traditionally have been very slow in recognizing emerging markets. When they see smaller companies making big gains, they are often motivated to jump into the market and quickly establish dominance.

- *Return on Assets (Net Income [after tax] / Total Assets).* Return on Assets (ROA) is a measure of how well a firm is using the assets it has and the larger this ratio, the better.

Additional areas that are worth investigating include:

- *Financial sources.* Seek out and analyze the corporate ratings issued by the financial community, such as Moody's or Dun and Bradstreet.

- *Industry contacts.* Contact references to verify what the vendor says about its performance. Go beyond superficial questions to find out such things as rate of turnover among salespeople and field technicians, the quality of customer service, the availability of spare components and subsystems, and the attitudes of executives toward customers.

- *Market direction.* Does the vendor have a strategic vision? Based on an assessment of the marketplace by industry analysts and market researchers, determine if the vendor seems to be moving in the right direction.

- *Current events.* Research current news stories about potential vendors to determine such things as lawsuits pending or in progress, financial irregularities, problems with environmental compliance, and employee layoffs or strikes. Any of these can indicate potential financial and legal problems that may disqualify the firm from further consideration.

As with any analysis, use all the sources available in an effort to get a clear picture of the vendor's present and future financial performance. Making the wrong purchase decision based only on a cursory analysis or incomplete information can result in such problems as unnecessary delays in installation, cost overruns, or poor service and support down the road.

4.10   Conclusion

Financial planning and financial management decisions are not made in a vacuum. The needs of the various departments and work groups must be considered, as well as top management's orientation to the corporate network, whether it is viewed as an asset or an expense. Moreover, new purchases must be justified in terms of their return on investment, which should not only include the point at which the purchase price can be recovered through cost savings over previous equipment and services, but less tangible benefits such as improving customer service and good will.

In order to offer intelligent advice on major purchase decisions, you, as a network manager, must do many things. You must track new technologies, adapt a structured approach to evaluating the need for new products and services, sell "soft-dollar" benefits to top management, implement pilot tests to objectively evaluate new technologies, focus on high-impact areas that will reduce expenditures while strengthening the business, formalize action plans to minimize risk, and take a skills inventory to fill knowledge and performance gaps. And if hiring (or staff cutbacks) must be done, hire or retain the people who can add depth to the minimally structured organization and who have the capability of reaching across boundaries to achieve results.

5

Managing, Evaluating, and Scheduling Technical Staff

5.1 Introduction

New and established companies in virtually any industry know that continuing success depends largely on technical expertise and marketing savvy—a powerful combination for attracting new business, retaining customers, and increasing market share. Even network managers are coming to appreciate the role of communications and computer resources in supporting the corporate mission.

But to reap the full advantages of technology, you, as a network manager, can no longer afford to rely on technical expertise alone to bring corporate projects to successful completion. Interpersonal communications skills are equally important because they facilitate the problem-solving process, contribute to better decision making, and enhance work group productivity. Both new and experienced managers tend to overlook the importance of finely honed communications skills, preferring instead to get by on technical know-how, business acumen, and plain old horse sense.

Managers in the technical environment must be concerned with interpersonal communications skills to avoid becoming part of the problem. A common complaint among technical professionals is that they start to discuss one problem, only to end up with another—such as the frustration of not being heard, of not being respected, or of not being taken seriously by their managers. When employees are weighted down with such problems with no way to relieve their anxiety or stress, the desire to excel is diminished and this affects performance.

A manager's inability to handle people problems can have disastrous consequences. Poor communications skills can create problems that stifle initiative and creativity, delay projects, and increase staff turnover—all of which can drive up costs. If the situation is allowed to continue, corporate objectives may even be jeopardized.

Generally, there is reluctance in the technical environment to come to grips with interpersonal relations. After all, human emotions do not lend themselves to neat, clear-cut rules. Nor do they lend themselves to easy diagnoses, fault isolation, and standard solutions. Compared to the challenges of circuit design, applications programming, or building an enterprise network, dealing with human emotions is a messy business that managers tend to avoid rather than meet head-on.

Despite the seeming "messiness" of interpersonal communications, the process can be broken down into discrete stages, which require specific skills to support. What follows is a model for the interpersonal communications process against which you, as a network manager, may evaluate your present comfort level and gauge your likelihood of success in dealing with present and future people problems.

5.2   Structuring the Setting

Problems usually surface in one of two ways: They are revealed by poor performance (lateness, sloppiness, errors, etc.) or brought to your attention after staff members have exhausted all other resources or reached an impasse. In such cases, problem resolution calls for private meetings with the parties involved so you can properly address all issues and viewpoints. There are other occasions when highly developed interpersonal communications skills may prove useful, such as in project development, where the cooperation of many individuals is required to implement a project on time and under budget.

In the meetings, the first element that requires attention is the setting. Ideally, most of the information during the initial meeting should come from staff members. Your responsibility at this point is to listen and ask probing questions, the answers to which help clarify the issues. It is best to avoid jumping to conclusions, questioning actions or motives, belittling others, or casting blame. Such actions not only confuse the issues but also create undercurrents of discontent that will be difficult to surmount later.

Since many people relate information in a casual, roundabout way, you must engage in active listening, note taking, and requests for clarification. It is your responsibility to analyze and develop the information that is provided. To facilitate this process, you can structure the setting in such a way as to make others feel comfortable about talking, and making them feel that their problem is also that of management.

5.2.1   Making time

With most employee meetings likely to occur in your office, there are several ways to structure the setting. Face-to-face meetings quite obviously require that enough time be set aside to develop a meaningful dialogue. When you give uninterrupted time, you are conveying to the other person that that individual is important and that the problem is serious enough to warrant your full attention.

5.2.2   Ensuring privacy

Privacy is another element that helps structure the setting. Since the other person may be divulging sensitive information about him- or herself, you must convey that the discussion will be treated as confidential. The person will not feel that he or she is at the center of the problem-solving process until assurances are conveyed that privacy will be protected. One way managers destroy their credibility is by dropping the names of present or past employees, and discussing how they were able to help them with similar problems.

Although such remarks are merely intended to gain the confidence of the employee so that he or she will open up, more often than not the tactic backfires. The employee is left wondering how the manager will talk about the present situation when he or she feels the need to impress others. Volunteer to keep information confidential and never use private information against that person at a later time, particularly in a staff meeting or in a performance review. Nothing destroys your credibility faster or more effectively than a violation of privacy.

5.2.3   Physical setting

Besides time and privacy, you should ensure that the physical setting supports the interpersonal communications process. This means being attentive to reaffirm the personal nature of the meeting. Such things as maintaining appropriate physical distance, eye contact, and body attention encourage the free flow of information. These affirmative actions also reinforce the employee's feeling that the meeting is taking place just for him or her.

Special care should be taken not to constantly check the time, interrupt the employee to take phone calls or check email, or stare as if preoccupied. These negative affirmations convey the message that the employee or the problem is just not important enough to spend much time on.

5.3   Managing the Process

It is not enough to structure the setting of the meeting to make employees or peers (or even superiors) feel at ease. The opportunity must be used for their benefit, and not to fulfill one's own needs. When your actions reflect your own needs, the whole interpersonal communications process is under-mined. Do not try to set the record straight, cut through the bull, or become preoccupied with establishing technical superiority. Avoid remarks about others' difficult circumstances, level of understanding, or lack of foresight. Above all, do not convey satisfaction—worse, enjoyment or glee—with another person's rebuke, demotion, or dismissal. Such state-ments and attitudes convey the idea that the interpersonal communica-tions process is being controlled by egotistical needs, rather than organizational needs.

If you find yourself in this situation, it is best to terminate the meeting immediately (but politely), analyze the situation, and open up new oppor-tunities for discussion later, but not too much later. In taking the initiative to continue the discussion, you will be demonstrating a sincere interest in gaining the acceptance and cooperation of staff members. This will go a long way toward healing wounds and reestablishing credibility.

Managing the interpersonal communications process involves bridging the inherent psychological distance that quite naturally exists between staff and management. The three qualities that help reduce that distance are *comfort, rapport,* and *trust.*

5.3.1   Comfort

You can first promote *comfort* by making sure that the setting is a good one. You can reinforce it with cordial greetings and appropriate physical contact, such as a handshake, or a friendly slap on the back. A skilled com-municator also displays good body attention while asking a question, such as nodding in agreement or furrowing the eyebrows, which conveys sincere interest.

5.3.2   Rapport

Rapport presumes some level of comfort, but goes deeper and tends to encourage the continued exchange of information. Rapport may be easier to achieve by drawing upon shared experiences—such as having gone to the same school, knowing the same people, or engaging in the same hobbies.

5.3.3   Trust

A third strategy for bridging psychological distance is the establishment of *trust.* From the employee's point of view, trust is the sense that he or she can believe in your competence and integrity. It carries a deeper level of commit-ment than does mere rapport. Alternatively, if the employee suspects that he or she is being manipulated, or has reason to doubt that you can be trusted, you may have yet another problem to compound the original problems.

5.4   A Closer Look at Bridging Skills

Nonverbal communications are tools that can be used to achieve a variety of objectives, such as bridging psychological distance.

5.4.1   Body language

Body language is one facet of nonverbal communications that is now widely recognized as a critical part of the total communications package. The manager who tells a staff member, "Go on, I'm listening," while writing, looking through files, or checking the time, presents the employee with physical behavior that contradicts verbal behavior. When verbal and nonverbal messages are delivered at the same time, but contradict each other, the listener tends to believe the nonverbal message.

There are definite actions that comprise positive body language: facial expressions and head movements that encourage, eye contact that is natural and continuous, body positioning, open posture, leaning forward, and maintaining appropriate physical distance. All of these skills, when used appropriately, can be very effective in clearing information bottlenecks.

5.4.2   Mental attention

While body language can be used to indicate that active listening is taking place, mental attention lets the employee know that you are using all of your senses to fully appreciate all parts of the message.

The message's verbal component is the simplest and most direct: the words. But the words themselves may not convey the whole message. The simple statement, "I'm tired of hearing excuses from marketing," might indicate despair, frustration, or anger. Here, what is included in the verbal package becomes the message's critical element.

Think of *verbal packaging* as the message's emotional content, which can be determined by observing how fast or loud someone speaks, and by listening for tone of voice. Virtually everything about people, including their competence, integrity, and well-being, is wrapped up in these emotional components.

The message's nonverbal components are particularly useful in meetings with employees because they provide real-time feedback that can be brought to bear on problem solving and decision making. Through the use of nonverbal signals, you can bridge the psychological distance between management and staff. If you greet the employee with a tense jaw, clenched fists, slumped shoulders, or avoid eye contact, subtle messages are being sent that may be interpreted by employees as the signal to stay away. When it comes to improving interpersonal communications skills, you must pay attention to the whole message, integrating its verbal, verbal packaging, and nonverbal elements into the big picture.

5.4.3 Respect

As a manager, you should do everything you can to show that you are interested in problem solving and to demonstrate respect for employees. Respect begins by managing a meeting's setting so that staff members know immediately that they are at the center of the problem-solving process. Without respect, there is little chance that anyone will want to continue the meeting, much less offer helpful information and participate in possible solutions that may improve individual or group performance. In fact, the only thought running through people's minds in such a setting is how to end the session or extricate themselves as quickly as possible.

5.4.4 Invitations

It is your responsibility to draw out answers, helping people to speak freely. This comes naturally to some managers, who use comments such as: "Where would you like to start?" or "How can I assist you?" Those are straightforward invitations to talk. But note that there is a danger in these otherwise valuable comments, especially when they are coupled with careless remarks such as, "After all, that's what I'm here for," "That's what I'm paid for," or "It's your nickel."

In such cases, staff members will not know if they are talking to a role or to a person. Role comments serve only to widen the psychological gap. Remember, until the psychological gap is bridged, solutions to problems will take longer, which may delay the successful completion of projects and cost the organization more money. In addition, if such problems persist, you could lose credibility among executives, perhaps giving them reason to reevaluate your ability to handle the job.

5.4.5 Acknowledgments

Look for opportunities to acknowledge what others are saying. This provides a clear indication that they are being understood and that they should continue speaking. Acknowledgment signals may be as simple as nodding in agreement or uttering a few "uh-huhs." But watch out for their overuse. Bobbing the head continually, for example, eventually loses its impact and might even offend some people.

These are some of the communications skills that managers should demonstrate during problem-solving sessions, especially when counseling individuals on performance-related matters. Such skills are important for enlisting the confidence and cooperation of staff, who may or may not forgive affronts to their intelligence, skills, knowledge, status, or abilities. But there's more. Let us turn now to a discussion of "integration" skills.

5.5   Integrating the Information

A separate step in the interpersonal communications process is integration. Here, the employee continues to talk about his or her needs with your encouragement and shared insights. You must not only bridge psychological distance but also encourage and stimulate the disclosure of more and more information until the employee gains new insights into the nature of the problem at hand. There are at least six skills that you should master to help staff members arrive at this point: reflection, self-disclosure, immediacy, probing, checking, and confrontation.

5.5.1   Reflection

Reflection is a response that lets others know they are being heard. Unlike mere acknowledgment, which is passive, reflection is an active process that helps the employee focus on areas that require further exploration. For example, a frustrated IT supervisor might talk about personnel turnover problems in her group, which she believes are responsible for low morale and missed deadlines on even routine equipment move-and-change requests. A skilled manager can use reflection to zero in on an unstated problem, as in the following scenario:

> IT SUPERVISOR: "I have so many problems with my staff, so much turnover. Now you give me responsibility for processing half a million call records per month to account for telecommunications costs."
>
> NETWORK MANAGER: "You're struggling to improve your operations, but you feel that taking on the additional burden of call accounting at this time may hamper your efforts?"

The manager should employ this strategy to examine certain aspects of a problem in more depth. If the staff member acknowledges the manager with an appreciative "yes" or merely a nod, and then continues, that is a clear sign that the strategy of reflection has worked effectively.

5.5.2   Self-disclosure

By using *self-disclosure,* you can encourage the speaker to continue by interjecting a relevant comment that indicates that you have run into a similar problem before. If appropriate, you can even relate the solution. But avoid the pitfalls of this technique. One is the tendency to monopolize valuable time, and the other is the tendency to change the subject.

Some managers are reluctant to use self-disclosure because it deals with here-and-now feelings and issues generated during face-to-face meetings. They are afraid to provoke or embarrass others, to appear intimidating, or just cannot bring themselves to confidently propose an appropriate course of action. Other managers resist self-disclosure, fearing that it will be interpreted as a sign of weakness. Notwithstanding these peripheral issues, self-disclosure reminds the employee that the manager is also a human being with a wide range of experience, which is what makes problem solving possible.

This skill should be used with discretion, however. Its injudicious use could interrupt the session entirely by drawing too much attention to yourself, which may be misinterpreted by listeners as your ego speaking. Managers who engage in this behavior are not only wasting organizational resources and widening the psychological gap but are also delaying a solution to the employee's problem.

5.5.3   Immediacy

Immediacy deals with the here-and-now behavior that staff may display among themselves or to others. The way to handle this situation may very well determine the outcome of a project. For example, a key staff member may appear anxious about an outside consultant's role. A staff member may think that top management is giving him or her a vote of no confidence by inviting the manager to bring in a consultant to deal with a particular problem and make a recommendation. If these concerns go unaddressed, those pent-up feelings may undermine these efforts by hampering the consultant's ability to obtain important information and staff participation.

A good manager will not let such problems fester for too long. At the most opportune time, the manager must defuse this time bomb, perhaps with the pointed observation, "I have the feeling that it is difficult for you to be as candid as you would like. Are you bothered because a consultant is looking into matters that you are responsible for?"

Applying the skill of immediacy is useful for dealing with other situations as well. For example, if any member of the staff has the tendency to ramble on aimlessly, you can use the skill of immediacy to help that person zero in on the problem. An appropriate remedy could come from a statement like: "You know, I'm feeling a bit confused right now. We seem to be covering a lot of ground, but not getting down to the real problem." As with self-disclosure, the skill of immediacy must be used carefully. Any hint of cleverness or of adopting a superior attitude could negate whatever benefits the strategy was intended to achieve in the first place.

5.5.4   Probing

Probing refers to the exploration of some area or issue that the employee has stated directly. It can help you, as a network manager. develop key points, define problems more completely, or identify patterns in nonconstructive thinking.

5.5.5   Checking

With checking, you can structure responses that confirm understanding of what the staff member has said. For example, you might begin with a phrase such as, "It seems to me that you have identified at least three network configurations worth looking into…," then repeat back the essence of the employee's message to verify understanding.

5.5.6   Confrontation

Unlike probing, confrontation deals with information that has only been hinted at and not directly stated. Using confrontation assumes that a satisfactory relationship has already been established. You should start with the least difficult or threatening point and move to the more difficult or threatening subject. Confrontation must be tentative, but specific, to be effective. Phrases that convey tentativeness include: "I wonder if…" or "Could we talk about…?"

The staff member should know that confrontation has taken place, but should not feel boxed in or under attack. A sign that confrontation has not registered is when the staff member ignores the statement, or becomes tight-lipped or defensive.

Confrontation can serve many purposes. It can help managers and staff look at problems from different points of view, or see possible consequences previously overlooked. It also aids in uncovering a fundamental need or an underlying problem. It may even help a staff member take ownership of a statement or feeling. For example, the person who felt threatened by the presence of a consultant may say something like, "Some of the staff are really going to be turned off by that vendor recommendation." You should be skilled in the use of confrontation, perhaps handling the problem in this manner: "It would help me a lot if I knew you were describing your own feelings. Do you know something about this vendor that should be taken into account by the consultant?"

5.6   Support

In the support phase of the interpersonal communications process, you must help define problems, deal with problem ownership, and develop action plans.

5.6.1   Problem definition

During problem definition, you will work with staff members to develop statements that accurately describe the problem(s) they intend to address. After appropriate discussions and research, you should test the best course(s) of action—through statements and restatements—until problems are specific and realistic. You should also make sure that staff members can "own the problem," at least temporarily, and that they have the appropriate experience, skill level, and resources to implement the action plan.

Typically, it will not be possible to develop an action plan until specific problem statements are formulated. This can be a difficult skill for some managers to master, particularly for those who are accustomed to acting on imprecise statements. "This is a lousy modem." "I'm hearing bad things about that new technician." "I can't see myself committing to that network design." All such statements beg for clarification. By themselves, they can never be translated into specific action plans that anyone will accept. The result will be that any solutions you come up with will be off target.

In helping others define the problem, you must deftly steer them from the general to the specific, as in the following progression of statements made by the IT supervisor in a previous illustration:

> "I don't like the idea of having responsibility for corporate telecommunications."
>
> "I don't have time for that sort of thing."
>
> "My staff is specialized and dedicated to other priorities."
>
> "I don't know anything about telecommunications; this assignment is an invitation to fail so they can get rid of me."

Until the manager can arrive at the crux of the problem—in this case, unfounded insecurities—any advice offered by the manager without this vital information may prove to be far off base. By moving toward specific problem statements, you can dramatically increase the employee's understanding of the problem's fundamental nature and scope.

5.6.2   Problem ownership

Problem ownership is a prerequisite to developing action plans. It is achieved through a common search for solutions. You can encourage others to own problems by presenting solutions with "we" comments. Staff members will assume ownership as they work with you to develop problem statements. Once those statements have been worked out and are accepted, both parties can turn the statements into goals, then agree on a strategy for reaching them.

5.6.3   Developing action plans

There are a variety of methods for developing action plans, but the minimum steps include:

- List several possible solutions and prioritize them.

- Reach a consensus on the alternatives that appear to be the most acceptable to the organization in terms of needs, budget, and personnel resources.

- Agree on a fallback position in case the preferred solution proves unworkable.

- Assign specific tasks to the individuals best qualified to carry them out.

- Develop milestones with mutually agreed-upon time frames for completion.

- Plan follow-up meetings to monitor satisfaction with progress and to make appropriate refinements in the action plan as new knowledge becomes available or as new developments in the implementation may warrant.

The action plan ensures that management and staff are working toward the same goal and that all know their role and what they must contribute to achieve a successful outcome.

5.7   Withdrawal

Having arrived at a solution with relevant input from staff members, you must initiate a withdrawal process that weans the staff members from ongoing dependence on management for support. Two skills are involved in this phase: centering and appreciation.

5.7.1   Centering

Centering involves identifying the staff members' strengths, especially those displayed in productive meetings, and making positive comments about them. This is not a public relations gimmick. Instead, it is an attempt to help the staff feel more secure and confident about implementing the action plan. This is especially important in cases where the results will not become immediately known. Employee strengths that may deserve mention include candor, analytical skills, commitment to problem solving, and progress at arriving at the most appropriate solution.

5.7.2   Appreciation

Other staff members may have played an important role in the problem-solving process by choice or by chance. A final skill in the withdrawal phase is for the manager to express appreciation to each staff member for the time and effort he or she has put into the sessions and into completing each task.

Why should you care about these skills, or even about the withdrawal phase in general? Quite simply, management's behavior at the conclusion of a project will determine the likely level of staff support in the future. If, for example, managers do not acknowledge the contributions of others in arriving at a solution, staff members are less likely to lend their cooperation in the future. After all, it is only human nature to want credit for one's efforts. Nothing is more demoralizing to staff members than to watch others repackage their ideas for top management's consumption and get rewarded for it. Engaging in such antics will not go unnoticed, and will reflect badly on the manager's credibility with staff, peers, and superiors.

5.8   Why Bother?

When interacting with technical professionals, be aware that, as a manager, you must have the skills necessary to deal with people problems while attempting to address technical issues. Recognizing the skills that are crucial to effective interpersonal communications will reduce the risk of making wrong decisions, which can cause staff turmoil, missed project deadlines, and cost overruns.

Mastering the elements of effective interpersonal communications is not easy. Unfortunately, it does not come natural to most people. Complicating the situation are differences in culture, background, experience, training, and a host of other intervening variables. Nevertheless, you have a responsibility to develop skills that facilitate rather than hinder interpersonal communications, mastering them as tools with which to improve problem solving and decision making.

Of course, each manager's style is unique. Moreover, style must be adjusted to the demands of particular situations. Interactions with subordinates, peers, superiors, and outside constituents call for slight adjustments in style and technique. Whatever the style, messages must be relevant, congruent, and comprehensible. After all, managers occupy a unique position in the organizational hierarchy. In a technical environment staffed with highly educated professionals, your credibility is your greatest asset. Credibility can be enhanced and overall performance improved when you set the example as a clear, direct, and honest communicator.

5.9   Staff Expectations

There is virtually universal consensus among network managers that they need to stay abreast of technologies to be effective at their jobs. Any technology manager who does not stay current probably will not succeed for long. This does not mean, however, that a working knowledge of all technologies is required. In fact, staff members typically do not want their managers taking a hands-on approach to technical tasks, including programming.

It is more important for you, as a network manager, to be able to provide staff with the resources they need and be able to remove any roadblocks that stand in the way of progress. It is most important to know the issues—organizational, business, and technical—so you can address them. Still, you should not rule out knowledge of technical details altogether. A working knowledge of various technologies can help move projects ahead, instead of allowing them to become bogged down by staff uncertainty and procrastination.

Although you should stay current on technical issues, you should not necessarily be required to perform technical duties, such as network reconfiguration, troubleshooting, or programming. Generally, nonmanagement staff members have the strongest technical skills. If the environment is such that the technical staff has the ability to empower themselves, you should trust that they know what they are doing from a technical standpoint. Your job becomes one of leading the staff, balancing workloads, and marshaling resources. To do this well, it is important that you have a basic understanding of the technologies they are working with and the communications skills that can be used to deal effectively with staff problems.

5.10   Employee Evaluations

Often evaluations are not performed for technical professionals. Many managers believe that technical staff is challenged every day and, consequently, enjoys a high degree of job satisfaction. To a large extent, this is true. But it does not follow that performance evaluations should be ignored. Technical professionals value feedback on their performance just as any other employee does. In fact, it can be argued that technical professionals value feedback *more* than anyone else. Because they are more apt to have ego involvement in their work—putting their knowledge and skills on the line every day—they tend to view performance evaluations in the same way. They need independent confirmation of their performance, and failure to receive it—good or bad—can be a serious blow to their ego.

5.10.1 Reasons to evaluate

There are a number of reasons for performing evaluations for technical professionals. These reasons do not differ significantly from the reasons for doing evaluations for any other type of professional. Performance evaluations can have one or more of the following objectives:

- Provide basis for performance feedback
- Set work goals
- Determine merit increases
- Identify training and development needs
- Identify candidates for promotion
- Document employment actions such as promotions, job assignments, discipline, and terminations
- Identify special skills, abilities, and interests

5.10.2 Information gathering

Relevant information should be gathered before starting any employee review. Some possible sources of information include:

- *The employee.* Interview the employee about the job and have the employee complete a self-review. Often, a self-review can yield insights that cannot be obtained in any other way.
- *The job description.* The original job description can be used to review and confirm the employee's essential duties and responsibilities. If the employee is doing more than expected, this is a good indicator of initiative and responsibility.
- *Previous reviews.* In comparing historical results, goals, and plans for improvement to the present, you can address the reasons for inferior performance or reward superior performance.

- *Personal notes.* An ongoing log of observations, critical incidents, and factual material is useful for obtaining a long-term view of employee performance, rather than relying on short-term memory, which does not present a fair and accurate assessment of employee accomplishments.

- *Others who work with the employee.* The observations of other managers, co-workers, subordinates, and even clients may help to validate your observations and opinions.

5.10.3   Common traps

In the process of performing employee evaluations and to avoid common evaluation traps, you, as a network manager, should ask yourself the following questions:

- *Is it job-related?* Make sure that the performance being evaluated relates to the essential duties of the job, rather than more peripheral assignments.

- *Can I support it?* Write evaluations that can be supported by objective observations or valid data such as response times and error rates.

- *Am I being fair and consistent?* Many management and labor law problems arise when employees are evaluated unfairly and inconsistently.

- *Am I being overly lenient?* If performance problems are not addressed immediately, they rarely go away by themselves. Leniency bias appears harmless but is actually very dangerous. When problems are addressed after a history of good reviews, far more work is required to support the corrective action that must be taken. Legal considerations can also become a factor.

5.10.4   Review methods

When it comes to the review itself, there are several ways to proceed, including:

- *Self review.* This method entails giving the employee a blank review form and asking him or her for a self-review. This increases employee involvement and commitment to the review process. When you meet, the observations are compared jointly and agreement is reached on the substantive points.

- *Pre-review.* This method entails giving the employee a copy of the completed review. During the meeting, observations of performance are compared. More attention should be spent on areas of agreement rather than on areas of disagreement.

- *No pre-review.* Without giving the employee a copy of the review, talk the employee through the performance review. This focuses the employee's attention on the discussion itself rather than on the written review.

- *Write the review together.* With this method you and the employee sit together to do the evaluation. With each element on the form, you can talk through the events of the review period. This can make the process a joint effort and facilitate agreement.

There are many other ways to conduct performance reviews, depending on the structure of the organization and the employee's position and job responsibilities. For example, there is the "360 review," which is designed to measure how well an employee performs throughout the organization. Typically, when conducting such an assessment, the employee's subordinates review this individual. Secondly, this person's co-worker, or perhaps someone in another department who works daily with the employee, conducts the evaluation. Finally, the employee's supervisor approves the evaluation. This process allows you to view how successfully certain employees conduct themselves within the company. For instance, after completing an evaluation in the 360-review manner, it may be discovered that help desk staff communicates extremely well with management and co-workers, but needs dramatic improvement when communicating with repeat callers, who are viewed as nuisances.

Another way of performing evaluations is the bottom-up method. In this case, a manager will receive a review from each of the employees that he or she supervises. Generally, the manager's supervisor or perhaps a co-worker conducts the evaluation. This option can provide a fair and accurate way to evaluate the performance of the management level within the organization.

Peer-to-peer reviews can be effective in determining how well a team performs together. To do this, each member of a team reviews every other member of the team. In turn, their supervisor generally conducts the evaluation.

5.10.5   Computerized evaluation tools

To make staff evaluations easier, systematic, and more accurate, a variety of software packages are available. Generally, they organize information into employee file folders that include the following fields:

- Employee name
- Gender
- Job title
- Job code

- Date of hire
- Date of job start
- Salary grade
- Reviewer title
- Salary
- Department
- Division
- Location
- Review period
- Date of last review
- Date of next review
- Reviewer name
- Type of review: introductory, merit, step, annual
- Previous review rating
- Date of last promotion
- Number of days absent during review period
- Number of days tardy during review period
- Date job description last reviewed

In addition to these fields, there may be provisions for an employee photograph and fields for free-text input that can be inserted into the file folder as notes (see Fig. 5.1). With this and other information entered into the computerized file folder, a database is created from which it is possible to call up historical views of employee performance based on past reviews plotted in graphical form. This allows you to easily track the progress of an employee from year to year, according to various standards of performance, which might include such things as communications skills, initiative, reliability, teamwork, and professionalism. It is also possible to view how an individual compares in these categories to other employees or to others with the same job description.

Some packages walk managers through every step of the evaluation process, ensuring consistency and completeness (see Fig. 5.2). Daily observations and coaching suggestions can be turned into valuable feedback with thorough documentation. A properly implemented appraisal program promotes a better understanding of expectations and work responsibilities among employees.

Each employee's computerized file folder can include a goals page that monitors individual performance with regard to the goals that he or she has attempted to attain. A complete list of the employee's goals—completed and not completed—appears. Any goals that are past their designated completion dates are highlighted. Completed goals are rated on a scale of 1 to 10, with 10 representing the highest rating.

When employee evaluations are computerized, other aspects of the evaluation process can be checked more easily. Take the time to view peer scores and comments to help prevent personal biases from affecting the results of an evaluation. Another way to guard against incorrect evaluation results is to compare evaluator ratings. If it is revealed that one evaluator consistently scores everyone far lower than any of the other reviewers on technical proficiency, for example, perhaps that person needs further instruction or coaching on how to perform evaluations in this area.

Software tools for employee evaluations provide managers with the ability to centrally control the review process from start to finish:

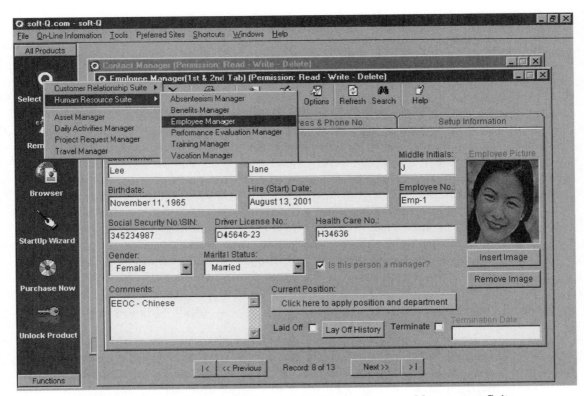

Figure 5.1   The performance evaluation module of soft-Q's Human Resource Management Suite.

- Set and measure goals.

- Encourage regular feedback between reviews.

- Extract performance data for analysis.

- Work on reviews at home or on the road.

- Choose from a library of performance elements.

- Assign weighting factors to goals and performance elements.

- Keep track of upcoming reviews with the built-in calendar.

Depending on the specific tool selected, you can customize forms, import employee data, extract review data for analysis and reporting, and even add your own performance criteria, review text, and coaching ideas.

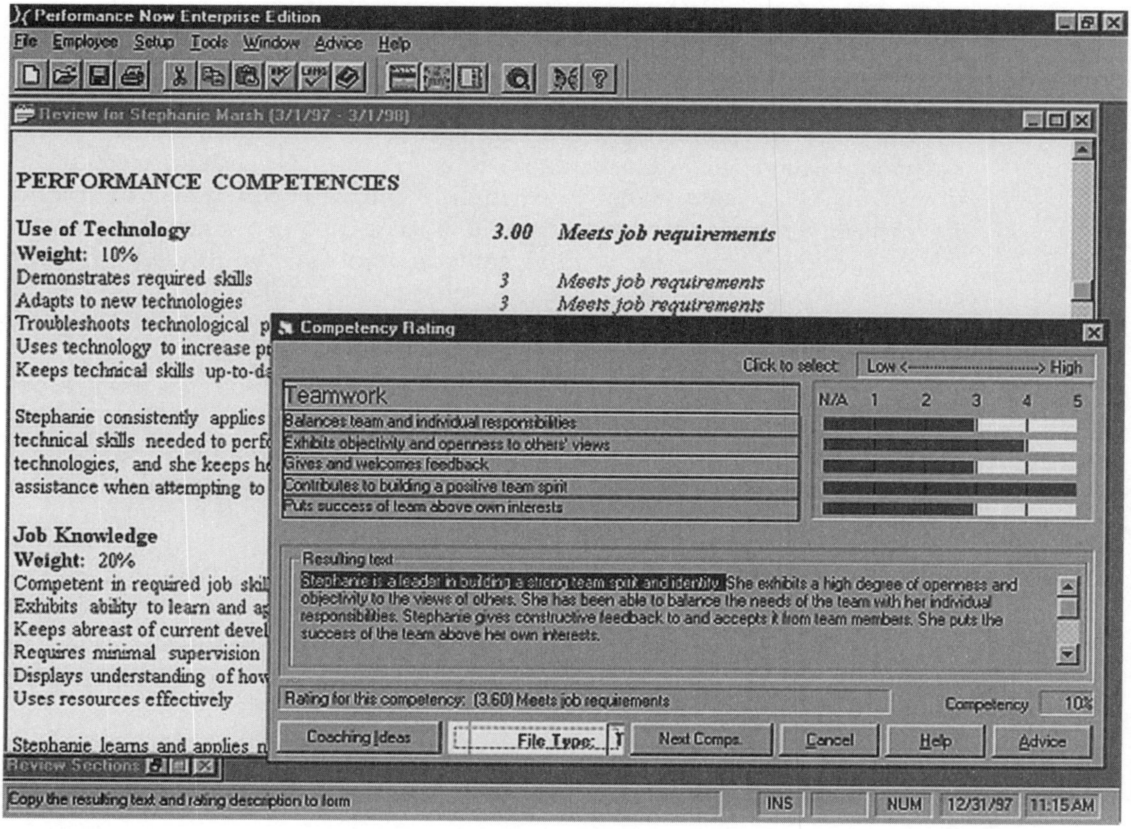

Figure 5.2   KnowledgePoint's Performance Appraiser walks managers through every step of the evaluation process.

The issue of security poses a dilemma when maintaining employee records. Many businesses keep several files for each employee in order to segregate sensitive information, or they strictly control the location of employee files, often placing them under lock and key. While these steps help ensure both the employees' and the employer's privacy, they usually hinder accurate maintenance, and make record auditing nearly impossible.

Computerized solutions provide security through password access to an encrypted database that allows for user-specific security. This restricts each evaluator to his or her appropriate area of responsibility. The computer screen is even modified to eliminate and deactivate inappropriate functions and views. This format provides the necessary wide access to the database, without permitting the display of sensitive information.

5.11 Computerized Scheduling Tools

Large telecom and IT departments can often benefit from software that tracks the schedules of staff members. Scheduling is both varied and complex, depending on company operations and management style. Scheduling software can be used to track such things as daily attendance, training, vacations, days off, and travel. It can also be used for work scheduling and to quickly determine which people are available for such things as special projects, overtime, and field visits. Finally, scheduling software can assist in finding replacements for people who are unable to work their assigned schedule.

For each employee, an annual calendar can be maintained, showing excused and unexcused absences, as well as notes that amplify on the reasons for absences such as lateness, sickness, training, vacation, medical appointment, or injury.

Not only can individuals be scheduled using this type of software but project teams, work groups, and skill groups can be scheduled as well. A work group might be help desk personnel, customer support staff, or data center staff. A skill group might be installers, test technicians, programmers, or trainers.

Some software packages allow existing employee database information to be imported, which eliminates manual data entry. Suitability lists make it easy to match employee qualifications with project needs. When scheduling conflicts or inconsistencies occur, a warning alerts the supervisor or manager. The use of such software can reduce the time spent on scheduling tasks by as much as 75 percent.

5.12 Conclusion

Managers and supervisors need to update employee records as part of their daily routine. Software packages that help manage, evaluate, and schedule employees make documenting employee activities easier. Such software can also be used to track benefit eligibility, job history, certificate expirations, and numerous other practical employee statistics, which even the smallest businesses need to keep current.

With proper documentation, a business can minimize legal exposure owing to wrongful termination and discrimination litigation. To further minimize legal exposure, businesses can use such software to keep pace with the frequent changes in government demands for employee record management. For example, employers with more than 50 workers need to keep records appropriate for compliance with the 1993 Family Leave Act, which requires detailed day-off accountability and benefit tracking. And when employers grow beyond this size or look to do business with the government, they need to prepare themselves for compliance with the Equal Employment Opportunity (EEO) regulations, which require detailed reporting about the ethnic and gender composition of the company laid out by positions of employment and advancement.

There are several bottom-line benefits that can accrue to the company as a result of being able to effectively manage, evaluate, and schedule technical professionals. First, there is the immediate payback on capital investments in computer and communication technologies, since they will be used to optimal advantage. Second, productivity will also be improved since outages by users and interruptions of telecom and IT staff will be minimized. Third, with increasing reliance on information systems and data networks to support mission-critical applications, their problem-free availability can improve end-user productivity and customer response, while enhancing corporate success in competitive markets.

Outsourcing Infrastructure

6.1 Introduction

With computing and networking environments becoming ever more powerful, complex, and expansive, an increasing amount of the corporate budget is being tied up to support them. Enormous amounts of capital are invested annually in bricks and mortar, hardware and software, lines and circuits, security, electrical power, backup systems, management tools, and technical staff—just to name a few. Instead of building and maintaining this infrastructure themselves, more and more companies are turning to outsourcing arrangements and hiring qualified third-party firms to take care of these essential functions. In offloading these burdens to outside firms and paying a fixed monthly charge, companies can contain operating costs, eliminate waste, use resources more effectively, and become far more competitive by funneling more of their resources to core business activities.

Outsourcing is not a new idea. It is part of a widespread trend that started in the mid-1980s called downsizing, whereby companies flattened organizational structure, streamlined business processes, distributed operations, and reduced staff to serve customers more efficiently, become more competitive, and improve financial performance. This trend continues today as companies seek to readjust themselves to new production technologies, market realities, and general economic conditions. What *is* new is the kind of activities that can be outsourced. In addition to the traditional management of systems and networks, companies can outsource entire data centers, security, content delivery, and enterprise applications. Even product development, marketing, billing, and customer service functions—still considered core business activities—can be outsourced. With outsourcing taken to the extreme, companies one day may not need to have a physical presence at all.

Instead, they may be just as effective in a virtual state of existence—merely a presence on the Web—that does not require the trappings of traditional infrastructure.

Outsourcing, once considered an arrangement of last resort for financially strapped businesses, has instead become part of the overall strategic vision of virtually all organizations, large and small. This arrangement is no longer just a cost-cutting measure for companies; many now view outsourcing as a way to acquire information technology, business processes, network capacity, global reach, and management expertise that will help them become more competitive and expand into new markets more quickly. While companies are still looking to reduce costs at every opportunity, they are also looking for a flexible infrastructure that will sustain them through alternating periods of expanding and contracting market demand without becoming bogged down by a standing army of people, fixed assets, and escalating expenses for data centers and networks. Outsourcing arrangements, properly structured, can meet all of these needs in a timely and economical manner.

6.2   Reasons to Outsource

The reasons a company may want to outsource are varied. They might include:

- Difficulty in using technical personnel efficiently or in upgrading their level of expertise

- Insulating management from day-to-day system problems and decisions, so it can focus more attention on core business issues

- Concern about buying expensive technology that could become obsolete shortly after purchase or before it is fully depreciated

- Greater flexibility to deal with fast-changing worldwide markets, government regulations, and differing standards

- Extending presence to new markets economically, without a heavy investment in basic infrastructure

- The need to improve the corporate balance sheet, extend available investment capital, improve efficiency of business operations, and contain long-term costs

Despite the many good reasons to outsource, there are still many concerns associated with putting critical resources in the hands of outsiders. Many of these concerns can be overcome with experience and knowledge of typical outsourcing arrangements.

Many corporate executives are concerned about giving up control when considering the move to outsource. However, control can increase when corporate management is better able to concentrate on issues that have potentially greater returns. Instead of consuming valuable resources in the nuts-and-bolts aspects of setting up an Automated Teller Machine (ATM) network, for example, bank executives can focus on developing the services customers will demand from such a network and devise test marketing strategies for potential new financial services.

A common refrain among corporate executives is that outsource firms do not know their companies' business. In any outsourcing arrangement, however, users continue to run their own applications as before; the service provider just keeps the data center or network running smoothly. In addition, outsourcing firms typically hire at least some members of the client staff who would have been let go upon the decision to outsource and who are familiar with the business.

Companies that are considering outsourcing should examine their current information system and network activities in competitive terms. Activities that are performed about the same way by everyone within a particular industry can be more safely farmed out than those that are unique or based on company-specific skills. Most important, the company must take precautions to remain in a position to recommend and champion strategic systems and new technologies, which may involve high initial payout and possible cross-functional applications.

As the company opts for external solutions, standards that were internally developed do not suddenly lose their relevance. Oversight of standards that address hardware, communications, and software should remain an internal responsibility to ensure compatibility of information systems and networks across the enterprise.

Outsourcing can increase service quality and decrease costs, but management control cannot just be handed over to a third party. The fact that work has been contracted out does not mean that you, as a network manager, can or should stop thinking about it. Typically, there is still a significant amount of advisory and supervisory overhead that consumes resources.

Someone within the company must ensure that contractual obligations are met, that the outsourcing firm is acting in the company's best interests, and that problems are not being covered up. Since the outsourcing firm will usually not have the detailed knowledge of network operations as in-house staff, you, as a network manager, should expect to play an advisory role in getting the outside firm up to speed. Just as important, considerable effort is usually required to establish and maintain a trusting relationship. To oversee such a relationship requires staff who are highly skilled in interpersonal communications and negotiation, and who are knowledgeable about business and finance.

6.3   Approaches to Outsourcing

There are two basic approaches to outsourcing—each of which can pose clear benefits. First, the outsourcing firm can buy existing information systems, networks, and other assets from the company and lease them back to the company for a fixed monthly fee. In this type of arrangement, the outsourcer can also take over the information systems payroll. This relieves the client of administrative costs, which can be applied to core business operations or used as a resource conservation mechanism to improve the corporate balance sheet.

Typically, the outsourcer maintains the communications equipment and information systems, upgrading them as necessary, but staying within the budgetary parameters and performance guidelines in the master agreement. Sometimes equipment leasing is part of the arrangement. Aside from its tax advantages, leasing can protect against premature equipment obsolescence and rid the company of the hassles of dealing with used equipment once it has been fully depreciated.

A second type of outsourcing arrangement entails the company selling off its equipment and migrating the applications and data to the outsourcer's data center and high-speed network. Often, the company's key personnel, who have been reassigned to the outsourcer's payroll and who receive a comparable benefits package, manage the data center and network. Employees can even find new career opportunities with the outsourcing firm, since it may provide a wide array of services to a broad base of clients worldwide.

In the first case, the outsourcing arrangement leaves the company's assets where they are, minimizing costs and simplifying administration. In the second case, the assets are moved from the company's location and reestablished at the outsourcing firm's location. This usually entails extra cost because duplicate systems and networks must be maintained to enable business processes to continue during the transition. Ultimately, this has the benefit of freeing up the company's office space for equipment and personnel for other uses or closing it down for cost savings. When network equipment is involved, this type of outsourcing arrangement is often referred to as *collocation*. When applications and data storage are involved, the arrangement is often referred to as *hosting*. Today, the distinction is blurred because outsourcing, more often than not, includes elements of each.

6.4   Outsourcing Trends

Outsourcing is not new in the information systems arena. Historically, service bureau activity has been associated with data center outsourcing, in which in-house data centers were turned into remote job-entry operations to support such applications as payroll, claims processing, credit card invoicing, and mailing lists for mass marketing operations. Under this arrangement, mainframes were owned and operated off-site by the computer service company and processing time was reserved for customers who paid an hourly fee.

In recent years, this type of arrangement has been extended to include facilities management. With this type of outsourcing, an outside firm takes over on-site management of a corporate data center. The service provider typically brings in some of its own people and keeps some of the client's staff, maintaining the data center as it is currently set up at the customer location.

As applied to networks, however, the outsourcing trend is more recent. An organization's Local Area Network (LAN) may be thought of as a computer system bus, providing an extension of data center resources to individual user's desktops. Through Wide Area Network (WAN) facilities, data center resources may be extended further to remote offices, telecommuters, and mobile professionals. Given the increasing complexity of current data networks and the expense of security to protect mission-critical data, it is not surprising that companies are seeking ways to offload management responsibility to those with more knowledge, experience, and hands-on expertise than they alone can afford.

Today's outsourcing vendors offer a mix of technology and management solutions, including consulting, client-server applications development, project management, and life-cycle services. A total outsourcing package typically includes an analysis of the company's business objectives, an assessment of current and future computing and networking needs, and a determination of performance parameters to support specific data transfer requirements. Responsibility for the resulting system or network design may encompass the local and interexchange facilities of any carrier and equipment from alternative vendors. Acting as the client's agent, the outsourcing firm coordinates the activities of equipment vendors and carriers to ensure efficient and timely installation and service activation.

In one type of outsourcing arrangement, an integrated control center—located at the outsourcing firm's premises or that of its client—serves as a single point of service support. Technicians are available 24 hours a day, 365 days a year to monitor network performance, contact the appropriate carrier or dispatch field service as needed, perform network reconfigurations, and take care of any necessary administrative chores.

Selective outsourcing—sometimes called *out-tasking*—is a growing trend. More companies want to outsource only a narrowly defined portion of their operation. For example, they may want to outsource only the help desk, application development for a particular platform, hardware procurement, network integration, or distributed computing services. Selective outsourcing includes transition outsourcing, which typically involves helping companies migrate from a legacy system to a networking or client-server system.

Rather than outsource responsibility for the entire network, many companies out-task specific planning, operations, or management functions to a contractor. This approach is less disruptive to daily business operations than a full-blown outsourcing deal because it eliminates the wholesale transfer of assets and personnel to an outside vendor. Some other advantages of out-tasking include:

- It is less risky than turning over responsibility for an entire network to an outsider.

- It can become the basis for establishing a partnership with a vendor that develops over time or is terminated as needed.

- It can result in greater control over the network, since only nonstrategic tasks are parceled out.

By reducing risks, out-tasking can result in immediate cost savings and faster productivity improvements. At the same time, short-term outsourcing is becoming more popular than long-term arrangements. More companies are willing to give up control of select applications or functions in the short term, either because they want to focus inwardly on employee training or retooling, or outwardly on adding value to business processes. Their desire to achieve short-term objectives is driving them to seek flexible outsourcing arrangements that last two or three years instead of seven to ten.

6.5   What to Outsource

Networks are growing rapidly in size, complexity, and cost; technical experts are expensive, hard to find, and hard to keep; new technologies and new vendors appear at an accelerating rate; and users clamor for more and better service while their bosses demand lower costs and increased work performance. Consequently, the question may no longer be whether to outsource but what to outsource.

Just about any aspect of systems and network operations can be outsourced in almost any combination. Start by looking at functions that are not critical to the company's core business, especially those that the telecom or Information Technology (IT) department cannot perform cost-effectively. The following network functions may be considered for outsourcing:

- Service and support operations:
  - Help desk
  - Customer support
  - Consulting: end user, workgroup, department, enterprise
  - Problem analysis and management
  - Billing inquiries and reconciliation
  - Repair and installation dispatch
  - Moves, adds, and changes
  - Management reporting

- Network monitoring and diagnostics

- Network design, installation, and management

- Local access and integrated services

- WAN voice and data services

- Network security

- Branch office communications management

- Traffic analysis and capacity planning

- Disaster planning and recovery

- New product testing and evaluation

- Technology assessment and migration

- Enterprise applications, such as email

A rule of thumb is to keep in-house any highly critical activity that is being performed cost-effectively, but outsource noncritical activities that are not being performed cost-effectively (see Fig. 6.1).

A company can let go of the commoditylike operational functions, sometimes called *tactical functions*. It is possible to both save money and eliminate headaches by letting someone else pull wires, set up circuits, move equipment, and troubleshoot problems. However, it may not be wise to farm out mission-critical or strategic functions. After all, if the outsourcing firm performs poorly, for whatever reason, the client company's competitive position could become irreparably damaged. On the other hand, resource-constrained companies may have no choice but to outsource an important function like network security. The cost for qualified staff such as Certified Security Engineers (CSE) or Certified Security Administrators (CSA) to manage firewalls on a 24×7 basis, stay updated on the latest security threats, and implement attack countermeasures is cost-prohibitive for all but the largest companies.

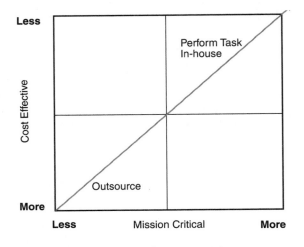

Figure 6.1   Outsourcing decision matrix.

One way for a company to take a first step down the outsourcing path is to give its network to a vendor during second and third shifts and weekends, while keeping control over the more critical prime time. As confidence in the vendor grows, more can be outsourced. Another way to ease into an outsourcing arrangement is to farm out only one or two discrete activities. Choose those that will provide the telecom or IT staff with some relief. If these activities are successful, consider adding more functions. Each additional function can be farmed out to new or existing vendors, and the desired amount of control can be exercised over each separate activity.

When considering network outsourcing, organizations typically expect to achieve one or more of the following:

- A single point of contact for the management of all voice and data communications

- A reduction in the number and severity of problems that affect service

- Improvements in network performance, availability, and reliability through proactive management

- Better response time for move, add, and change requests

- A reduction in the number of personnel involved in day-to-day operations, with the possibility of redeploying these resources to core business activities

- Better management of equipment inventories, vendor warranties, license agreements, and maintenance schedules

- Obtaining all of the above more efficiently and more economically, while providing higher-quality services to end users without compromising security

With network management costs increasing by more than 30 percent a year, large companies are being forced to reengineer their networks more often. Among the reasons for skyrocketing costs are the increased emphasis on network security, the pressure to deploy e-commerce and hosted applications on the Internet, the need to distribute storage and mission-critical data to safeguard them from loss due to disaster, continued implementation of client-server architectures, and development of enterprise applications. These activities entail enormous hidden costs in the form of skilled people. Many of these costs can be trimmed or eliminated by outsourcing.

Companies with multinational locations are not likely to have a significant level of in-house expertise in more than a few national markets. For example, it is difficult to stay informed of the latest standards and compliance issues of each Post Telephone & Telegraph (PTT) administration. This aspect of international operations alone could require dedicated staff. From logistics, billing, and network management standpoints, an outsourcing firm with an international presence can shave as much as 30 percent off the cost of global network operations.

6.6 Typical Outsourcing Services

The specific activities performed by the outsourcing firm may include:

- Moves, adds, changes

- Systems and network integration

- Project management

- Trouble ticket administration

- Management of vendor-carrier relations

- Maintenance, repair, replacement

- Disaster recovery

- Technology migration planning

- Training

- Equipment leasing

- LAN administration and management

- WAN administration and management
- Network security
- Enterprise applications

The list of activities that can be outsourced is continually expanding. The list above represents the most commonly outsourced activities and is discussed in the following sections.

6.6.1   Moves, adds, changes

Even with the right tools, handling Move, Add, Change (MAC) requests constitutes a daily grind that can consume enormous corporate resources if handled by in-house network administrators. For both LANs and WANs, the MAC process typically includes the following activities:

- Processing move, add, and change orders
- Assigning due dates for completion
- Providing information required by technicians
- Monitoring move, add, and change service requests, scheduling, and completions
- Updating the directory database
- Handling such database modifications as features, ports, and user profiles
- Creating equipment orders upon direction
- Maintaining order and receiving logs
- Preparing monthly move, add, and change summary and detail reports
- Managing and accounting for all hardware and software assets

In assigning these activities to an outside firm, the company can realize substantial cost savings in staff and overhead, without sacrificing efficiency and timeliness.

6.6.2   Systems and network integration

Current information systems and communications networks consist of a number of different intelligent elements: host systems and servers, LAN hubs and switches, routers and gateways, and local and WAN facilities from various carriers, to name a few. The selection, installation, integration, and maintenance of these elements require a broad range of expertise that is not usually found within a single organization. Consequently, many companies are increasingly turning to outsourcing firms for systems and network integration services.

Briefly, the integration function is concerned with unifying disparate computer systems and transport facilities into a coherent, manageable utility, a major part of which is reconciling different physical connections and protocols. The outsourcing firm also ties in additional features and services offered through a public-switched network. The objective is to provide compatibility and interoperability among different products and services so that they are transparent to the users.

The evaluation of various integration firms should reveal a well-organized and staffed infrastructure that is enthusiastic about helping to reach the customer's networking objectives. This includes having the methodologies already in place, the planning tools already available, and the required expertise already on staff. Beyond that, the integrator should be able to show that its resources have been successfully deployed in previous projects of a similar nature and scope. And since most integrators are not staffed and equipped to handle every conceivable type of project, they should be able to tap the expertise they need through strategic relationships with partners.

6.6.3   Project management

Project management entails the coordination of many discrete activities, starting with the outsourcing firm's development of a customized project plan based on the client's organizational needs. For each major and minor task, critical requirements are identified, lines of responsibility are drawn, delivery milestones are established, dependencies are identified, and problem resolution and escalation procedures are agreed on.

Line and equipment ordering, for example, might be included in a network upgrade project. Acting as the client's agent, the outsourcing firm negotiates with multiple suppliers and carriers to economically upgrade or expand the network in keeping with predefined performance requirements and budget constraints. Before new systems are installed at client locations, the outsourcing firm performs site survey coordination and preparation, ensuring that all power requirements, air conditioning, ventilation, and fire protection systems are properly installed and in working order.

When an entire node must be added to the network or a new host must be brought into the data center, the outsourcing firm will stage all equipment for acceptance testing before bringing it online, thus minimizing potential disruption to daily business operations. When new lines are ordered from various carriers, the outsourcing firm conducts the necessary performance testing before turning the lines over to user traffic.

6.6.4   Trouble ticket administration

A trouble ticket is an electronic record that describes a problem on the network and its resolution. In assuming responsibility for daily network operations, a key service performed by the outsourcing firm is *trouble ticket processing,* which typically involves a high degree of automation. The sequence of events in trouble ticket processing is as follows:

- An alarm with a severity indication is received at the network control center operated by the outsourcing firm.

- Alternatively, a user may call the outsourcing firm's customer support team or help desk to report a problem that has not been discovered through automated monitoring capabilities.

- A trouble ticket is opened, which describes the problem, and it is given a tracking number.

- Restoral mechanisms are initiated (manually or automatically) to bypass the affected equipment, network node, or transmission line until the faulty component can be brought back into service.

- An attempt is made to determine the cause of the problem and, if necessary, duplicate the problem in an effort to find the cause:
  - If the problem is with hardware, a technician is dispatched to swap out the appropriate board or subsystem.
  - If the problem is with software, analysis and correction may be performed remotely.
  - If the problem is with a particular line, the appropriate carrier is notified.

- The client's help desk is kept informed of the problem's status so that the help desk operator can assist local users.

- If the problem cannot be solved within a reasonable time, it is escalated to a higher level. This escalation is noted on the trouble ticket.

- Before closing out the trouble ticket, the repair is verified with an end-to-end test by the outsourcing firm.

- Upon successful end-to-end testing, the primary Customer Premises Equipment (CPE) or facility is turned back over to user traffic and the trouble ticket is closed.

- A record of the transaction is filed in the trouble database to aid future problem solving and staff performance evaluation.

- A message is sent to the affected user or group of users, informing them of the problem's resolution.

Not all problems go through such an elaborate resolution process. For self-service-related PC tasks that do not require help desk assistance, for example, some vendors offer tools that automatically open, resolve, and close trouble tickets, completely eliminating help desk involvement. Help desk personnel even receive credit for successful trouble tickets, even though the open-resolve-close process is completely automated.

6.6.5   Management of vendor-carrier relations

Another benefit of the outsourcing arrangement comes in the form of improved vendor-carrier relations. Instead of having to consume staff resources managing multiple relationships with vendors and carriers, the client needs to manage only one: the outsourcing firm. Dealing with only one firm has several other advantages, including:

- Elimination of delays caused by vendor and carrier finger-pointing

- Improvement of response time to alarms and trouble calls

- Expedited order processing and move, add, and change requests

- Reduced time spent in reconciling multiple invoices

- Freed-up staff time for planning, prototyping, and pilot testing applications

- Containment of the long-term cost of network ownership

Short of a traditional outsourcing arrangement, reliance on an Integrated Communications Provider (ICP) might be a viable alternative. An ICP is a Competitive Local Exchange Carrier (CLEC) that provides multiple services over the same access connection and has its own data backbone network. The ICP provides end-to-end management of customer equipment, lines, and services. In addition to local and long-distance voice services, some of these providers offer Asynchronous Transfer Mode (ATM), frame relay, or Internet Protocol (IP) services through the same multiprotocol platform under centralized 24 × 7 surveillance and proactive management from a network operations center.

ICPs give customers the opportunity to take advantage of the most appropriate technology, or facilitate migration between them as their needs change, without having to deal with multiple service providers and equipment vendors. And with the latest generation of Integrated Access Devices (IADs), which support all of these traffic types, there is no need for customers to add or change the access equipment to take advantage of any or all of these technologies. The ICP takes care of reconfiguring the equipment to handle any combination of services, based on the customer's applications and bandwidth needs.

6.6.6   Maintenance, repair, and replacement

Some outsourcing arrangements include maintenance, repair, and replacement services. Relying on the outsourcing firm for maintenance services minimizes a company's dependence on in-house personnel for specific knowledge about system design, troubleshooting procedures, and the proper use of test equipment. Not only does this arrangement eliminate the need for ongoing technical training and vendor certification but the company is also buffered from the effects of technical staff turnover, which can be a persistent problem. Repair and replacement services can increase the availability of systems and networks, while eliminating the cost of maintaining a spare components inventory.

Contracting with a third-party depot service firm is an alternative to a traditional outsourcing firm for maintenance, repair, and replacement. Many service firms are authorized to work only on equipment from specific vendors so several such firms may be needed to take care of all the systems a company might have. The service firm should be staffed with technicians who have undergone vendor training and passed a certification test. A large firm will offer both on-site and in-house repair services and have a hefty parts inventory. A variety of plans are usually offered, including manufacturer's warranties in which there are no charges, customized warranties in which fixed pricing applies, and no-warranty plans in which time and materials pricing applies for repairs.

Typically, a third-party maintenance and repair firm works with customers in the following manner:

- Customer calls the firm to request service.

- The firm calls its logistics partner to arrange for pickup of the item at the customer site.

- The item is delivered to the depot center overnight where a qualified technician assesses the problem and makes necessary repairs.

- The repaired item is packed and shipped overnight back to the customer site.

The kind of equipment that can be maintained, repaired, or replaced by a depot service center varies. This arrangement is common with printers, computers, and servers, but there are firms that specialize in network equipment such as modems, channel service units, routers, hubs, switches, and multiplexers—just about any device that attaches to a LAN or WAN. Depending on the device, turnaround time averages three to five days, or optional expedited repairs are available. All repairs should be accompanied by at least a 90-day warranty.

6.6.7   Disaster recovery

Disaster planning has always been a mission-critical concern of businesses. And with increasing reliance on information systems and telecommunications services, companies of all types and sizes are reevaluating their plans to accommodate previously unforeseen events in the wake of the terrorist attacks in 2001. As part of this reevaluation process, companies are assessing options that will ensure business continuance under a variety of disaster scenarios.

Disaster recovery includes several components that may be outsourced to ensure the maximum performance, availability, and reliability of computer systems and data networks:

- Disaster impact assessment
- Network recovery objectives
- Evaluation of equipment redundancy and dial backup
- Network inventory and design, including circuit allocation and channel assignments
- Vital records recovery
- Procedure for initiating the recovery process
- Location of a hot site, if necessary
- Distribution of mission-critical applications and data among multiple, geographically separate locations
- Acceptance test guidelines
- Escalation procedures
- Recommendations to prevent loss of systems and networks in the face of new threats
- Security assessment and recommendations to prevent unauthorized access and hacker intrusions to corporate resources

Many of these activities are outsourced to multiple sources such as local and interexchange carriers, Application Service Providers (ASPs), managed service providers, and security service providers. Rarely will there be one source that can handle such a diverse and growing array of corporate disaster recovery needs. And for smaller companies to try to take responsibility for all these activities in-house would be cost-prohibitive.

6.6.8 Technology migration planning

A qualified outsourcing firm should be able to provide numerous services that can assist the client with strategic planning. Specifically, the outsourcing firm can assist the client company in determining the impact of the following:

- Proposed national and international standards
- Emerging products and services
- Regulatory and tariff trends
- International political and economic developments on service availability
- Strategic alliances among vendors and service providers
- Mergers and acquisitions among key vendors and service providers
- Competitive aspects of industry deregulation

With experience drawn from a broad customer base as well as its daily interactions with hardware and software vendors, service firms, carriers, and application developers, the outsourcing firm has much to contribute to clients in the way of assisting in strategic planning.

6.6.9 Training

Outsourcing firms can fulfill the varied training requirements of users, including:

- Basic and advanced communications concepts
- Product-specific training
- Custom training
- Resource management
- Security planning and implementation
- First-level testing and diagnostic procedures
- Help desk operator training

The last type of training is particularly important, since 80 percent of reported problems are applications oriented and can be solved without the outsourcing firm's involvement. This can speed up problem resolution and reduce the cost of outsourcing. For this to be effective, however, the help desk operator must know how to differentiate between applications problems, system problems, and network problems. This basic knowledge may be gained by training and improved with experience.

The outsourcing firm may have an arrangement with one or more strategic partners to actually design and deliver basic, advanced, and custom training to its corporate customers. Training may be provided on-site or at a facility leased from a hotel or conference center and supplemented with self-paced materials on a CD-ROM or on the Web.

6.6.10   Equipment leasing

Many times an outsourcing arrangement will include equipment leasing. There are a number of financial reasons for including leasing in the outsourcing agreement, depending on the company's financial situation. Because costs are spread over a period of years, leasing can improve a company's cash position by freeing up capital for other uses. It also makes it easier to cost-justify technology acquisitions that would normally prove too expensive to purchase.

Leasing makes it possible to procure equipment that has not been planned or budgeted for. Leasing, rather than buying equipment, can also reduce balance sheet debt, because the lease or rental obligation is not reported as a liability. At the least, leasing represents an additional source of capital, while preserving credit lines and investor confidence—the importance of both are magnified during periods of slow economic growth and the consequent scarcity of investment capital.

With new technology becoming available every 12 to 18 months, leasing can prevent the user from becoming saddled with obsolete equipment. This means that the potential for losses associated with replacing equipment that has not been fully depreciated can be minimized. With rapid advancements in technology and shortened product life cycles, it is becoming even more difficult to sell used equipment on the gray market. Leasing eliminates such problems, since the leasing company assumes responsibility for disposing of used equipment.

6.6.11   LAN administration and management

As the size, complexity, and expense of LAN management become too big to handle, companies may want to consider outsourcing the job. Certain LAN functions such as moves, adds, and changes can be safely outsourced because they cause little or no disruption to daily business operations. Such functions as planning, design, implementation, operation, management, and remedial maintenance are commonly outsourced, and

LAN operation and management are the services increasingly needed by many firms.

Providers of LAN-related services can be categorized into three groups: computer makers, management and systems integration firms, and regional and long-distance carriers.

The strengths of the computer firms include a sound service and support infrastructure, knowledge of technology, and a diverse installed base. Their weaknesses include an orientation toward their own products and skills that are limited to certain technologies or platforms.

Systems integration and facilities management firms approach LAN outsourcing from the time-sharing, data center, and mainframe environments. Their strengths include experience in data applications, familiarity with multivendor environments, and a professional service delivery infrastructure. Their weakness is that they often lack international capabilities, although this is changing.

Carriers, including interexchange carriers, Local Exchange Carriers (LECs), and Value Added Network (VAN) providers, have expanded their network integration services to include LAN management outsourcing. They use the remote monitoring functions of Hewlett-Packard's OpenView, for example, to manage the LAN and WAN components of their clients. Some offer different levels of service to suit customer needs and budgets. Basic service might include fault, performance, and operations management. Enhanced service might include all elements in the basic service package, plus network planning and security management. Comprehensive service might include configuration management along with all the features of the enhanced service package.

Among the strengths of carriers is that they typically have a large service and support infrastructure, significant experience in physical cabling and communications, network integration expertise, and remote management capabilities. In addition, they have available considerable investment capital as well as strategic partnerships and alliances worldwide.

The strength of the value-added network providers is their ability to manage an internetwork of LANs as part of their Internet and frame relay services. Their chief weakness is that they do not have extensive service infrastructures and staff, and rely on local authorized contractors.

6.6.12 WAN administration and management

WAN management is the strength of regional and interexchange carriers. Carrier-provided outsourcing services include:

- Help desks for network administrators
- Router-to-router monitoring

- Coordinated maintenance of network services
- Analysis of WAN performance
- Network capacity planning
- Equipment vendor and carrier coordination
- Network monitoring and problem resolution
- Network efficiency analysis
- Network performance and utilization analysis

Pricing for such services depends on the number and location of network nodes, as well as the level of customization required. Some of these activities, such as network monitoring and problem resolution, may come with performance guarantees that are written into a Service Level Agreement (SLA). If the carrier exceeds a given threshold of noncompliance, the customer may be entitled to credits.

An outsourcing agreement can include the monitoring and management of wiring hubs, routers, multiplexers, Channel Service Unit/Data Service Units (CSU/DSUs), switches, and gateways. It can also include support for a variety of WAN protocols such as Systems Network Architecture (SNA), Transmission Control Protocol/Internet Protocol (TCP/IP), Internet Packet Exchange (IPX), frame relay, and ATM. As part of the outsourcing agreement, the company also can have the carrier arrange for all of the necessary lines and equipment, including local access connections with the various local exchange carriers and CLECs, and international carriers.

With regard to lines, the carrier that provides the outsourcing service can manage the access circuits provided by local carriers and monitor their performance from a network operations center. When problems arise, a trouble ticket is generated, and the local carrier is notified. The status of the trouble ticket is monitored until service is restored.

The pricing for WAN management services is determined by the types of services customers need and the size and complexity of their networks. The goal of some carriers is to charge companies 15 to 20 percent less than the estimated cost of performing WAN management tasks in-house.

The demand for WAN management services is especially high when international services are involved. Few multinational companies have a significant level of expertise in more than a few national markets. The creation of strategic alliances between U.S. and international service providers virtually guarantees dense network coverage and local support in many countries, making the outsourcing arrangement less risky today than in years past.

6.6.13   Network security

With hacker attacks and viruses becoming more frequent and malicious, companies are examining Internet security issues more seriously than ever before and taking steps to close potential breach points on their networks. The solution most commonly applied to meet these threats is the firewall, which enforces security by regulating access to and from the Internet, holding back unidentifiable content, and implementing countermeasures to thwart suspected break-in attempts. However, with large enterprises now spending lavishly on security, many hackers are turning their attention to small and midsize businesses. They know such firms cannot afford the expertise required to configure and manage their own firewalls, much less take effective countermeasures to stop attacks in progress at any time of the day or night.

For smaller companies, outsourcing the responsibility for security to an experienced service provider represents a cost-effective, no-hassles alternative to firewall ownership. In fact, acquiring security expertise is the biggest hidden cost of the do-it-yourself approach. The dearth of knowledgeable security personnel and the high salaries they command put seasoned talent out of reach for many smaller companies. An outsourced firewall solution, on the other hand, allows small and medium-size firms to implement best-of-breed security solutions at a fixed monthly cost, and without the ongoing challenges of recruiting and retaining quality staff.

After submitting the network to a battery of vulnerability tests, the managed firewall service provider presents the customer with recommendations for fixing the problems that have been identified. The recommendations are codified in the form of rule sets that are loaded into the firewall. Rule sets can be defined to regulate passage of traffic according to its source and destination, specific applications and types of files, users or groups of users, and even limit access to resources by time of day. At this point, the service provider must be prepared to take full responsibility for configuring and fine-tuning the firewall and for ongoing management.

The managed firewall service provider should be able to generate performance reports that can be accessed by the customer on a secure Web site using a browser that supports 128-bit key encryption. By entering a username and password, the customer can view high-level charts and graphs that summarize the quality of network and application resources. Comparative performance data on specific network resources and groups of resources should also be available on the Web.

The managed firewall service should come with an SLA that includes performance guarantees for the following parameters:

- Network Security Operations Center (NSOC) availability.

- NSOC answers call from customer within the predefined number of rings.

- Callback from NSOC (if required by customer) within 30 minutes of contact.

- Status reports of outage provided to customer every four hours until resolved (unless customer requests otherwise).

- NSOC complies with established escalation requirements.

- Firewall configuration changes accomplished within predefined number of business hours.

- Mean Time to Response (MTTR) for troubleshooting is two hours or less.

- Firewall uptime, assuming firewall CPE is functional, as negotiated at time of contract.

6.6.14 Enterprise applications

Many businesses find it difficult and time-consuming to run their own enterprise applications. Not only are they expensive to customize and maintain, but skilled personnel are hard to find and keep as well. Even email systems, which are taken for granted because they are so ubiquitous, are actually difficult for most organizations to run problem-free for any length of time. Businesses with up to 1000 employees may find outsourcing their email a more attractive option. By subscribing to a carrier-provided email service, these companies can save from 50 to 75 percent on the cost of buying and maintaining an in-house system.

Carriers even offer guaranteed levels of service that will minimize downtime and maintain 24 × 7 support for this mission-critical application. An interface allows the subscribing company to partition and administer accounts. Companies can even create multiple domains for divisions or contractors. Administration is done through the service provider's Web site. Changes are easy and intuitive—no programming experience is necessary—and the changes take place immediately. Many services even offer spam controls to virtually eliminate junk mail.

Outsourcing can level the playing field for companies competing in the Internet economy. Businesses no longer need to budget for endless rounds of software upgrades. A carrier-provided service can also include value-added applications such as fax, collaboration, calendaring, and unified messaging services.

For companies that are not ready to entrust their entire email operation to a third-party service provider, there are customized solutions that allow them to selectively outsource certain aspects and functions. For example, a company can choose to have its email system hosted on the service provider's server. Such "midsourcing" can reduce the costs of administration and support, and provide improved performance. In addition, the company can add new functionality easily, without incurring a costly upgrade.

6.7   Decision to Outsource

Strategic, business-oriented issues play a significant role in the decision to outsource. It is essential that potential users take stock of their operations before making this decision. Arriving at the correct solution requires an examination of the company's unique characteristics, including its human resources and technological infrastructure. For example, it is advisable to compare the costs of in-house operations with the services that will be performed by the outside firm. This entails performing an audit of internal computing and networking operations to determine all present and planned costs for hardware, software, services, and overhead. These costs should cover a minimum of three years and a maximum of seven years and should include specifics on major expenses that may be incurred within that time frame. This establishes a baseline figure from which to evaluate more effectively the bids of potential outsourcing firms and monitor performance after the contract is signed.

A detailed description of the operating environment should be prepared, starting with computer and network resources. This description should include:

- Hardware configurations
- Storage requirements
- Backup media and devices
- Operating systems
- Applications software
- Bandwidth requirements
- Communications facilities and services
- Protocols in use
- Quality of Service (QoS) requirements
- Locations of spare bandwidth and redundant lines
- Restoral methods
- Applications at remote locations
- Critical processing periods
- Peak traffic loads

The next step is to identify potential outsourcing firms. These vendors can then be invited to visit corporate locations to view the various internal operations, thereby gaining an opportunity to understand the company's requirements so that these can properly be addressed in a formal proposal.

A company that turns to outsourcing to alleviate problems in managing information systems or networks should realize that transferring management to a third party may not turn out to be the hoped-for panacea. Although outsourcing represents an opportunity for companies to lower costs and enhance core business activities, before such an arrangement is considered, it should be determined how well internal staff, vendors, consultants, and contract programmers are managed. If there are already difficulties in this area, chances are that the situation will not improve under an outsourcing arrangement. In this case, perhaps some changes in staff responsibilities or organizational structure are warranted.

6.8   Vendor Evaluation Criteria

Most vendors are flexible and will negotiate contract issues. Each outsourcing arrangement is different and requires essentially a custom contract. It is important to identify all the issues that should be written into the contract. This can be a long list, depending on the particular situation. The following criteria, however, should be included in any rating scheme applied to potential outsourcing firms:

- Financial strength and stability over the last three to five years
- Demonstrated ability to manage domestic and multinational computer systems and data networks
- Number of employees, their skills, and years of experience
- Ability to tailor computer and network management tools to client needs
- History of implementing the most advanced technologies and best-of-breed solutions
- Reputation in the industry, news media, investment community, and customers
- Fair employee transfer policies and benefits packages
- Quality and scope of strategic partnerships with equipment vendors, carriers, and application developers

The company, in keeping with its unique short- and long-term requirements, should set the weights of these criteria.

When it comes to software, suppliers may impose hefty transfer fees on licensed software if an outsourcing vendor takes over internal operations. This is often a hidden and potentially costly surprise. The common assumption among software users is that they can just move software around as they please from site to site. For the most part, software firms do not allow third parties to provide use to customers without a new license or significant transfer fees. Interestingly, this is the basis of a new business model called applications outsourcing, whereby Application Service Providers provide the infrastructure for allowing companies to access enterprise applications that they would not normally be able to afford. The ASP pays the software vendor a license fee based on anticipated usage.

The outsourcing firm should be required to submit a detailed plan—with time frames—describing the transition of management responsibilities. Although time frames can and often do change, setting them gives the company a better idea of how well the outsourcing firm understands the company's unique requirements. Time frames also provide a structure that imposes project discipline.

Performance guarantees that mirror current internal performance commitments should be agreed upon—along with appropriate financial penalties for substandard performance. The requirements should not exceed what is currently provided, unless that performance is insufficient, in which case the company should review its motives for outsourcing in the first place.

Satisfactory contractual performance guarantees for data center and/or network operations can be developed if sufficient information on current performance exists. Although it is possible to develop such guarantees in the applications development arena, the open-endedness of such projects makes it more difficult, which is the reason why many companies do not outsource this function.

A detailed plan for migrating management responsibilities back to the company at a future date should also be required. Despite the widely held belief that outsourcing is a one-way street, proper planning and management of the outsourcing firm will keep the option open of bringing the management function back into the corporate mainstream should it become necessary. Despite this option, the company may decide that, after the three or five years of the contract are up, the outsourcing arrangement should be made permanent.

6.9   Structuring the Relationship

Companies that outsource face a number of critical decisions about how to structure the relationship. Without proper safeguards, entering into a long-term partnership with the outsourcing firm can be risky. As previously noted, poor performance on the part of the service provider could jeopardize the client company's customer relationships and, ultimately, its competitive position.

It must be determined at the outset which party will respond to application, computer system, and network failures, and the degree to which each party will be responsible for restoring these elements back to proper operation. This includes spelling out what measures the outside firm must take to ensure the security and integrity of the data, financial penalties for inadequate performance, and what amount of insurance must be maintained to provide adequate protection against losses.

The outsourcing relationship must make explicit provisions for maintaining the integrity of critical business operations and the confidentiality of proprietary information. The firm must ensure that the outside firm will not compromise any aspect of the company's business.

The typical outsourcing contract covers a lengthy period of time—perhaps three to five years. Some contracts can go to seven years. Outsourcing firms justify this by citing their need to spread the initial costs of consolidating the client's data processing or network operations over a long period of time. This also allows them to offer clients reasonable monthly rates for services.

The relationship must provide for the possibility that the client's needs will expand or contract substantially as a response to customer demand and general economic conditions. The outsourcing firm's ability to meet changing needs (e.g., from the addition of a new division or acquisition of another company) should be evaluated and covered under the existing contract.

Companies entering into outsourcing relationships must also establish what rights they have to bring some or all of the management responsibilities back in-house without terminating the contract or paying an exorbitant penalty. However, this should not be done lightly, since it can take a long time to hire appropriate staff and bring them up to an acceptable level of performance.

To avoid getting locked into the outsourcing arrangement, organizations should minimize the sharing of data centers, networks, and application software and from relying too heavily on customized software, applications, and networks. It can be very difficult for a company to extricate itself from outsourcing arrangements when its operations are too tightly woven into those of other companies operating under similar arrangements. The contract should be structured so that upon expiration of the term, it can be put up for bid by other parties.

Contracts should provide an escape clause that allows the user to migrate operations to an alternative service provider should the original firm fail to meet performance objectives or other contract stipulations. Because it is difficult to rebuild in-house systems or network staff from scratch, it is imperative that users not outsource anything that cannot be immediately taken over by another firm. In fact, having another firm on standby should be an essential element of the company's disaster recovery plan.

The outsourcing agreement can provide major long-term benefits to the organization, if structured properly. The following ingredients make for successful outsourcing agreements:

- Prepare a separate, detailed Service Level Agreement that specifies financial penalties for missing performance targets on such things as network and circuit availability, mean time to repair, outage notifications, and response time to trouble calls.

- Refer to the SLA in the contract, but do not make it part of the contract. This allows minor changes to the agreement without affecting the contract and possibly causing implementation delays.

- At the outset, obtain separate pricing for each service provided by the outsourcing firm. This makes adding and deleting services a more straightforward process.

- If the network is international, obtain monthly detailed and summary billing reports in one currency.

- Require a single point of contact for sales, technical, and management issues. This minimizes opportunities for errors and finger-pointing between the outsourcing firm's partners.

6.10   Negotiations

When it comes to face-to-face bargaining, be aware that the outsourcing firm negotiates contracts every day, while a CIO may do it only a few times a year. This imbalance can be overcome by hiring an experienced consulting firm or specialized technology advisor to take part in the negotiations. These experts know just how far outsourcing firms are willing to go in winning new business. Regardless of who is actually negotiating the outsourcing agreement, there are several pitfalls that should be avoided.

For example, conventional wisdom contends that everything is negotiable when dealing with outsourcing vendors. However, a careful examination of the assumptions that lie hidden beneath the surface of this claim may go a long way toward ensuring the long-term success of the arrangement.

The first assumption that merits attention is that the buyer has properly defined the objective. This may sound rather rudimentary, but it is easy to get sidetracked in the heat of negotiations. For example, in negotiating a network management outsourcing agreement, is the objective to save money? Or is it to improve the availability of information systems and communication networks? The former objective is certainly more attractive, but it might come at the expense of the latter. Bottom line: If competitive advantage is compromised, what has been gained?

Second, there is the assumption that the buyer knows what outsourcing alternatives exist. In the case of network management outsourcing, the buyer must become familiar with the comparable offerings of carriers, vendors, and traditional service companies, as well as the large accounting and/or management firms. Such knowledge provides the buyer with negotiating leverage on such matters as response times, service options, training, maintenance, and equipment repairs and upgrades. But the more emphasis one puts on cost savings at the outset, the more inflexible the outsourcing firm is likely to become on other matters as one tries to negotiate terms and conditions.

The third assumption is that the buyer is prepared to ask the right questions. Related to this assumption is the fourth assumption, which is that the buyer will aim these questions at the right person. For example, asking a carrier how much it will cost to protect the entire T1 network against every possible link failure will elicit a few smiles from the account representative, plus a dollar figure that approximately doubles the cost of the network.

Pose the same question to a network designer, and the answer will turn out to be quite different. In eliminating redundant T1 lines in favor of Fractional T1 and/or multirate ISDN, the T1 network is not only totally protected, but at much less cost despite the addition of extra circuits. When confronting the carrier's representative with this information, suddenly he or she is ready to deal. Without the ability to aim the right questions at the right people, the buyer is at a severe disadvantage during the negotiations.

The fifth assumption is that the buyer understands technical terminology. While plain language is the preferred style of communicating with others, the use of technical (as well as legal and financial) jargon serves critical social functions. It is a form of ritual initiation, and it can be used to expose what one knows or does not know before serious negotiations begin.

If technical jargon is used correctly, one is treated as a peer and invited into the inner circle, where candor and fair dealing are the norm. If one does not talk the same language or—worse—pretends to talk the same language by relying only on buzzwords, not only is that person exposed but is never treated as a peer. The inner circle will be forever closed, which is a severe handicap during negotiations.

Boldly going into the negotiation process without adequate preparation is fraught with risks. Careful examination of the hidden assumptions behind this and other broad claims can save buyers from making serious mistakes, which can result in lost credibility among senior management and an outsourcing arrangement that is both costly and inflexible over the long term.

6.11   Conclusion

The pressures for third-party outsourcing are considerable and on the increase, especially as companies in virtually every industry look for ways to become more efficient and profitable without compromising the integrity of business operations. Requirements to service large amounts of debt have made every corporate department the target of close budgetary scrutiny, the data center and corporate network included. In addition, competition from around the world is forcing businesses to scale back the ranks of middle management, streamline operations, and distribute workflows for greater efficiency and economy. Outsourcing allows businesses to meet these objectives.

Although outsourcing promises bottom-line benefits, deciding whether such an arrangement makes sense is a difficult process that requires considerable analysis of a range of factors. In addition to calculating the baseline cost of managing in-house information systems and communications systems and networks, and determining their strategic value, the decision to outsource often hinges on the company's business direction, the state of its present data center and network architecture, the internal political situation, and the company's readiness to deal with the culture shock that inevitably occurs when two firms must work closely together on a daily basis.

Even if the outsourcing contract specifies financial penalties, discounts, or the withholding of payments in cases of subpar vendor performance, many companies have discovered too late that these provisions do not begin to compensate them for related business losses. All things must be considered and if doubt persists, outsourcing should be limited to a narrowly focused activity and expanded to other activities as the comfort level with the arrangement rises.

Organizing Technology Assets

Downsizing and Distributing Information Resources

7.1 Introduction

Downsizing involves easing the load of the data center by distributing applications and databases to the Local Area Network (LAN) and Wide Area Network (WAN), specifically, the clients and servers located throughout the organization, including telecommuters and mobile clients with dial-up or virtual connections. This arrangement can yield optimal results to the organization. It provides users at all levels with ready access to the information they need in a friendly format, while permitting mission-critical applications and databases to remain within the data center where security and access privileges can best be applied.

Distributing resources over the WAN through collocation arrangements with carriers and service providers represents another way to downsize. Collocation represents a key component of disaster recovery planning, offering opportunities for power protection, network redundancy, carrier diversity, physical security, and distributed operations that can help businesses withstand a natural or man-made calamity in one city by having duplicate resources fully available in other cities. Users in the distressed location can be serviced from resources available from nonaffected locations by such means as traffic rerouting and node rehoming under control of the collocation provider's Network Operations Center (NOC).

Downsizing entails more than just moving applications and databases from a data center to a network of clients and servers, or distributing these resources for improving security, availability, and response time. Many companies are discovering that downsizing constitutes a major modification in business philosophy and in the way the organization exploits the vast amount of information at its disposal. Before seeking to downsize operations, corporate managers should examine their attitudes toward the data center and client-server environments to ensure that they do not have unrealistic expectations.

This chapter describes the benefits of downsizing information resources and making the transition from a highly centralized operation to a distributed one, and the pitfalls you should avoid when planning and implementing such a transition.

7.2   Benefits of Downsizing

In the distributed computing environment, information can be accessed by individuals who are most capable of exploiting it. In moving applications and information closer to departments and individuals, knowledge workers at all levels in the organization are empowered, improving the quality and timeliness of decision making and allowing a more effective corporate response in dynamic business environments.

Downsizing applications from the data center to users on LANs and WANs holds many other potential advantages:

- Downsized applications may run as fast or faster on laptops and workstations than on mainframes in the data center, at only a fraction of the cost.

- Even if downsizing does not improve response time, it can improve response-time consistency.

- In a distributed environment, the organization need not be locked into a particular vendor for much of the hardware and software, as is typically the case with centralized host-based architectures.

- In the process of rewriting applications to run in the downsized environment, there is the opportunity to realize greater functionality and efficiency from the software.

- In the downsized environment, the heavier reliance on off-the-shelf application programs reduces the number of programmers and analysts needed to support the organization.

- The lag time between applications development and business processes can be substantially shortened, enabling the corporation to realize significant competitive advantage.

7.3   Opportunity Assessment

To assess the feasibility of downsizing and determine where to focus initial efforts, an evaluation of the opportunities should be performed from different perspectives: senior management, systems designers, systems developers, and Information Technology (IT) management.

7.3.1   Senior management perspective

Senior management must consider the data processing and networking needs of the entire organization. Toward this end, resources must be distributed appropriately, that is, in keeping with each department's mission. There must be a balance between the needs, wishes, and preferences of department managers. Striking this balance requires that the following questions be considered to determine the nature and scope of the downsizing effort:

- To what extent does each department currently rely on the data center for applications, database maintenance, customization, and problem resolution?

- Should this reliance be maintained at the current level, or be increased or decreased?

- What is the strategic value of each department's data?

- What is the degree of sensitivity for each department's data, and what levels of security must be employed to protect it?

- How will the organization's anticipated growth and evolution affect its need for information from each department?

- How reasonable is the current IT budget and is the staffing sufficient to meet these needs?

- If additional resources are needed to implement the downsizing effort, are they available?

7.3.2   Systems designer perspective

One of the key responsibilities of the systems designer is to stay informed of the company's business plan or marketing strategy so that information systems and networks can support rather than hinder corporate growth or entrance into new markets. Since downsizing often plays a role in such efforts, the systems designer should consider the following questions:

- What are the true costs of operating the current application and database platforms?

- What is the true cost of operating each alternative platform, including maintenance, support, and disaster recovery?

- What course of action offers a natural migration path toward emerging technologies?

- What are the costs of retraining, redevelopment, and software conversion?

- What incentives can be devised to ensure cooperative participation in downsizing?

7.3.3   Systems developer perspective

The systems developer is responsible for developing the applications that will carry out various business processes as efficiently and economically as possible. When evaluating the feasibility of downsizing, the systems developer should consider the following questions:

- What has to be done to allow applications to take full advantage of the new platform?

- What documentation provisions must be adopted to allow the applications to be enhanced by future staff who will have played no part in the original development?

- What provisions must be made for user retraining?

- How will user feedback be gathered and incorporated to enhance existing applications or develop new ones?

7.3.4   Systems designers and developers

Together, the systems designers and systems developers work out the methodology for achieving corporate objectives. They must consider the following questions:

- What is the best application development choice (e.g., 4GL or object-oriented programming), from both technical and cost-benefit viewpoints?

- What is the best operating system choice from both technical and cost-benefit viewpoints?

- What is the best LAN interconnectivity choice [e.g., Ethernet, Fast Ethernet, Gigabit Ethernet, Frame Relay, Asynchronous Transfer Mode (ATM)] from both technical and cost-benefit viewpoints?

- What is the best hardware platform choice from both technical and cost-benefit viewpoints?

- What is the best systems management platform choice from both technical and cost-benefit viewpoints?

- What can possibly go wrong, and what corrective measures are available or need to become available to handle the various disaster scenarios?

7.3.5   IT management perspective

IT management is responsible for data center operations. Traditionally, this meant only the mainframe. As companies decentralized their data resources over the LAN, between LANs, and over the Internet, IT management has been charged with additional responsibilities, such as administering security, maintaining corporate directories, implementing email and other messaging services, providing dial-in access and VPNs, operating the help desk, and managing software licenses. In a downsizing project, IT management should consider the following questions:

- How appropriate are the current IT platforms, given the continuing trend toward distributed information resources?

- What are the realistic alternatives for implementing the various strategic applications?

- What is the maximum time an application can be out of service without causing major disruption to key business operations?

- How much is it worth to protect against this possibility?

- Who needs access to the various applications and the databases?

- What applications and databases can be open to everyone, and what applications and databases will require appropriate levels of security?

7.4   Making the Transition

The transition from a centralized to a decentralized information systems environment need not entail sudden and dramatic changes to existing business operations. In fact, most downsizing efforts are gradual and initially limited to a few select applications. In some cases, it can be counterproductive to shift all applications to the LAN. Assuming this is even desirable, putting the extra load onto an existing network could overwhelm the backbone. Not only can this slow or deny access to the distributed computing resources, but it can force the company to spend even more money on new facility installation and network expansion.

With homegrown applications, a pilot test is usually conducted to ascertain the feasibility of the downsizing concept. The pilot test usually involves a lot of software writing, testing, debugging, and network performance monitoring. With the first downsized application, the goal is to develop confidence among IT staff and demonstrate the benefits of a successful implementation to the entire user community. When there is success with one application, another may by added. An application-by-application approach to downsizing is the safest way to proceed because IT staff will gain experience as they progress. Eventually, the circle of users is widened to include the entire enterprise, even those at branch offices, telecommuters, and mobile professionals. Packaged applications do not require as much work. Many companies migrate their accounting systems first because virtually all commercial mainframe accounting packages are now available for use on LANs.

Although concentrating most data processing and information storage at the mainframe has become inefficient, eliminating the mainframe entirely should not necessarily be the objective of downsizing. The mainframe can be retained to perform tedious number-crunching tasks, for example, which can free up users to perform other tasks. Upon completion of processing, the user can then simply retrieve the results from the mainframe. In fact, this is the idea behind client/server computing, which is one way to implement downsizing. Some mainframe manufacturers, including IBM, have capitalized on this trend by adding server capabilities and LAN interfaces to their mainframes.

Another factor that must be entered into the cost-benefits equation is that all PCs and workstations are not necessarily built to do everything mainframes can. For example, applications that require the sorting of millions of items accumulated every hour or so can take too long to execute on a PC, server, or special-purpose workstation. Another concern is that PCs and workstations lack the reliability of the mainframe, and that the systems administration, management, and security functions that organizations have come to expect on the mainframe are not yet widely available in the distributed environment.

7.5 Distributed Processing Environment

The distributed processing environment can take many forms. There is no universal solution; what works for one organization may not work for another. There are several commonly used computing architectures from which to select. Multiple solutions can be implemented to carry out different business operations of the organization.

7.5.1 Dedicated application servers

Dedicated application servers can be used on the LAN to control access to mission-critical and basic business software and to prevent users from modifying or deleting certain types of data files. Several file servers can be deployed throughout the network, each supporting a single application (e.g., email, word processing, spreadsheet, or specific types of databases). This multiserver arrangement has the added benefit of keeping some applications available even if one server should go down. Metering tools can be included on the servers to regulate usage according to various software licenses and to prevent unauthorized software copying.

7.5.2 Minicomputers

Using a minicomputer as an application server might entail using equipment such as an IBM AS/400 or an HP 9000 connected to an Ethernet LAN. The drawbacks of this type of solution are cost, amount of set-up time needed, limited expansion possibilities, and the need for continual administrative vigilance to keep things working properly.

7.5.3 Super servers

With so many organizations looking to downsize applications from mainframes to local area networks, the super server concept has become a key strategic tool. *Super servers* are high-end workstations specifically equipped to act as network servers. These systems typically come with multiple high-speed processors and redundant subsystems, offer data storage in the hundreds-of-gigabytes range, and use mainframelike restoral and security techniques. Although a super server may be used to support multiple applications, it is ideal for processing-intensive applications [e.g., Computer-Aided Design/Computer-Aided Manufacturing (CAD/CAM)], relational database applications, and applications based on expert systems or neural networking.

7.5.4 Server clusters

A variation in the super server concept is the *server cluster,* which is a collection of servers on a network that can function as a single computing resource. This is accomplished by tying together server nodes using Input/Output (I/O) channels and using special software that controls job and system resources, while exploiting idle processing cycles by load-balancing tasks between machines. *Load balancing* is the process of spreading work throughout the cluster to prevent the saturation of one host. Most cluster systems implement load balancing by collecting system activity information on each computer in the cluster and passing this information to a master load-balancing server. The master server uses this information to place batch jobs on hosts in the cluster.

Organizations favor clustering, especially for applications that run on the Internet, because it provides high availability and a means to handle growth by scaling hardware to very large configurations. Virtually every UNIX system manufacturer has some facility for coupling at least two machines to attain higher availability by what is called *failover.* The machines listen to "heartbeat" signals from each other, and when one stops, the other takes over its work. Most manufacturers allow failover to occur between machines of different sizes. So a large system can be the primary and the second can be a smaller, less expensive machine.

7.5.5 Mainframes

The mainframe can continue supporting traditional applications that require access to voluminous customer records or financial data, but also act as a server. In the IBM environment, for example, the addition of the following software can allow the mainframe to act as a server:

- *LAN Resource Extension and Services (LANRES).* A software server-based product that is used with Net Ware LANs.

- *Data Facility Storage Management Subsystem (DFSMS).* A suite of software programs that automate storage management.

- *Network File Server (NFS).* Originally designed to operate on LANs, this software can now operate with Multiple Virtual System (MVS) on the mainframe.

- *File Transfer Protocol (FTP).* The FTP server application can be used as part of the native Transmission Control Protocol/Internet Protocol (TCP/IP) stack under Virtual Telecommunications Access Method (VTAM) or as a single third-party FTP server application also running as a VTAM application.

The mainframe can also play the role of master server in a hierarchical arrangement, backing up, restoring, and archiving data from multiple LAN servers. This is an important role for the mainframe, considering that many PC-based LANs lack a strong central backup capability. The mainframe can also provide additional space for users with large storage requirements.

7.5.6 Client-server

With the client-server approach, an application program is divided into two parts on the network. The client portion of the program, or front end, is run by individual users at their desks, performing such tasks as querying databases, producing printed reports, or entering new records. These functions can be carried out through a common access language, which operates in conjunction with existing application programs. The front-end part of the program executes on the user's workstation, drawing on its Random Access Memory (RAM) and Central Processing Unit (CPU).

The server portion of the program, or back end, is resident on a computer that is configured to support multiple clients. This setup offers users shared access to numerous application programs as well as to printers, file storage, database management, communications, and other capabilities. Consequently, the server is typically configured with more RAM and higher-speed CPUs (or multiple CPUs) than are the clients connected to it over the network.

The design of client-server software is so different from mainframe software that the applications usually must be developed from scratch—not merely recompiled—in order to fully exploit the true potential of the client-server architecture. This calls for new development methodologies as well as new application development tools (discussed in Sec. 7.6).

7.5.7   Thin-client architecture

The thin-client computing model is a lower-cost variation of the traditional client-server model in which applications are deployed, managed, supported, and executed on a server. This allows organizations to deploy low-cost client devices on the desktop and, in the process, to overcome the critical application deployment challenges of management, access, performance, and security.

The operation of thin clients is fairly simple: They are dependent on servers for boot-up, applications, processing, and storage. Since most thin clients may not have a hard drive, the server provides booting service to the network computers when they are turned on. The server can be a suitably equipped PC, a RISC-based workstation, a midrange host like the IBM AS/400, or even a mainframe. The server typically connects to the LAN with an Ethernet or token ring adapter and supports TCP/IP for WAN connections to the public Internet or a private intranet.

Since all applications reside on the server, installation is done only once—not hundreds or thousands of times at individual desktops. Periodic updates to applications are conducted on the server. This ensures that every network computer uses the same version of the application every time it is accessed.

Network computers can access both Java and Windows applications on the server, as well as various terminal emulations for access to legacy data. Users accessing Java applications do so through a Java-enabled Web browser, which also gives them access to applications on the Internet or intranet.

The promises of the thin-client computing model are that it allows organizations to more quickly realize value from the applications and data required to run their businesses, receive the greatest return on computing investment, and accommodate both current and future enterprise computing needs. This does not mean thin clients will replace PCs. The two are really complementary architectures that can be centrally managed. In some cases, it may even be difficult to distinguish between the two—the line of demarcation seems to be quite fluid. Overall, thin clients could provide the economies and efficiencies many organizations are looking for, but these advantages will primarily come from among a user population that is task oriented.

7.5.8 Cooperative processing

Cooperative processing is another variation of the client-server architecture in which one or more clients are used to offload some of the work usually done by the central server. Among the most popular techniques for cooperative processing applications are Advanced Program-to-Program Communications (APPC) in the Systems Network Architecture (SNA) environment and Internet Protocol (IP) sockets in the UNIX environment. Emulation is most often used in the Windows environment. The most basic cooperative processing approach converts a 3270 application to a more user-interactive application, making it mouse-based with colors, pull-down menus, and other Windows-like features. Other options include integrating different applications via front ends that merge the results from several dissimilar systems and display them on a single screen. The arrangement reduces overall processing costs.

There are tools available that bring together the power of APPC with the graphical user interface of Microsoft Windows. With such tools, Windows-based APPC transaction programs can be created which communicate with partner programs running on other computer systems. Such APPC conversations occur on a direct peer-to-peer basis in an efficient cooperative processing environment, eliminating the overhead associated with traditional terminal emulation communications.

7.5.9 Peer-to-peer data sharing

Perhaps the simplest and most economical method of distributed computing is peer-to-peer data sharing. Each computer is an equal or "peer" and can share the files and peripherals of the others. For a small business doing routine word processing, spreadsheets, and accounting, this type of network is the low-cost solution to sharing resources such as files, applications, and peripherals. Multiple computers can even share an external cable, Digital Subscriber Line (DSL) modem, or wireless access point, allowing them to connect to the Internet at the same time.

Although this approach is economical in that it does not rely on a dedicated file server, it has several disadvantages that can outweigh cost savings. Unlike servers, this data-sharing scheme does not have a central administration facility to enforce database integrity, perform backups, or maintain security. In addition, the performance of each user's hard disk may degrade significantly when continually accessed by other users.

7.6   Transition Aids

Standardization of reporting tools such as Structured Query Language (SQL) for database access, operating systems, and Graphical User Interfaces (GUIs) can ease the transition from centralized to distributed computing. This transition allows companies to maintain a smooth information flow throughout the organization and realize a substantial reduction in end-user training requirements while reaping the economic and performance benefits of hardware diversity.

The X Window System, for example, permits the creation of a uniform presentation interface that shields users from having to learn multiple ways of accessing data across different computing platforms. It is the de facto standard graphical engine for the UNIX and Linux operating systems and provides the only common windowing environment bridging the heterogeneous platforms in today's enterprise computing.

Under X Windows, there is a single GUI front end that allows users to access multiple mainframe or server sessions simultaneously. These sessions can be displayed in separate windows on the same screen. X Windows also allows users to cut and paste data among the applications displayed in different windows.

All major hardware vendors support the X Window System. The inherent independence of the X Window System from any operating system or hardware, its network transparency, and its support for a wide range of popular desktops are responsible for its continuing and growing popularity.

Concurrent with the downsizing trend is the growing acceptance of the client-server architecture as the optimal way to share information and resources. Client-server applications typically are developed using Object Oriented Programming (OOP) or Computer Aided Software Engineering (CASE), or a blend of the two.

7.6.1   Working with objects

The basic premise of object-oriented programming is that business functions and applications can be broken up into classes of objects that can be reused. This greatly reduces application development time, simplifies maintenance, and increases reliability.

Objects provide functionality by tightly coupling the traditionally separate domains of programming code and data. With separate domains, it is difficult to maintain systems over time. Eventually the point is reached when the entire system must be scrapped and a new one put into place at great expense and with serious disruption to business processes. In the object-oriented approach, data structures are more closely coupled with the code, which is allowed to modify that structure. This permits more frequent enhancements of applications, while resulting in less disruption to end users' work habits.

With each object viewed as a separate functional entity, reliability is improved because there is less chance that a change will produce new bugs in previously stable sections of code. The object-oriented approach also improves the productivity of programmers in that the various objects are reusable. Each instance of an object draws on the same piece of error-free code, resulting in less application development time. Once the object method of programming is learned, developers can bring applications and enhancements to users more quickly, thus realizing the full potential of client-server networks. This approach also makes it easier to maintain program integrity with changes in personnel.

The object-oriented paradigm signals a fundamental shift in the way distributed applications and databases are put together as well as how they are used, upgraded, and managed. The ability to create new objects from existing objects, change them to suit specific needs, and otherwise reuse them across different applications promises compelling new efficiencies and economies, especially in the client-server environment.

7.6.2 Working with CASE tools

CASE tools, Fourth-Generation Languages (4GLs), and various code generators have been used over the years to help improve applications development, but they have yet to offer the breakthrough improvements demanded by an increasingly competitive business environment. In fact, although these tools have been around for years, the applications development backlog has not diminished appreciably.

This is because many CASE tools are too confining, forcing programmers to build applications in one structured way. This can result in redesign efforts that often fall behind schedule and go over budget. Furthermore CASE tools are not usually "plug-and-play." This would require a CASE framework with widely disclosed integration interfaces. Complicating matters is the growing number of government and industry standards organizations that are offering proposals that supersede and overlap one another.

Object-oriented technologies, however, have delivered on the breakthrough promise. In some instances, OOP technology has brought an order-of-magnitude improvement in productivity and system development time over that of CASE tools. Some IT staffs can compress five or six months of applications development time to only five or six weeks.

There are few technologies available to the applications development community that hold as much promise as the object orientation.

7.6.3   Distributed network management

Although it is becoming increasingly popular to migrate mainframe applications to a client-server environment, thought must be given to the network management tools for measuring and monitoring distributed resources. Without a centralized view of the network, it may take specialists in databases, communications, systems network architectures, gateways, servers, and client servers to provide a complete picture of what is happening on the network. Fortunately, there are tools available to address these issues, such as intelligent network agents.

Intelligent agents are autonomous and adaptive software programs that accomplish their tasks by executing commands remotely. System administrators, network managers, and software developers can create and use intelligent agents to execute critical processes including performance monitoring, fault detection and restoration, hardware and software asset management, virus protection, and information search and retrieval. One of the primary uses of intelligent agents is in continuously monitoring the performance and availability of applications. A just-in-time applications performance management capability captures detailed diagnostic information at the precise moment when a problem or performance degradation occurs, pinpointing the source of the problem so it can be resolved immediately.

Such agents are usually installed on client as well as application servers. They monitor every transaction that crosses the user desktop, traversing networks, application servers, and database servers. They monitor all distributed applications and environmental conditions in real time, comparing actual availability and performance with service-level thresholds. Further analysis performed by the agents enables network and application managers to understand the source of application response time problems by breaking down response times into network, application, and server components. Troubleshooting that once took weeks can now be accomplished in a matter of minutes with intelligent agents.

7.6.4   Role of SNMP

The most widely implemented standard for managing the network infrastructure is the Simple Network Management Protocol (SNMP), which serves as a polling mechanism for reporting the status of workstations, hubs, servers, and other hardware-based devices. There is even a database Management Information Base (MIB) that extends SNMP polling to distributed client-server databases.

Although database vendors employ radically different architectures, at the very least, LAN managers are able to discover databases in a heterogeneous environment using SNMP-based management platforms. Depending on the extent to which each database vendor supports the database MIB, LAN managers may also be able to use their SNMP-based management software to obtain statistics on the size of the database, available disk space, reads, writes, transactions, and network activity levels. By adding SNMP MIB capabilities to relational databases, system administrators can manage remote hosts and networks and their associated databases through a single management application.

With the database MIB, network managers can go across a LAN or WAN, including the Internet, to identify and characterize databases from any vendor via a third-party SNMP tool. When queried via the SNMP protocol, a database uses a proscribed format to return information to the MIB identifying itself and its vendor. The following parameters about database status and activity level are also provided:

- Databases installed on a host and/or system
- Actively opened databases
- Database configuration parameters
- Database limited resources
- Database servers installed on a system
- Active database servers
- Configuration parameters for a server
- Server limited resources
- Relationship of servers and databases on a host

Depending on the feedback, the MIB management tool has a command set for carrying out several types of management actions, such as resetting the configuration parameters of a server. Although this information has broad applicability among database systems and is enough for many monitoring tasks, it is far from adequate for detailed management or performance monitoring of specific database products. This gap is filled with vendor- and product-specific MIBs addressing information that has not been codified in the database MIB.

7.7   Organizational Issues

The promised benefits of downsizing are often too compelling to ignore: improved productivity, increased flexibility, and cost savings. The impetus for change may come from as many as three directions at once:

- *Senior management,* who is continuously looking for ways to streamline operations to improve financial performance. This usually translates into doing more with fewer people.

- *End users,* who are becoming more technically proficient, resent the gatekeeper function of data center staff, and want immediate access to data that they perceive as belonging to them. In the process they benefit from being more productive and from being able to make better and faster decisions, which comes from increased job autonomy.

- *IT management,* who is responding to budget cutbacks or scarce resources and is looking for ways to do more with less-powerful computers, by, for example, adopting open source operating systems, such as Linux, for enterprise applications.

From their own narrow perspectives, each camp looks on downsizing as the most feasible solution. The feasibility of downsizing, however, does not always revolve around technical issues alone. There may be some long-held political concerns to deal with as well. If these issues are ignored, the promised benefits of downsizing may never be realized.

Such problems are often at least partially the fault of differences between two opposing mindsets—mainframe and PC—within the organization, which results in competition for control of critical resources. Consequently, downsizing in an atmosphere of enmity can quickly erupt into a high-stakes game, supercharged by heated bickering and political maneuvering that come from perceived ownership of information. Managing such challenges is often more difficult than dealing with the technical issues.

7.7.1   Mainframe mindset

Data center managers are often forced into adopting a defensive posture during the downsizing effort because it usually means that they must give up something: resources (in the form of budget and staff), power and prestige, or control of mission-critical applications. In addition—and perhaps as important—data center managers regard downsizing as the sacrifice of an operating philosophy in which they have invested considerable time and effort throughout their careers.

In the past, data center managers have frequently been insufficiently attuned to the needs of users. Whereas users originally wanted access to the host, data center managers often seemed to focus on other things, such as upgrading a particular application, adding a certain controller, or waiting for a programmer to finish work on a specific piece of code. These managers clearly had their own agenda, which did not always include serving the needs of others. Unable to wait, users took matters into their own hands. Today, in large corporations, virtually all desktop systems are networked.

Users' rapid adoption of desktop computers did not occur because of the data center's failure to provide fast applications; rather, it occurred because the data center failed to provide applications quickly. By focusing on their own needs and answering only to senior executives, IT personnel tended to lose sight of the real business issues.

7.7.2   PC mindset

With PCs now well entrenched in corporate offices, individuals, work groups, and departments have become acutely aware of the benefits of controlling information resources and of the need for data coordination. In becoming informationally self-sufficient and being able to share resources via LANs, users can better control their own destinies within the organization. For instance, they can increase the quality and timeliness of their decision making, execute transactions faster, and become more responsive to internal and external constituencies—all without the need to confront a gatekeeper in the data center.

In many cases, this arrangement has the potential of moving accountability to the lowest common denominator in the organization, where many end users think it properly belongs. This scenario also has the potential of peeling back layers of bureaucracy that have traditionally stood between users and centralized resources.

7.7.3   Addressing the "soft" issues

Change can be very threatening to those most affected by it. This is especially true with downsizing, because it essentially involves a redistribution of responsibilities and, consequently, of accumulated power and influence. Therefore, the "soft" issues of feelings and perceptions must be addressed first to assure success with the issues that address technical requirements.

The best way to defuse emotional and political time bombs that can jeopardize the success of downsizing is to include all affected parties in the planning process. The planning process should be participative and start with the articulation of the organizational goals targeted by the downsizing effort, outlining anticipated costs and benefits. This stage of the planning process should also address the most critical concern of the participants—how they will be affected. Once the organizational goals are known, these become the new parameters within which the participants can shape their futures.

7.8   Implementation

The technologies involved in downsizing are neither simple nor straightforward. An organization cannot afford to have each individual department research, implement, and support various complex technologies. Instead, a single group—Information Technologies—must be responsible for this. This department, however, must be responsive to other departments' business needs and must be technically competent and capable of proving that it can perform these tasks for the good of the entire organization.

7.8.1 Success factors

As with any corporatewide project, there are actions an organization should take when downsizing which, if followed, can increase the chances of yielding a successful outcome:

- Form an oversight committee of IT staff members, corporate management, departmental management, and representative end users to explore, propose, define, review, and monitor the progress of the project. This approach is usually effective for both short- and long-range planning.

- Identify the applications that are appropriate for downsizing. The initial project should be of limited scope that is easy to define and control, and its success should be easy to determine. The urge to accomplish too much too quickly should be resisted; instead, small successes should be encouraged and built upon.

- Identify the work and information flows that are currently in place for the existing system and determine the effect the project will have on those processes.

- Determine which staff members will be the owners of the data and which will be responsible for maintaining that information.

- Identify clearly the project's objectives and quantify the benefits these objectives will provide the company.

- Obtain the support and involvement of senior management from the start and secure its commitment to the project's objectives and articulated benefits.

- Ensure that the rationale for downsizing is based on strategic business goals rather than on political ambition or some other private agenda that could easily derail the whole project.

- Review on a regular basis the progress of the project with the multidepartmental committee, modifying the plan as the committee deems appropriate.

A well-defined and well-implemented training program can also help ensure the success of a downsizing project. This includes preparing documentation and developing training courses designed to acquaint users with how to work in the distributed operating environment and the specific procedures governing such things as logging on, backup and restoral procedures, file synchronization, security mechanisms, and use of the help desk.

This participative approach to downsizing not only facilitates cooperation but also has the effect of spreading ownership of the solution among all departments and work groups. Instead of a solution dictated by senior management, which often engenders resistance through emotional responses and political maneuvering, the participative approach provides each party with a stake in the outcome of the project. With success comes the rewards associated with a stable work environment and shared vision of the future; with failure comes the liabilities associated with a chaotic work environment and uncertainty about the future. Although participative planning takes more time, its effects are often more immediate and longer lasting than imposed solutions, which are frequently resisted and short-lived.

7.8.2   Risk factors

There are certain applications that will resist most attempts at downsizing until hardware and software are upgraded. Other applications should wait until changes in infrastructure are made, which will facilitate downsizing. Depending on the organization, the following applications should be thoroughly evaluated before the commitment is made to downsize:

- Applications that are closely dependent on other mainframe applications

- Applications of a mission-critical nature that require strong, centrally managed security; accounting; backup; and recovery services

- Applications that issue voluminous customized reports on a daily basis that must be distributed throughout the organization

- Applications that span multiple time zones and require around-the-clock availability

Another factor that merits consideration is that the downsizing effort may have to be accompanied by organizational reengineering so that the power of distributed computing can genuinely improve business processes. The value of downsizing will be limited if old business practices based on old technologies are merely transferred to new systems. The value of downsizing can be increased if there is a comprehensive analysis and restructuring of work flow.

7.8.3   Setting objective criteria

Downsizing may not be the best solution for certain applications, such as those that span multiple divisions or departments. Therefore, it makes sense to use the participative planning process, which can be helpful in reaching a consensus on which applications qualify for possible off loading from the mainframe to the local area network. This minimizes disputes and further removes emotion and politics from the downsizing effort. Issues that should be examined in the participatory planning process are discussed in the following sections.

Mission criticality.   There is currently some controversy about whether mission-critical applications should be moved from mainframes to Local Area Networks. Any application that serves a large number of users across organizational boundaries, requires ongoing changes, and conveys great competitive advantage to the company can be deemed mission-critical. On the other hand, any application that helps users accomplish the mission of a department, is highly specialized, and requires few ongoing changes to be propagated through various corporatewide databases is an ideal candidate for downsizing.

Response-time requirements.   Sometimes information must be decentralized to ensure a quick response to competitive pressures. This decentralization permits multiple users to instantly apply their knowledge and expertise to the resolution of a given problem and to share information across microcomputer and LAN systems. Users would not have to wait for access to mainframe-resident databases, assuming that these are available and up to date when a mainframe or communications controller port is finally accessed.

A marketing department, for example, might want control of product pricing and availability information so that it can quickly reposition its offerings against those of its competitors. Other candidates for decentralization are applications that are highly aligned with product design and service development, such as CAD/CAM and CASE.

Control.   On the mainframe, data reliability is virtually guaranteed, because there is only one place where all data resides. In addition to providing better control, this structure facilitates keeping the data organized, current, and more secure. However, new tools are emerging that provide LAN managers with sophisticated control capabilities that go well beyond simple file transfer.

This higher level of control is provided by change management tools that allow managers to centrally control the distribution and installation of operating systems, application programs, and data files. Verification reports are provided on each transmission to confirm the contents' arrival at the proper destination, the success or failure of the installation, and the discovery of any errors. Changes to software and the addition of unauthorized software

are reported automatically. Data compression and decompression can be performed on large files to prevent network throughput problems during heavy use. Data files may be collected for consolidation or backup.

Security.   An ongoing concern of mainframe managers is security—the protection of sensitive information from unauthorized access as well as from accidental corruption and damage. In the centralized computing environment, security is maintained by such methods as passwords, port assignments, work space partitioning, and access levels. In addition, database maintenance and periodic backups, which ensure the integrity of the data, are easy to perform. Although possible with new tools such as firewalls, security is much harder to implement in the decentralized computing environment. Staff must be dedicated to the security function and stay abreast of the latest viruses, hacker attacks, and software patches. Programming the firewall is a complex undertaking and should be entrusted only to certified security administrators and engineers.

Interconnectivity.   Applications that involve a high degree of interconnectivity among users are best implemented in a decentralized environment of microcomputer and LAN systems. A publishing operation, for example, typically entails the movement of information among multiple users in a work group. At a higher level, text and graphics are gathered from multiple work groups for document assembly. The assembled document is then transmitted to a printing facility, where it is output in paper form. If attempted in a centralized environment, this kind of application would tie up mainframe resources, denying other users access to the limited number of ports, for example. Therefore, any specialized application that requires a high degree of interconnectivity with local or remote users is a prime candidate for downsizing.

End-user expertise.   The success of a proposed downsizing effort may hinge on how knowledgeable end users are about their equipment and applications. If users are not capable of determining whether a problem is hardware- or software-oriented, for example, they may still require a high degree of support from the IT department. This is an important concern, considering that the primary motivation behind many downsizing efforts is to greatly reduce IT staff.

Downsizing is not simply a matter of buying a desktop computer for each person, connecting these computers together, handing out applications packages, and hoping for the best. Instead of eliminating staff, downsizing may instead result in the redeployment of current staff to supply extensive assistance to inexperienced users, provide training and technical support, set up and enforce security procedures, operate a help desk, evaluate equipment for acquisition, and deal with unexpected crises (e.g., limiting the corruption of data from computer viruses).

Attitudes of IT personnel.   A concern related to users' expertise is the attitudes of IT staff. The crucial question to be answered is whether current IT staff members are willing and able to change with the times; specifically, do they have the interpersonal communications skills necessary to deal effectively with end users in all of these areas? If key IT staff stick doggedly to a mainframe-only attitude, they may have to be replaced with personnel who are more knowledgeable about LANs and microcomputers, and more sensitive to end-user needs. This extreme scenario is probably unnecessary in most cases, however.

During the participative planning process, other department heads can become sensitive to the anxiety that IT staff are possibly experiencing. Other department managers can make the idea of downsizing more palatable to IT staff by pointing out that this endeavor is a formidable challenge that represents a vital, experience-building opportunity. In becoming a partner in the project's success, IT staff will have contributed immeasurably toward a revitalized organizational structure that is better equipped to respond to global competitive pressures.

Hidden costs.   The primary motivation for downsizing has ostensibly been to cut costs. Organizations frequently think that the switch to cheap microcomputers connected over LANs from expensive mainframes will cut their budgets and help them realize productivity benefits in the bargain. This attitude ignores the hidden costs associated with desktop assets.

In fact, many times, little or no money is saved immediately as a result of downsizing. According to some industry estimates, the cost of operating and supporting a single workstation can reach $40,000 over five years. About 90 percent of this amount consists of hidden support costs that remain unmanaged and unaccountable. A hidden cost can be anything a LAN administrator or systems manager does not know about, such as having too many software licenses or overconfigured workstations that have more disk capacity or memory than is required for particular applications.

Another hidden cost is training. Nonstandardized configurations mean training cannot be implemented on a corporatewide basis, which means possible economies of scale are missed. A well-planned asset management program can reduce costs in these and other areas, including software licensing, help desk operations, network management, technology migrations, and planning a reengineering strategy. It also can ease the record-keeping burden of routine moves, adds, and changes.

Moving applications to PCs and LANs requires specialized management, technical staff, and testing equipment. The money saved by trimming IT staff is inevitably spent in the many and varied hidden costs and in trying to maintain a problem-free distributed computing environment.

As mentioned earlier, downsizing frequently entails a redeployment rather than a reduction of IT staff. One of the key roles of IT staff, once downsizing is approved, is to create the architecture on which dispersed systems will be built. IT staff may also be charged with controlling and managing the connectivity of the company's networks and setting guidelines and a methodology for using the network. Consequently, even though downsizing may initially have been motivated by the desire to trim IT staff and reduce costs, the head count may not change significantly—if at all.

7.9   Assuming a Leadership Role

With users already installing their own applications on company computers, bringing Linux into the workplace, or setting up wireless access points to the office LAN in order to increase mobility, it is incumbent upon IT to assert a leadership role; after all, users are not network architects, and they may be violating software license agreements or creating security problems. You, as a network manager, must not become content with the role of spectator rather than of participant. When users attempt to overcome problems with their own quick fix solutions, many IT managers prefer to take no action rather than attempt to explain basic concepts to users. This can be a serious mistake, since the problems caused by rogue employees are ultimately the responsibility of the company. Inevitably, top management will want to know how these problems arose in the first place and how they were allowed to continue. The person who must answer these questions is the network manager.

As more equipment is added to expand the distributed environment—such as bridges, routers, and multiplexers—new support capabilities must be added, usually in the form of specialized personnel. This too is an excellent opportunity for IT to assume a leadership role. The peripheral activities of operating networks provide ample justification for maintaining existing IT staff—and perhaps even augmenting staff. With more vendors to deal with, personnel are required to administer such things as contracts, responses to trouble calls, inventory, and invoice and payment reconciliation. Some companies are trying to rid themselves of such encumbrances by outsourcing these responsibilities to an outside firm.

Outsourcing is generally defined as contracting out the design, implementation, and management of information systems and networks. Even this arrangement provides opportunities for IT staff, because it often entails their transfer to the outsourcing vendor with comparable pay and benefits. They not only continue working at their company but also become exposed to other opportunities as well.

When it comes to the question of downsizing, it is incumbent upon the IT department to assume leadership for another reason. If downsizing is forced on the organization from senior management or is planned by another department, IT will face an even greater loss of credibility and influence and will have to accept whatever decisions are imposed upon it. Leadership, however, enhances credibility and influence, which in turn often leads to control.

7.10   Collocation Arrangements

Collocation refers to an arrangement whereby a company installs its application, database, and storage servers in the building of a carrier or service provider, saving on the cost of infrastructure and obtaining access to bandwidth and services at reduced rates.

Distributing resources via collocation arrangements is particularly well suited for application and content delivery business models, previously based on data center deployment over highly meshed networks, which are burdened with high operations costs. Instead of building data centers and installing high-speed networks to reach subscribers in selected markets, much can be gained through collocation arrangements with carriers that provide the necessary infrastructure at a fixed monthly cost.

Collocation arrangements provide secure, environmentally controlled space that is leased in the form of cages, cabinets, and racks into which service providers place their own routers, servers, and switches. This type of arrangement also offers access to a local fiber ring as well as access to a national fiber backbone for interconnecting systems in multiple markets. The collocation provider monitors the entire network infrastructure on a 24×7 basis, freeing application and content providers to focus more of their resources on core business operations.

More than just rack space for equipment and cross-connects between multiple carriers, collocation is a platform for distributed operations, market entry, value creation, and revenue generation—activities that can make or break Application Service Providers (ASPs), Content Delivery Providers (CDPs), Internet Service Providers (ISPs), and e-commerce firms in competitive markets.

Collocation sites can also provide performance advantages. CDPs, for example, can use these sites to expedite content delivery to new markets. The arrangement allows caching servers and content switches to be positioned closest to points of demand, while the resulting geographical diversity protects information assets from potential loss due to unforeseen events. Multiple paths between the node switches in the collocation provider's backbone network offer protection against congestion and link failures, while the built-in protection mechanisms of its metro Synchronous Optical Network (SONET) rings offer fail-safe transport of local data.

ASPs can engage in collocation arrangements that can very economically extend enterprise applications to new markets. This arrangement allows powerful multiprocessor application servers, for example, to be positioned closest to customers without having to build a data center in every market. This arrangement also improves application response time for customers. All customization and upgrades of the enterprise applications can be done at the ASP's data center and thoroughly tested there before being loaded to the remote application servers for customer access.

With the right collocation service provider, even the data center capabilities can be leased. If an ASP needs data center capacity and resources, even on a standby basis, there are collocation service providers that offer "hardened" data centers that provide critical amenities, including:

- Multiprocessor servers for the applications
- Carrier-class routers
- Robotic tape library (terabytes capacity)
- Supercomputer processing capacity
- Redundant fiber links to Internet Network Access Points (NAPs)
- High-speed links to remote collocation facilities
- Multilevel power protection
- Environmental controls
- On-site technicians and application experts

The ASP can contract for data center capacity and resources to incrementally supplement its own data center, as a backup data center for disaster recovery, or as its virtual data center. In the virtual data center scenario, the ASP's programmers, application maintenance staff, and other support personnel can work from any location via secure links over a VPN. This allows the ASP to recruit specialists from any location to further reduce its infrastructure costs, while drawing from the talent pool in any city. A project manager coordinates the activities of programmers and quality assurance and software maintenance staff.

VPNs can be built with IP for economy, or with frame relay and ATM for performance. The right collocation service provider can support all three, allowing ISPs, ASPs, and content providers to choose an appropriate transport for service delivery to customers and another for internal business processes—or move from one to another, as business needs change.

In not having to duplicate infrastructure, there is less risk of running out of capital when it is needed most—for improving service to customers, expanding local markets, and adding features to further differentiate services. Risk is further minimized by leveraging the collocation provider's infrastructure for use as a test bed for new services using pilot subscribers on a localized basis before a full rollout.

As a key component of disaster recovery planning, collocation also offers opportunities for power protection, network redundancy, carrier diversity, physical security, and distributed operations that can help service providers withstand a natural or human-caused calamity in one city by having resources fully available in other cities. Users in the distressed location can be serviced from resources available from nonaffected locations by such means as traffic rerouting and node rehoming under control of the collocation provider's network operations center.

From a central operations center, the collocation provider can monitor network equipment and lines for impending fault conditions and take corrective action before a failure actually occurs and disrupts service. This capability translates into high network availability and throughput. These capabilities save collocation customers from having to recruit and retain qualified network operations staff of their own, plus the need to acquire network management platforms and tools.

The collocation provider also monitors the environment to ensure proper temperature and humidity. Battery banks are monitored to ensure that they are properly charged and ready to kick in during a commercial power outage to protect network systems. After 10 seconds or so, a diesel-fuel generator kicks in to supply power.

Managed servers for Web site and e-commerce hosting, firewalls, and routers can also be included in this power protection arrangement. Collocation customers may install their own Uninterruptible Power Supply (UPS) equipment to protect their own systems against momentary power interruptions. When monitored systems reach predefined thresholds, alarms go off at the network operations center so corrective action can be taken before the condition results in an outage.

The collocation provider's on-site technicians provide physical management of the site, performing such tasks as installing new racks and cable, upgrading network systems, replacing faulty hardware components, and escorting visitors to their space. Visiting privileges allow technicians from customer organizations to enter the collocation site to install or upgrade equipment, swap out faulty components, and do extensive hardware troubleshooting. If the customer organization does not have a field technician available, the collocation provider's technicians can provide first-level support.

The more a company can geographically distribute its IT assets, the greater protection it will have against disruption to business processes and service delivery to customers. Collocation arrangements in each market can provide this protection. With a managed security service, for example, rule sets can be defined for a firewall that filters traffic according to various parameters.

An ASP might want to specify the domain names that are allowed access to its application servers. This would prevent access from any other source. An e-commerce firm might want a firewall configuration to protect its internal network where back-office processes reside to support the online activities of its customers. The firewall is fully managed by the collocation provider and includes an SLA that guarantees the proper functioning of the various rule sets.

Collocation arrangements address a wide range of strategic business needs and protect against virtually any disaster scenario. With collocation, service providers can outsource facilities, connections, hardware, security, and network management. This leaves them with more resources to focus on product development, testing, feature enhancement, marketing, billing, and customer service—which differentiate them in competitive markets. The payoff comes in the form of increased sales, low overhead, market share, and competitive advantage.

7.11 Conclusion

Claims that downsizing is the wave of the future and that the mainframe is dead are generalizations that have little relevance to the real world. Each organization must ignore such generalizations when considering whether or not downsizing offers an economical solution to its particular problems of information access and resource sharing.

The results of downsizing, however, are potentially too compelling to ignore. Among these results is the wide range of benefits over the existing centralized mainframe environment, including lower processing costs, configuration flexibility, a more responsive applications development environment, and systems that are easier to use. In addition, networked, distributed processing systems may provide the most cost-effective way for users to access, transfer, and share corporate information.

The cost benefits of downsizing can be near-term, as in the more effective utilization of internal IT staff, or long-term, as in the investment in newer applications and technologies. Determining the benefits of downsizing, the extent to which it will be implemented, who should take responsibility for it, and the justification of its up-front costs, are highly subjective matters. The answers will not appear in the form of solutions borrowed from another company. In fact, the likelihood of success will be improved greatly if the downsizing effort is approached as a custom reengineering project, with reliance on outsourcing arrangements as new needs arise.

Network Service and Facility Selection

8.1 Introduction

Companies are under increasing pressure to get all the value possible out of private network facilities and carrier-provided services, as well as to extend services to branch offices, telecommuters, and mobile professionals, while controlling communications expenditures. At the same time, companies continue to make investments in information systems and networks to gain competitive advantage. In implementing various combinations of services, facilities, media, and equipment, a balance of efficiency and economy can be achieved to obtain or sustain that advantage.

A variety of communications services and facilities are available over such transmission media as optical fiber, twisted-pair copper, and wireless. In conjunction with such Customer Premises Equipment (CPE) as Private Branch Exchanges (PBXs), multiplexers, channel banks, routers, and modems, these media support communications services that provide access to virtually any location in the world. The selection of services and facilities depends on such factors as response time requirements, performance characteristics, and cost.

8.2 Transmission Media

A variety of transmission media are available for both voice and data communications. Among the commonly used media are twisted-pair wiring (shielded and unshielded), coaxial cable (thick and thin), and optical fiber (single- and multimode). In addition, there is a variety of wireless technology available, including laser, infrared, and microwave.

8.2.1 Twisted-pair wiring

Twisted-pair wiring is the most common transmission medium; it is currently installed in most office buildings and residences. Twisted-pair wiring consists of one or more pairs of copper wires. To reduce cross talk or electromagnetic induction between pairs of wires, two insulated copper wires are twisted around each other. Ordinary wire to the home is Unshielded Twisted Pair (UTP). Only one pair of wires is used in the local loop to support ordinary phone calls. In the transport portion of the network, beyond the central office switch, two pairs of wires are used to support the call.

Local Area Networks (LANs).   Many businesses use unshielded twisted-pair wiring for LANs. In this case, two twisted pairs of wire are used for the connections between computers, printers, and other devices on the network. Each color-coded wire is solid conductor, 24 American Wire Gauge (AWG). In some corporate data centers, however, the twisted pairs or wires are enclosed in a shield that protects the signal against interference, including cross talk. This is known as Shielded Twisted Pair (STP).

Unshielded twisted-pair wiring, also called Category 5 cable, has become the most popular transmission medium for local area networking (see Fig. 8.1)

Figure 8.1   A typical Category 5 UTP cable contains 4 pairs of wire identified with a solid color and the same solid color striped onto a white background.

because of its low cost and ease of installation. The pairs of wires in UTP cable are colored-coded so that they can be easily identified at each end. The most common color scheme is the one that corresponds to the Electronic Industry Association/Telecommunications Industry Association's (EIA/TIA) Standard 568B, which is summarized in Table 8.1.

The cable connectors and jacks that are most commonly used with Category 5 UTP cables are RJ45. The RJ simply means Registered Jack and the number 45 designates the pin numbering scheme. The connector is attached to the cable, and the jack is the device that the connector plugs into, whether it is in the wall, the network interface card (NIC) in the computer, or in the hub.

In response to the growing demand for data applications, cable has been categorized into various levels of transmission performance, as summarized in Table 8.2. The levels are hierarchical in that a higher category can be substituted for any lower category.

The use of UTP wiring has several advantages. The technology and standards are mature and stable for voice and data communications. Telephone systems, which use twisted pair wiring, are present in most buildings, and unused pairs usually are available for LAN connections. When required, additional twisted pair can be installed relatively easily and the cost of Category 5 cabling is relatively inexpensive.

Of course, unshielded twisted-pair wiring has a few disadvantages as well. It is sensitive to Electromechanical Interference (EMI), so new installations must be planned to route around sources of EMI. Because signals have a tendency to radiate outward from the medium, unshielded twisted pair is also more susceptible to eavesdropping, which makes encryption and other security precautions necessary to safeguard sensitive information. An additional requirement is a wiring hub. Although this involves another expense, the hub actually facilitates the installation of new wiring and keeps all wires organized, while making it easier to implement moves,

TABLE 8.1   The Color Scheme Specified by Electronic Industry Association/Telecommunications Industry Association's Standard 568B for Category 5 UTP Cable

Wire pair	Color code
#1	White/Blue Blue
#2	White/Orange Orange
#3	White/Green Green
#4	White/Brown Brown

TABLE 8.2   Categories of UTP Cable

Category	Maximum bandwidth, in MHz	Application	Standards
7	600	1000BaseT and faster	Standard under development*
6	250	1000BaseT	TIA/EIA 568-B (Category 6)
5E	100	Same as CAT 5 plus 1000BaseT	ANSI/TIA/EIA-568-A-5 (Category 5E)
5	100	10/100BaseT 100 Mbps TPDDI (ANSI X 319.5) 155 Mbps ATM	TIA/EIA 568-A (Category 5) NEMA (Extended Frequency) ANSI/ICEA S-91-661
4	20	10 Mbps Ethernet (IEEE 802.3) 16 Mbps Token Ring (IEEE 802.5)	TIA/EIA 568-A (Category 4) NEMA (Extended Distance) ANSI/ICEA S-91-661
3	16	10 Mbps Ethernet (IEEE 802.3)	TIA/EIA 568-A (Category 3) NEMA (Standard Loss) ANSI/ICEA S-91-661
2	4	IBM Type 3 1.544 Mbps T1 1 Base 5 (IEEE 802.3) 4 Mbps Token Ring (IEEE 802.5)	IBM Type 3 ANSI/ICEA S-91-661 ANSI/ICEA S-80-576
1	Less than 1	POTS (Plain Old Telephone Service) RS 232 and RS 422 ISDN Basic Rate	ANSI/ICEA S-80-576 ANSI/ICEA S-91-661

* In new installations, fiber to the desk may be less expensive than installing Category 7 cable.

adds, and changes to the network. Over the long term, a hub saves much more than it costs.

UTP cable has evolved over the years, and different varieties are available for different needs. Improvements such as variations in the twists or in individual wire sheaths or overall cable jackets have led to the development of EIA/TIA-568 standard-compliant categories of cable that have different specifications for supporting high-speed signals. Because UTP cable is lightweight, thin, and flexible, as well as versatile, reliable, and inexpensive, millions of nodes have been and continue to be wired with UTP cable, even for high-data-rate applications. For the best performance, UTP cable should be used as part of a well-engineered structured cabling system. However, businesses that require reliable gigabit-per-second data transmission speeds should give serious consideration to moving to optical fiber, rather than Category 7 UTP.

Local loop.   In the local loop, hundreds of insulated wires are bundled into larger cables, with each pair color-coded for easy identification. Most telephone lines between central offices and local subscribers consist of this type of cabling, which is mounted on poles or laid underground. Such bundling facilitates installation and reduces costs. Special sheathing offers protection from natural elements. For businesses that want digital connections, such as T1 at 1.536 Mbps from a PBX to the central office, two pairs of wires are used to achieve full-duplex transmission—one pair for the send path and one pair for the receive path.

Copper wiring has bandwidth and distance limitations. While UTP is adequate for a LAN, it may not be adequate for linking multiple LANs between the floors of an office building or for linking multiple LANs between buildings in a campus environment. While T1 lines can link multiple LANs over great distances, the bandwidth is limited to 1.536 Mbps (channelized) or 1.544 Mbps (unchannelized). Since today's LANs operate at 10 to 100 Mbps and higher, the use of T1 lines for LAN interconnectivity can be a source of bottlenecks that slows application response time. Of course, the use of T3 lines at 44.736 Mbps can ease the bottleneck somewhat, but their monthly cost may be prohibitive for most companies.

Wide Area Networks (WANs).   When companies require high-speed connectivity between geographically dispersed locations, cost becomes a key issue, especially as distance increases. Maintaining high performance over long distances can be dealt with by turning to alternative transmission media, such as optical fiber or a wireless technology that uses licensed or unlicensed spectrum. But building and operating a private network is expensive, and the costs escalate dramatically as distance increases.

If there are multiple corporate locations that need to be interconnected within a metropolitan area, a carrier-provided service such as Ethernet may offer the best value, even if there are only two locations that need to be connected. If the corporate locations are in different cities, a frame relay or Asynchronous Transfer Mode (ATM) service may offer the best value, more so as distance increases. This is because frame relay and ATM services are not priced according to distance, but on the bandwidth of the virtual circuits and the capacity of the port into the network. And with all three services—Ethernet, frame relay, and ATM—the underlying transport technology is optical fiber, making the services very scalable to meet customer needs quickly.

8.2.2   Coaxial cable

Until the 10Base-T standard (10 Mbps over unshielded twisted pair), coaxial cable was the preferred media for LANs. This type of cable contains a conductive cylinder with insulation around a wire in the center. Coaxial cable is typically shielded to reduce interference from external sources. Coaxial cable can transmit at a much higher frequency than a wire pair, allowing more data to be transmitted in a given period. By providing a "wider"" channel, coaxial cable allows multiple LAN users to share the same medium for communicating with host computers, servers, front-end processors (FEPs), peripheral devices, and personal computers.

LANs started out using thick coaxial cabling that was heavy, rigid, and difficult to install. The maximum cable length of Ethernet is 500 m, which is referred to as 10Base-5. This technical shorthand means "10 Mbps data rate, baseband signaling, 500-m maximum cable segment length." The initial Ethernet implementations used 50-ohm coaxial cable with a diameter of 10 mm, which is referred to as "thick Ethernet cable." Another cable standard, 10Base-2 (meaning "10 Mbps data rate, baseband signaling, 200-m maximum cable segment length") uses ordinary CATV-type coaxial cable, called "thin Ethernet cable." Today, UTP is the most common cable type used for LANs, although optical fiber increasingly is being used for the backbone between floors and between buildings.

Coaxial cable networks, once used only to deliver television programming, have been upgraded to support a whole array of new business services. Traditional coaxial cable systems typically operate with 330 or 450 MHz of capacity, whereas modern Hybrid Fiber Coax (HFC) systems operate to 750 MHz or more. Each standard television channel occupies 6 MHz of Radio Frequency (RF) spectrum. A traditional cable system with 400 MHz of downstream bandwidth can carry the equivalent of 60 analog TV channels and a modern HFC system with 750 MHz of downstream bandwidth has the capacity for over 200 channels when digital compression is applied. Newer cable systems operate at 860 MHz, offering capacity for value-added services.

Many cable companies are migrating their networks from analog to digital. This provides customers with greater programming diversity, better picture quality, improved reliability, and enhanced services, such as Internet access, telephone service, and video-on-demand. Advanced compression techniques can be applied to digital signals, allowing up to 12 digital channels to be inserted into the space of only one traditional 6-MHz analog channel, enabling cable companies to greatly increase the capacity of their networks. The larger cable companies now offer 250 channels including enhanced pay-per-view service, digital music channels, new networks grouped by genre, and an interactive program guide.

While cable TV makes up 99 percent of cable operators' revenues, some operators have advanced HFC networks that can be leveraged for new revenue streams. In addition to interactive television and video on demand (VOD), cable operators offer business-class broadband Internet access and telecommunications services that compete with Internet service providers and telephone companies. One cable company, Cox, offers a diversified portfolio of services, illustrating the new service direction of the cable industry, as well as new choices for corporate network managers.

In early 2002, Cox completed the development of its own nationwide Internet Protocol (IP) network to deliver high-speed data and Internet services to its subscribers. This network utilizes Cox's HFC infrastructure to connect subscribers to the Internet and other on-line services, and includes standard Internet service provider functionality such as web page hosting for subscribers, access to Internet news groups, multiple e-mail accounts, and remote access.

Cox delivers telecommunications services to businesses through its competitive local exchange carrier operation, Cox Business Services. Through both its dedicated fiber-optic networks and its hybrid fiber coaxial cable networks, Cox Business Services provides business customers with video, telephony, and high-speed Internet access services.

The company's Network Operations Center (NOC) provides 24×7 active monitoring, diagnostics, and remote control. The center continually tests all customer lines and troubleshoots any problems before they become severe enough to disrupt service. When a service error occurs, technicians immediately reroute network traffic and dispatch local crews to ensure minimal or no network downtime. The NOC supports all of the metro fiber and HFC networks, including the management of building connections.

Not all cable operators offer such an extensive range of business-class services. But in areas where these services are available, corporate network managers have another alternative to consider, which might lower the cost of voice and data services. The medium of coaxial cable offers the added value of scalability. With the operator's coax cable installed at each company location, adding a business trunk for handling more voice traffic or adding more bandwidth for Internet access is just a matter of placing the order to start the billing process. There is no 30- to 45-day wait for an additional trunk or line to be provisioned by the telephone company.

Since it is still too early to draw conclusions on how successful the cable companies will be in attracting enterprise accounts in significant numbers, network managers are advised to weigh this option carefully.

8.2.3   Optical fiber

Optical fiber offers several advantages over copper-based transmission media, and even microwave radio systems, including:

- Higher bandwidth capacity
- The ability to increase capacity by upgrading the light sources and receivers, without changing the cable itself
- Immunity to EMI and Radio Frequency Interference (RFI)
- Protection from unauthorized, clandestine wire tapping, which is extremely difficult over a fiber-optic system

Types of fiber.   There are two types of optical fibers in common use today: single-mode and multimode. Single-mode fibers allow only one light wave to be transmitted along the core, whereas multimode fibers enable many light waves to be transmitted along the core. The difference is important. Single-mode fiber entails lower signal loss and supports higher transmission rates over longer distances than multimode fiber. For this reason, single-mode optical fiber is the preferred medium.

Multimode fibers have relatively large cores. Light entering a step-index or graded-index multimode fiber will take many paths. Light pulses simultaneously entering a multimode fiber may exit at slightly different times. This phenomenon, called *intermodal pulse dispersion,* creates minor signal distortion that limits the data rate and the distance that the optical signal can be sent without using repeaters. Initially, multimode fiber was used on LAN backbones over short distances, but now single-mode fiber is the preferred solution, regardless of distance.

Fiber Distributed Data Interface.   One of the first standards for fiber-based LANs and campus area networks was the Fiber Distributed Data Interface (FDDI), an American National Standards Institute (ANSI) specification, which provides 100-Mbps transmission. FDDI uses a timed token-passing access protocol for passing frames of up to 4500 bytes. It supports up to 500 stations over a maximum fiber path of 200 km (124 mi) in length with a maximum of 2 km (1.2 mi) between adjacent stations. The FDDI standard also includes built-in management capabilities, which detect failures and automatically reconfigure the network to bypass faulty nodes or LAN segments.

An extension to FDDI, called FDDI-2, supports the transmission of voice and video information as well as data. Another variation of FDDI, called FDDI Full Duplex Technology (FFDT) uses the same network infrastructure, but can potentially support data rates up to 200 Mbps. Although FDDI networks are still in operation and are valued for their self-healing capability, they are very expensive compared to Ethernet and do not match Ethernet in terms of scalability and transmission speed.

Fibre Channel.   Another ANSI standard is Fibre Channel, a high-performance interconnect standard designed for bidirectional, point-to-point serial data channels between desktop workstations, mass storage subsystems, peripherals, and host systems. Serialization of the data permits much greater distances to be covered than parallel communications. Unlike networks where each node must share the bandwidth capacity of the media, Fibre Channel devices are connected through a flexible circuit and/or packet switch capable of providing the full bandwidth to all connections simultaneously.

The key advantage of Fibre Channel is speed—it is 10 to 250 times faster than typical LAN speeds. Fibre Channel started out by offering a transmission rate of 100 MB/s (200 MB/s in full-duplex mode), the equivalent of 60,000 pages of text per second. Such speeds are achieved simply by transferring data between one buffer at the source device and another buffer at the destination device without regard for how it is formatted—cells, packets, or frames. It is inconsequential what the individual protocols do with the data before or after it is in the buffer—Fibre Channel only provides complete control over the transfer and offers simple error checking.

Unlike many of today's interfaces—including the Small Computer Systems Interface (SCSI)—Fibre Channel is bidirectional, achieving 100 MB/s in both directions simultaneously. Thus, it provides a 200 MB/s channel if usage is balanced in both directions. Fibre Channel also overcomes the restrictions on the number of devices that can be connected—up to 126, versus 15 for SCSI.

Fibre Channel also overcomes the distance limitations of today's interfaces. A fast SCSI parallel link from a disk drive to a workstation, for example, can transmit data at 20 MB/s, but it is restricted in length to about 20 m. In contrast, a quarter-speed Fibre Channel link transmits information at 25 MB/s over a single, compact optical cable pair at up to 10 km in length. This allows disk drives to be placed almost anywhere and enables more flexible site planning.

The high-speed, low-latency connections that can be established using Fibre Channel make it ideal for a variety of data-intensive applications, including:

- *Backbones.* Fibre Channel provides the parallelism, high bandwidth, and fault tolerance needed for high-speed backbones. It is the ideal solution for mission-critical internetworking. The scalability of Fibre Channel makes it practical to create backbones that grow as one's needs grow—from a few servers to an entire enterprise network.

- *Workstation clusters.* Fibre Channel is a natural choice to enable super-computer-power processing at workstation costs.

- *Imaging.* Fibre Channel provides the "bandwidth on demand" needed for high-resolution medical, scientific, and prepress imaging applications, among others.

- *Scientific/engineering.* Fibre Channel delivers the needed throughput for today's new breed of visualization, simulation, CAD/CAM, and other scientific, engineering, and manufacturing applications, which demand megabytes of bandwidth per node.

- *Mass storage.* Current mass storage access is limited in rate, distance, and addressability. Fibre Channel provides high-speed links between mass storage systems and servers at distances of up to several kilometers, making possible Storage Area Networks (SANs).

- *Multimedia.* Fibre channel's bandwidth supports real-time videoconferencing and document collaboration between several workstation users, and is capable of delivering multimedia applications containing voice, music, animation, and video.

The demands on networks and systems for moving and managing data are increasing exponentially, and improvements in performance across the infrastructure are required to enable users to move and manage their data efficiently and reliably. The current specification for Fibre Channel allows for speeds of 1 Gbps and higher. The FC-SW-2 Open Fabric standard boosts the speed of Fibre Channel to 2 Gbps and establishes the foundation for building interoperable, multivendor switch fabrics. Users can connect existing 1-Gbps products with the newer 2-Gbps technology and, through standards-based auto-negotiation, extend their current SAN installations instead of having to replace them. Not only does this technology provide a clear path for the industry but it also holds great promise for new and revolutionary products that greatly extend the capabilities of storage area networks.

The Fibre Channel Industry Association (FCIA) has introduced a proposal for 10-Gbps Fibre Channel that supports LAN and WAN devices over distances ranging from 15 m (about 50 ft) to 10 km (about 6 mi). The standard also supports bridging SANs over metropolitan-area networks through Dense Wavelength Division Multiplexing (DWDM) and SONET. The 10-Gbps draft specification requires backward compatibility with 1- and 2-Gbps devices. Devices of 10 Gbps will also be able to use the same cable, connectors, and transceivers used in Ethernet and InfiniBand.

InfiniBand.   Short for "infinite bandwidth," InfiniBand is a bus technology that provides the basis for an Input/Output (I/O) fabric designed to increase the aggregate data rate between servers and storage devices. The point-to-point linking technology allows server vendors to replace outmoded system buses with InfiniBand to greatly multiply total I/O traffic compared with legacy system buses such as PCI and its successor PCI-X.

The current PCI bus standard supports up to 133 Mbps across the installed PCI slots, providing shared bandwidth of up to 566 Mbps, whereas PCI-X permits a maximum bandwidth of 1 Gbps. Fibre Channel offers bandwidth up to 2 Gbps and is used to build storage area networks. In contrast, InfiniBand utilizes a 2.5-Gbps wire speed connection with multiwire link widths. With a four-wire link width, InfiniBand offers 10 Gbps of bandwidth. The InfiniBand specification supports both copper and fiber implementations.

The I/O fabric of the InfiniBand architecture takes on a role similar to that of the traditional mainframe-based channel architecture, which uses point-to-point cabling to maximize overall I/O throughput by handling multiple I/O streams simultaneously. The move to InfiniBand means that I/O subsystems need no longer be the bottleneck to improving overall data throughput for server systems.

In addition to performance, InfiniBand promises other benefits such as lower latency, easier and faster sharing of data, built-in security and quality of service, and improved usability through a form factor that makes components much easier to add, remove, or upgrade than today's shared-bus I/O cards.

InfiniBand technology works by connecting host-channel adapters to target channel adapters. The host-channel adapters tend to be located near the servers' CPUs and memory, while the target channel adapters tend to be located near the systems' storage and peripherals. A switch located between the two types of adapters directs data packets to the appropriate destination based on information that is bundled into the data packets themselves.

The connection between the host-channel and target-channel adapters is the InfiniBand switch, which allows the links to create a uniform fabric environment. One of the key features of this switch is that it allows data to be managed based on variables such as Service Level Agreements (SLAs) and a destination identifier. In addition, InfiniBand devices support both packet and connection protocols to provide a seamless transition between the system area network and external networks.

InfiniBand will coexist with the wide variety of existing I/O standards that are already widely deployed in user sites. Existing architectures include PCI, Ethernet, and Fibre Channel. Likewise, InfiniBand fabrics can be expected to coexist with newer I/O standards, including PCI-X, Gigabit Ethernet, and 10X Gigabit Ethernet. The key advantage of the InfiniBand architecture, however, is that it offers a new approach to I/O efficiency. Specifically, it replaces the traditional system bus with an I/O fabric that supports parallel data transfers along multiple I/O links. Furthermore, the InfiniBand architecture offloads CPU cycles for I/O processing, delivers faster memory pipes and higher aggregate data-transfer rates, and reduces management overhead for the server system.

Synchronous Optical Network.   Synchronous Optical Network (SONET) is an industry standard for high-speed, Time Division Multiplexer (TDM) transmission over optical fiber. Carriers and large companies use SONET facilities for fault-tolerant backbone networks and fiber rings around major metropolitan areas. SONET-based services have a performance objective of 99.9975 percent error-free seconds and an availability rate of at least 99.999

percent. SONET combines bandwidth and multiplexing capabilities, allowing users to fully integrate voice, data, and video over the fiber-optic facility.

The SONET standards were developed by the Alliance for Telecommunications Industry Solutions (ATIS), formerly known as the Exchange Carriers Standards Association (ECSA), with input from Bellcore (now known as Telcordia Technologies), the former research and development arm of the original seven Regional Bell Operating Companies (RBOCs) that were created in 1984 as a result of the breakup of American Telephone and Telegraph (AT&T). The standards for North America are published and distributed by the American National Standards Institute (ANSI). In 1989, the International Telecommunication Union (ITU) published the Synchronous Digital Hierarchy (SDH) standard, which is what the rest of the world uses.

SONET provides a highly reliable transport infrastructure, offering numerous benefits to carriers and users. For example, the enormous amounts of bandwidth available with SONET and its integral management capability permit carriers to create global intelligent networks capable of supporting the next generation of services. SONET-based information superhighways support bit-intensive applications such as three-dimensional Computer Aided Design (CAD), medical imaging, collaborative computing, interactive virtual reality programs, and multipoint videoconferences, as well as new consumer services such as video-on-demand and interactive entertainment. Under current SONET standards, bandwidth is scalable from about 52 Mbps to about 13 Gbps, with the potential to go much higher. The SONET standard specifies a hierarchy of electrical and optical rates as summarized in Table 8.3.

Even though the SONET standards only go up to OC-256, some equipment vendors support SONET-compliant OC-768 and OC-1536. Their systems offer a switch fabric and backplane that is OC-768/OC-1536 ready. When the optics become available and viable for 40–80 Gbps, carriers will only need to change the line cards in the optical transport switching systems to take advantage of these speeds.

The base signal rate of SONET on both the electrical side and optical side is 51.84 Mbps. On the electrical side, the Synchronous Transport Signal (STS) is what goes to customer premises equipment, which is electrical. On the optical side, the Optical Carrier (OC) is what goes to the network. Both the electrical and optical signals can be multiplexed in a hierarchical fashion to form higher rate signals.

A key feature of SONET framing is its ability to accommodate existing synchronous and asynchronous signal formats. The SONET payload can be subdivided into smaller "envelopes" called Virtual Tributaries (VTs) to transport lower-capacity signals. Because VTs can be placed anywhere on higher-speed SONET payloads, they provide effective transport for existing

TABLE 8.3 The Hierarchy of Electrical and Optical
Rates Specified in SONET Standards

Electrical level	Optical level	Line rate
STS-1	OC-1	51.84 Mbps
STS-3	OC-3	155.520 Mbps
STS-9	OC-9	466.560 Mbps
STS-12	OC-12	622.080 Mbps
STS-18	OC-18	933.120 Mbps
STS-24	OC-24	1.244 Gbps
STS-36	OC-36	1.866 Gbps
STS-48	OC-48	2.488 Gbps
STS-192	OC-192	9.95 Gbps
STS-256	OC-256	13.271 Gbps
STS-768*	OC-768*	39.813 Gbps
STS-1536*	OC-1536*	79.626 Gbps

*Not yet an official SONET standard.

North American and international formats. Table 8.4 highlights some of
these VTs.

SONET permits sophisticated self-diagnostics and fault analysis to be
performed in real time, making it possible to identify problems before they
disrupt service. Intelligent network elements, specifically the SONET
Add-Drop Multiplexer (ADM), can automatically restore service in the
event of failure with a variety of restoral mechanisms.

SONET's embedded control channels enable the tracking of end-to-end
performance and identification of elements that cause errors. With this
capability, carriers can guarantee transmission performance, and users
can readily verify it without having to go off-line to implement various
test procedures. For you, as a network manager, these capabilities allow
a proactive approach to problem identification, which can prevent service
disruptions. Along with the self-healing capabilities of ADMs, these diagnostic capabilities ensure that properly configured SONET-compliant
networks experience virtually no downtime. SONET offers multiple ways
to recover from network failures, including:

- *Automatic protection switching.* The capability of a transmission system
  to detect a failure on a working facility and to switch to a standby facility to recover the traffic. One-to-one protection switching and "one-to-n"
  protection switching are supported.

TABLE 8.4   Select Virtual Tributary (VT) Payload Envelopes Specified in SONET Standards

VT level	Line rate (Mbps)	Standard
VT1.5	1.728	DS1
VT2	2.304	CEPT1
VT3	3.456	DS1C
VT6	6.912	DS2
VT6-N	$N \times 6.9$	Future
Async DS3	44.736	DS3

- *Bidirectional line switching.* Requires two fiber pairs between each recoverable node. A given signal is transmitted across one pair of fibers. In response to a fiber facility failure, the node preceding the break will loop the signal back toward the originating node, where the data traverses a different fiber pair to its destination.

- *Unidirectional path switching.* Requires one fiber pair between each recoverable node. A given signal is transmitted in two different paths around the ring. At the receiving end, the network determines and uses the best path. In response to a fiber facility failure, the destination node switches traffic to the alternative receive path.

Wavelength Division Multiplexing. Wavelength Division Multiplexing (WDM) is a technology that increases the bandwidth of existing optical fiber by splitting light into constituent wavelengths (i.e., colors), each of which is a high-speed data channel that can be individually routed through the network (see Fig. 8.2). This technology has been in use by long-distance carriers in recent years to expand the capacity of their trunks by allowing a greater number of signals to be carried on a single fiber, but there are systems available for use on enterprise networks as well. The main advantage of WDM is that it allows huge amounts of bandwidth to be added to existing backbone networks without the need to install new fiber.

WDM systems are compatible with SONET rings. WDM systems are installed on the front end of the existing fiber. The SONET equipment is plugged into the WDM systems at each network node, thereby increasing the bandwidth capacity of the fiber and supporting the SONET payloads as well. The WDM systems even pass through SONET's embedded overhead channels, which perform link supervision and gather performance statistics, and allow SONET's fault recovery procedures to operate as normal to ensure network availability.

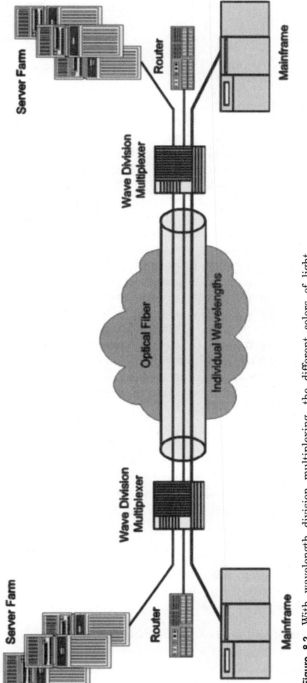

Figure 8.2 With wavelength division multiplexing, the different colors of light constitute channels for the transmission of data at high speeds, regardless of traffic type or protocol.

In terms of multivendor interoperability, SONET has the clear advantage over WDM/DWDM. This is because SONET supports standard mappings for a wide variety of payloads, including T1 and T3, ATM, and Internet Protocol (IP), to mention only a few. This permits true multivendor interconnection of a variety of equipment. By contrast, WDM/DWDM technologies are still relatively new with standards still being hammered out. Currently, multivendor interoperability of WDM/DWDM systems is achieved through cooperative arrangements negotiated among equipment manufacturers.

SONET is less efficient than WDM in the use of the available bandwidth because it is TDM-oriented. This means the offered traffic must fit into time slots, which wastes bandwidth if they are not filled. With WDM, there are no time slots. Instead there are wavelengths, which can be dedicated to specific traffic types such as IP, Gigabit Ethernet, Fibre Channel, or even SONET, over the same fiber. These wavelengths can also be shared by multiple applications, making for very efficient use of the available bandwidth.

SONET is known for its sub-50 ms path protection mechanism to sustain ring availability and prevent traffic loss in case of fiber cuts or node failures. Much of the installed base of WDM systems cannot do this. Newer systems, however, can be configured with or without this type of protection.

SONET does not scale nearly as well as WDM systems. SONET bandwidth tops out at OC-256 (13 Gbps). Nonstandard extensions at OC-768 (40 Gbps) and OC-1536 (80 Gbps) are becoming available as well. But commercial WDM systems currently scale to 320 Gbps and higher without the need to install additional fiber.

8.2.4   Wireless communications

A variety of wireless technologies have become available in recent years to support the voice and data communications needs of businesses. There are wireless adjunct systems that can connect to an existing PBX to economically extend the range of communication to mobile employees within a building. There are wireless LANs that eliminate the need to install new wiring to accommodate growth or keep up with changes in business needs. One technology, Bluetooth, supports both voice and data over short distances of up to 30 ft. And in the local loop, the Local Exchange Carrier (LEC) can be bypassed entirely with the use of various broadband wireless technologies. All of these wireless technologies give network managers more flexibility in meeting corporate communications needs and can offer opportunities for cost savings.

Narrowband voice communications systems. Narrowband is generally defined as a family of mobile or portable radio services that operate in the 900- and 1900-MHz bands of the electromagnetic spectrum to provide telephone, data, advanced paging, and other services to individuals and businesses. Unlicensed spectrum may be used to provide wireless communication services that can be integrated with a variety of competing networks and communications systems, including corporate PBX systems, giving network managers more choices in meeting corporate communications needs.

With office workers spending increasing amounts of time away from their desks—supervising various projects, working at temporary assignments, attending meetings, and just walking corridors—there is a growing need for wireless technology to help them stay in touch with colleagues, customers, and suppliers. According to various industry estimates, as many as 75 percent of all workers are mobile, spending significant amounts of time away from their desks. This lack of communication can adversely affect personal and organizational performance. The idea behind the wireless PBX is to facilitate communication within the office environment, enabling employees to be as productive while moving about as they would be if they were sitting at their desks.

Almost any organization can benefit from improved communications offered by a wireless PBX system, especially those engaged in:

- *Manufacturing.* Roving plant managers or factory supervisors do not have to leave their inspection or supervisory tasks to take important calls.

- *Retail.* Customers can contact in-store managers directly, eliminating noisy paging systems.

- *Hospitality.* Hotel event staff can stay informed of guests' needs and respond immediately.

- *Security.* Guards can relay emergency information quickly and clearly, directly to the control room or police department without needing to reach a desktop phone.

- *Business.* Visiting vendors or customers have immediate use of preassigned phones without having to borrow employee desktop phones.

- *Government.* In-demand office managers can be available at all times for instant decision making.

Wireless PBX technology differs from other forms of mobile communications that are provided by carriers. Unlike cellular phone service, for example, there is no charge for airtime. While cellular is a high-powered system designed primarily for high-speed use in cars, wireless PBX is used indoors and the mobile phones get better reception and longer battery life—up to six hours of talk-time or 60 hours of standby time on a single charge.

Via an adjunct switch, the wireless capability can be added to virtually any key system, PBX, or Centrex system to provide an integrated facility running both wired and wireless extensions. Mobile users have access to the same functions and features of the host system to which they are ultimately connected, including caller identification, call screening, extension dialing, speed dialing, conferencing, hold, transfer, programmable buttons, and voice messaging.

Under ideal conditions, mobile users have the same digital voice quality as conventional PBX users. Even the charges for outgoing calls are integrated into the same billing system as desktop phones. Via strategically placed base stations and distribution hubs, the wireless system also provides seamless communication throughout the building, including hard-to-reach places like elevators, tunnels, or parking garages.

As noted, corporate wireless communication systems are designed to operate in unlicensed frequency bands. The term *unlicensed* refers to the spectrum that is used with equipment, which can be purchased and deployed without Federal Communications Commission (FCC) approval because it is not part of the public radio spectrum. In other words, since corporate wireless systems operate over a dedicated frequency within a very narrow geographical area, the signals have little chance of interfering with other wireless services in the surrounding area. Consequently, there is no need for frequency coordination or FCC licensing. The individual channels supported by the wireless communication system are spaced far enough apart to prevent interference with each other. Privacy is enhanced through the use of digital speech encoding and spread spectrum frequency hopping to make eavesdropping nearly impossible.

The capacity of wireless communication systems is easily expanded—portable telephones and base stations are added as needed—up to the maximum capacity offered by the vendor's particular system. Likewise, the coverage of wireless systems can be expanded through the strategic placement of base stations and distribution hubs. While single-zone coverage up to 500,000 square feet effectively covers most office buildings, the addition of base stations and distribution hubs can achieve multizone coverage of up to 12 million square feet for a campus environment.

Network managers can achieve substantial savings over time through the elimination of traditional phone moves, adds, and changes. There are also significant savings in cabling, since there is less need to rewire offices and other locations for desktop telephones. This is significant because companies typically spend between 10 and 20 percent of the original cost of their PBX annually on reconfiguring the system.

Wireless LANs.   A Wireless Local Area Network (WLAN) is a data communications system implemented as an extension—or as an alternative—to a wired LAN. Using a variety of technologies including narrowband radio, spread spectrum, and infrared, wireless LANs transmit and receive data through the air, minimizing the need for wired connections. They are especially useful in health care, retail, manufacturing, and warehousing environments where employees move about and use hand-held terminals and notebook computers to transmit real-time information. Warehouse workers, for example, can use wireless LANs to exchange information with a database server, thereby increasing productivity, without having to walk great distances to a desktop computer.

While the initial investment required for wireless LAN hardware can be higher than the cost of conventional LAN hardware, overall installation expenses and life-cycle costs can be significantly lower. Long-term cost savings are greatest in dynamic environments requiring frequent moves, adds, and changes. Wireless LANs can be configured in a variety of topologies to meet the needs of specific applications and installations. They can grow by adding access points and extension points to accommodate virtually any number of users.

There are several technologies to choose from when selecting a wireless LAN solution, and each has its advantages and limitations. Most wireless LANs use spread spectrum, a wideband radio frequency technique developed by the military for use in reliable, secure, mission-critical communications systems. To achieve these advantages, the signal is spread out over the available bandwidth and resembles background noise that is virtually immune from interception.

There are two types of spread spectrum radio: frequency hopping and direct sequence. Frequency-Hopping Spread-Spectrum (FHSS) uses a narrowband carrier that changes frequency in a pattern known only to the transmitter and receiver. Properly synchronized, the net effect is to maintain a single logical channel. To an unintended receiver, FHSS appears to be short-duration impulse noise.

Direct-Sequence Spread-Spectrum (DSSS) generates a redundant bit pattern for each bit to be transmitted and requires more bandwidth for implementation. This bit pattern is called a *chip* (or chipping code), which is used by the receiver to recover the original signal. Even if one or more bits in the chip are damaged during transmission, statistical techniques embedded in the radio can recover the original data without the need for retransmission. To an unintended receiver, DSSS appears as low-power wideband noise.

Another technology used for wireless LANs is Infrared (IR), which uses very high frequencies that are just below visible light in the electromagnetic spectrum. Like light, IR cannot penetrate opaque objects. To reach the target system, the waves carrying data are sent in either directed (line-of-sight) or diffuse (reflected) fashion. Inexpensive directed systems provide a very limited range of not more than 3 ft and typically are used for personal area networks but occasionally are used in specific wireless LAN applications. High-performance directed IR is impractical for mobile users and is therefore used only to implement fixed subnetworks. Diffuse IR wireless LAN systems do not require line-of-sight, but cells are limited to individual rooms. As with spread spectrum LANs, IR LANs can be extended by connecting the wireless access points to a conventional wired LAN.

In a typical wireless LAN configuration, a transmitter-receiver (transceiver) device, called an *access point,* connects to the wired network from a fixed location using standard cabling. At a minimum, the access point receives, buffers, and transmits data between the wireless LAN and the wired network infrastructure. A single access point can support a small group of users and can function within a range of fewer than one hundred to several hundred feet. The access point (or the antenna attached to the access point) is usually mounted high but may be mounted essentially anywhere that is practical as long as the desired radio coverage is obtained.

Users access the wireless LAN through wireless LAN adapters with a built-in antenna. These adapters provide an interface between the client Network Operating System (NOS) and the airwaves via an antenna. The nature of the wireless connection is transparent to the NOS.

Wireless LANs can be simple or complex. The simplest configuration consists of two PCs equipped with wireless adapter cards, which form a network whenever they are within range of one another (see Fig. 8.3). This peer-to-peer network requires no administration. In this case, each client would only have access to the resources of the other client and not to a central server.

Installing an access point can extend the operating range of the wireless network, effectively doubling the range at which the devices can communicate. Since the access point is connected to the wired network, each client would have access to the server's resources as well as to other clients (see Fig. 8.4). Each access point can support many clients—the specific

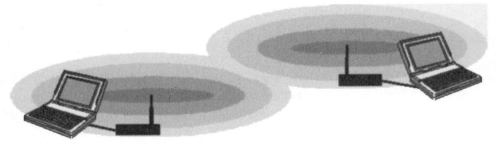

Figure 8.3 A wireless peer-to-peer network created between two laptops equipped with wireless adapter cards.

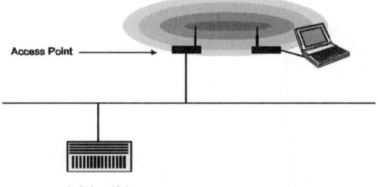

Access Point

Switch or Hub

Figure 8.4   A wireless client connected to the wired LAN via an access point.

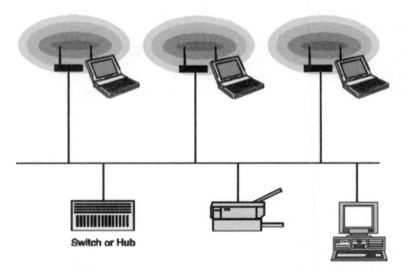

Switch or Hub

Figure 8.5   Multiple access points extend coverage and enable roaming.

number depends on the nature of the transmissions involved. In some cases, a single access point can support up to 50 clients.

Access points have an operating range of about 500 ft indoors and 1000 ft outdoors. In a very large facility such as a warehouse, or on a college campus, it will probably be necessary to install more than one access point (see Fig. 8.5). Access point positioning is determined by a site survey. The goal is to blanket the coverage area with overlapping coverage cells so that clients can roam throughout the area without ever losing network contact. Access points hand the client off from one to another in a way that is invisible to the client, ensuring uninterrupted connectivity.

To solve particular problems of topology, the network designer might choose to use Extension Points (EPs) to augment the network of access points (see Fig. 8.6). These devices look and function like Access Points (APs), but they are not tethered to the wired network, as are APs. EPs function as repeaters by boosting signal strength to extend the range of the network by relaying signals from a client to an AP or another EP.

Another component of wireless LANs is the directional antenna. If a wireless LAN in one building must be connected to a wireless LAN in another building a mile away, one solution might be to install a directional antenna on the two buildings—each antenna targeting the other and connected to its own wired network via an access point (see Fig. 8.7).

There are several wireless LAN standards, each suited for a particular environment: IEEE 802.11b, HomeRF, and Bluetooth.

For the corporate environment, the IEEE 802.11b (also called Wireless Fidelity, or simply Wi-Fi) offers a data transfer rate of up to 11 Mbps at a range of up to 300 ft from the access point. It operates in the 2.4-GHz band and transmits via the DSSS method. Multiple repeaters can be linked to increase that distance as needed, with support for multiple clients per access point. IEEE 802.11a, known as Wi-Fi5, offers a data transfer rate of up to 54 Mbps at a range of about 150 ft from the access point. It operates in the 5-GHz band and transmits via the DSSS method. While Wi-Fi5 does not penetrate walls as well as Wi-Fi, it does have a compelling speed advantage and operates in a less congested spectrum.

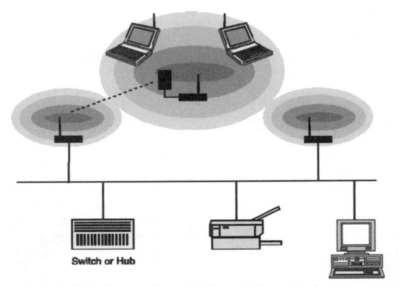

Switch or Hub

Figure 8.6   Use of an extension point in a wireless network.

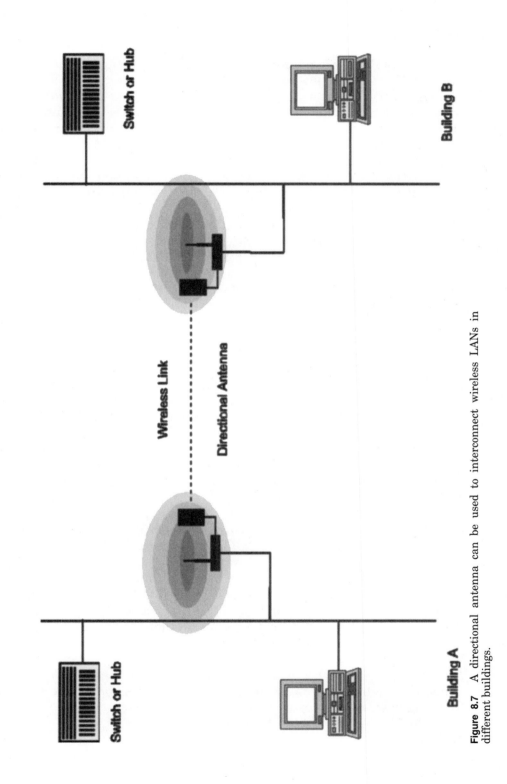

Figure 8.7 A directional antenna can be used to interconnect wireless LANs in different buildings.

The HomeRF 2.0 standard draws from 802.11b and Digital Enhanced Cordless Telecommunication (DECT), a popular standard for portable phones worldwide. Operating in the 2.4-GHz band, HomeRF was designed from the ground up for the home market for both voice and data. It offers throughput rates comparable to 802.11b, and supports the same kinds of terminal devices in both point-to-point and multipoint configurations. HomeRF transmits at up to 10 Mbps over a range of about 150 ft from the base station, which makes it suitable for the average home. HomeRF transmits using frequency hopping; that is, it hops around constantly within its proscribed bandwidth. When it encounters interference, like a microwave oven or the wireless LAN in the next apartment, it adapts by moving to another frequency.

The key advantage that HomeRF has over 802.11b in the home environment is its superior ability to adapt to interference from devices like portable phones, microwave ovens, and the latest generation of fluorescent lighting. As a frequency hopper, it coexists well with other frequency-hopping devices that proliferate in the home. Another advantage of HomeRF is that it continuously reserves a chunk of bandwidth via "isochronous channels" for voice services. Speech quality is high; there is no clipping while the protocol deals with interference.

The 802.11b standard for Wi-Fi does not include frequency hopping. In response to interference, 802.11b simply retransmits, or waits for the higher-level TCP/IP protocol to sort out signal from noise. This works well for data but can result in choppy-sounding voice transmissions. More often, however, the connection shuts down. Moreover, voice and data are treated the same way, converting voice into data packets, but offering no priority to voice. This results in unacceptable voice quality. Another problem with 802.11b is that its Wired Equivalent Privacy (WEP) encryption, designed to safeguard privacy, has had problems living up to its claim. Hackers have been known to break 64-bit WEP encryption within 15 minutes.

Bluetooth also operates in the 2.4-GHz band, but was not originally created to support wireless LANs; it was intended as a replacement for cable between desktop computers, peripherals, and hand-held devices. Operating at the comparatively slow rate of 30 to 400 kbps across a range of only 30 ft, Bluetooth supports "piconets" that link laptops, personal digital assistants, mobile phones, and other portable devices on an as-needed basis. It can also provide a wireless link to conventional LANs through an access point that acts as a gateway. Bluetooth improves upon infrared in that it does not require a line-of-sight between devices and has greater range than infrared's 3 to 10 ft. Bluetooth also supports voice channels, allowing push-to-talk radio communication.

While Bluetooth does not have the power and range of a full-fledged LAN, its master-slave architecture does permit the devices to face different piconets. In effect this extends the range of the signals beyond 30 ft. Like HomeRF, Bluetooth is a frequency hopper, so devices that use these two standards can coexist by hopping out of each other's way. Bluetooth has the faster hop rate, so it will be the first to sense problems and act to steer clear of interference from HomeRF devices.

Each of the three standards has particular strengths that make them ideal for certain situations. However, they also have specific shortcomings that render them inadequate for use beyond their intended purpose:

- While suited for the office environment, 802.11b is not designed to provide adequate interference adaptation and voice quality for the home. Data collisions force packet retransmissions, which are fine for file transfers and print jobs, but not for voice or multimedia that cannot tolerate the resulting delay.

- HomeRF delivers an adequate range for the home market, but not for many small businesses. It is better suited than 802.11b for streaming multimedia and telephony, applications that may become more important for home users as convergence devices become popular.

- Bluetooth does not provide the bandwidth and range required for wireless LAN applications, but instead is suited for desktop cable replacement and ad hoc networking for both voice and data within the narrow 30-ft range of a piconet.

Wireless LAN technology is continually improving. Standards developers seek to improve encryption (802.11i) and make the standard more multimedia-friendly (802.11e). Dozens of vendors are shipping 802.11b products, and the standard's proliferation in corporate and public environments is a distinct advantage. An office worker who already has an 802.11b-equipped notebook will not likely want to invest in a different network for the home.

Furthermore, the multimedia and telephony applications HomeRF advocates have not yet arrived to make the technology a compelling choice. As Wi-Fi gains popularity in the workplace, HomeRF will lose popularity in the home because telecommuters will not want to invest in different types of equipment. For these and other reasons, 802.11b will soon overtake HomeRF in the consumer marketplace as well, especially since the price difference between the two has reached parity.

8.2.5 Broadband wireless

Microwave systems use the high end of the radio frequency spectrum and are used by carriers for long-haul communications and by large companies that wish to bypass the local exchange carrier's local loop. Microwave is advantageous in that it does not require stringing wire over long distances, it is immune from the impairments that affect copper wire, and it provides greater bandwidth capacity. However, microwave transmission systems require a line-of-sight path and, typically, permission to site a dish antenna on a building rooftop. When properly deployed in the right environment, broadband wireless systems can provide substantial cost savings in bandwidth for voice, data, and Internet access.

Local Multipoint Distribution Service. LMDS is a two-way millimeter microwave technology that operates in the 27- to 31-GHz range. It offers enormous bandwidth, but has a maximum signal range of only 7.5 mi from the service provider's hub to the customer premises. New technology allows LMDS operation without having a direct line-of-sight with the receiver. This feature, highly desirable in built-up urban areas, may be achieved by bouncing signals off buildings so that they get around obstructions. At the receiving location, the data packets arriving at different times are held in queue for resequencing before they are passed to the application. This scheme does not work well for voice, however, because the delay resulting from queuing and resequencing disrupts two-way conversation.

Competitive Local Exchange Carriers (CLECs) can deploy LMDS to completely bypass the local loops of the Incumbent Local Exchange Carriers (ILECs), eliminating access charges and avoiding service-provisioning delays. Since the service entails setting up equipment between the provider's hub location and customer buildings for the microwave link, LMDS costs far less to deploy than installing new fiber. This allows CLECs to very economically bring customer traffic onto their existing metropolitan fiber networks and, from there, to a national backbone network.

The strategy among many CLECs is to offer LMDS to owners of multitenant office buildings and then install cable to each tenant who subscribes to the service. The cabling goes to an on-premises switch, which is cabled to the antenna of the building's roof. That antenna is aimed at the service provider's antenna at its hub location. Subscribers can use LMDS for a variety of high-bandwidth applications, including television broadcast, videoconferencing, LAN interconnection, broadband Internet access, and telemedicine.

At the carrier's hub location there is a roof-mounted multisectored antenna (see Fig. 8.8). Each sector of the antenna receives and/or transmits signals between itself and a specific customer location. This antenna is very small, some measuring only 12 in in diameter. The hub antenna brings the multiplexed traffic down to an indoor switch (see Fig. 8.9), which processes the data into 53-byte ATM "cells" for transmission over

Figure 8.8 A multisectored antenna at the carrier's hub location transmits and/or receives traffic between the antennas at each customer location.

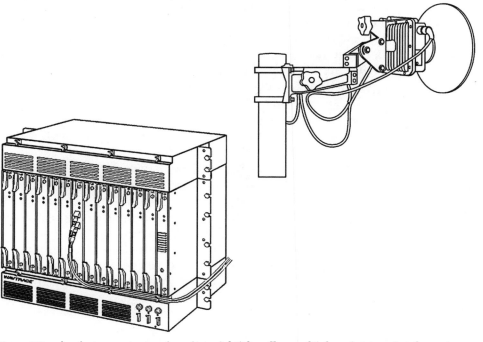

Figure 8.9 A microwave transceiver (top right) handles multiple point-to-point downstream and upstream channels to customers. The transceiver is connected via coaxial cables to an indoor switch (bottom left), which provides the connectivity to the carrier's fiber network. The traffic is conveyed over the fiber network in the form of 53-byte ATM cells. (*Source: Wavtrace, Inc.*)

the carrier's fiber network. These individually addressed cells are converted back to their native format before going off the carrier's network to their proper destinations—the Internet, Public Switched Telephone Network (PSTN), or to the customer's remote location.

At each customer's location, there is a rooftop antenna that sends and/or receives multiplexed traffic. This traffic passes through an indoor Network Interface Unit (NIU), which provides the gateway between the RF components and the in-building equipment, such as a LAN hub, PBX, or videoconferencing system. The NIU includes an up/down converter that changes the frequency of the microwave signals to a lower Intermediate Frequency (IF) that the electronics in the office equipment can more easily (and inexpensively) manipulate.

A potential problem for LMDS users is that heavy rainfall and dense fog can disrupt the microwave signals—even foliage can block a signal. In metropolitan areas where new construction is a fact of life, a line-of-sight transmission path can disappear virtually overnight. For these reasons, many network managers are leery of trusting mission-critical applications to this wireless technology. Service providers downplay this situation by claiming that LMDS is just one local access option and that fiber links are the way to go for mission-critical applications. In fact, some LMDS providers offer fiber as a backup in case the microwave links experience interference.

With an overabundance of fiber in the ground and metropolitan-area Gigabit Ethernet services coming on-line at a competitive price, the time for LMDS may have come and gone. In addition, newer wireless technologies like free-air laser hold a significant speed advantage over LMDS, as does submillimeter microwave in the 60- and 95-GHz bands.

Multichannel Multipoint Distribution Service. MMDS is a microwave technology that traces its origins to 1972 when it was introduced to provide an analog service called Multipoint Distribution Service (MDS). For many years, MMDS was used for one-way broadcasting of television programming, but in early 1999 the FCC opened up this spectrum to allow for two-way transmissions, making it useful for delivering telecommunication services, including high-speed Internet access to homes and businesses.

This technology, which has now been updated to digital, operates in the 2- to 3-GHz range, enabling large amounts of data to be carried over the air from the operator's antenna towers to small receiving dishes installed at each customer location. Operating at a lower frequency range means that the signals are not as susceptible to interference as those using LMDS technology. MMDS can be used for LAN interconnectivity between buildings, high-speed Internet access, videoconferencing, and television programs. The useful signal range of MMDS is about 30 mi, which beats LMDS at 7.5 mi.

Most of the time the operator receives TV programming via a satellite downlink. Large satellite antennas installed at the head-end collect these signals and feed them into encoders that compress and encrypt the programming. The encoded video and audio signals are modulated, via Amplitude Modulation (AM) and Frequency Modulation (FM), respectively, to an IF signal. These IF signals are up-converted to MMDS frequencies and then amplified and combined for delivery to a coax cable, which is connected to the transmitting antenna. The antenna can have an omnidirectional or sectional pattern.

The small antennas at each subscriber location receive the signals and pass them via a cable to a set-top box connected to the television. If the service also supports high-speed Internet access, a cable also goes to a special modem connected to the subscriber's PC. MMDS sends data as fast as 10 Mbps downstream (toward the computer). Typically, service providers offer downstream rates of 512 kbps to 2.0 Mbps, with burst rates up to 5 Mbps whenever spare bandwidth becomes available.

Originally, there was a line-of-sight limitation with MMDS technology. But this has been overcome with a complementary technology called Vector Orthogonal Frequency Division Multiplexing (VOFDM). Because MMDS does not require an unobstructed line of sight between antennas, signals bouncing off objects en route to their destination require a mechanism for being reassembled in their proper order at the receiving site. VOFDM handles this function by leveraging multipath signals, which normally degrade transmissions. It does this by combining multiple signals at the receiving end to enhance or recreate the transmitted signals. This increases the overall wireless system performance, link quality, and availability. It also increases service providers' market coverage through non–line-of-sight transmission.

MMDS equipment can be categorized into two types based on the duplexing technology used: Frequency Division Duplexing (FDD) or Time Division Duplexing (TDD). Systems based on FDD are a good solution for voice and bidirectional data because forward and reverse use separate and equally large frequency bands. However, the fixed nature of this scheme limits overall efficiency when used for Internet access. This is because Internet traffic tends to be "bursty" and asymmetrical. Instead of preassigning bandwidth with FDD, Internet traffic is best supported by a more flexible bandwidth allocation scheme.

TDD is more efficient because each radio channel is divided into multiple time slots through Time Division Multiple Access (TDMA) technology, which enables multiple channels to be supported. Because TDD has flexible time-slot allocations, it is better suited for data delivery—specifically, Internet traffic. TDD enables service providers to vary uplink and downlink ratios as they add customers and services. Many more users can be supported by the allocation of bandwidth on a nonpredefined basis.

MMDS is being used to fill the gaps in market segments where cable modems and DSL cannot be deployed due to distance limitations and cost concerns. MMDS will be another access method to complement a carrier's existing cable and DSL infrastructure, or it can be used alone for direct competition. With VOFDM technology, MMDS is becoming a workable option that can be deployed cost-effectively to reach urban businesses that do have line-of-sight access, and in suburban and rural markets for small businesses and telecommuters.

8.2.6 Satellite

Satellite transmission is a variation of microwave. Satellites circle the earth in geostationary orbit, which allows them to maintain the same position relative to the earth. From this position, satellites act as relay stations for earthbound communications links. Because altitude precludes interference caused by the earth's curvature and other geophysical obstructions (e.g., mountains and atmospheric conditions), satellites are ideal for broadcast applications and long-distance domestic and international communications.

Satellite services are available to customer sites equipped with Very Small Aperture Terminals (VSATs) that integrate transmission and switching functions to provide preassigned and on-demand links for packet transmission on point-to-point and broadcast networks. VSAT services are particularly well-suited for far-flung, transaction-oriented applications, such as automotive dealership support; retail Point-of-Sale (POS) environments; stock, bond, and commodity transactions; travel and lodging reservation systems; remote utility monitoring and control applications; remote data acquisition activities; video broadcasting; and local exchange bypass services.

The components associated with VSAT earth stations include the antenna, which is typically a parabolic reflector (dish) with a diameter of 0.3 to 2.4 m, and an RF power unit that supplies the 1 to 5 W necessary to support communications at up to 1.544 Mbps over the C-band or Ku-band—the two bands most commonly used for VSAT transmission. (Although most VSAT networks use either the C-band or the Ku-band, it is possible to build hybrid networks that use both bands.) The company's VSAT earth stations are linked via satellite to a Master Earth Station (MES), which provides bandwidth assignment, routing, and management functions (see Fig. 8.10).

VSAT networks offer advantages over terrestrial networks in terms of network expansion. Instead of interacting with multiple carriers for a new line installation, for example, capacity can be added to a VSAT network in a matter of minutes by simply allocating additional transponder bandwidth (usually by requesting it from the service provider). Also, with proper planning, additional VSAT locations can be brought on-line within a single business day.

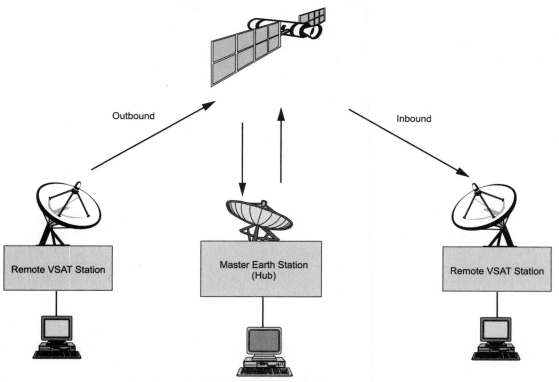

Figure 8.10   Satellite/VSAT network configuration.

VSAT networks are also a viable alternative to terrestrial networks in other ways. With regard to equipment failures, for example, each VSAT node operates independently of the others so that the failure of one does not affect the performance of the others. In contrast, a failure on shared terrestrial facilities can bring down a major portion of the network—and cost the companies that rely on them millions of dollars.

To further ensure the reliability of VSAT networks, some VSAT companies offer a complete package of equipment, data recovery services, and emergency procedures that are implemented in the event of a data communications failure. Included in the subscription fee for such a package are all of the necessary transponders, receivers, and dishes.

Much of the competition between private leased lines and satellite services is cost-related. Advances in technology have combined to bring about smaller, more powerful, and more economical satellite dishes that in many cases offer considerable advantages in efficiency, convenience, and flexibility over terrestrial private leased lines—including fiber-optic transmission systems. There is continual competition between the cost of a satellite system with its VSAT, transponders, and hub (MES), and the equivalent private line tariffs.

A relatively recent innovation in satellite technology is the Low Earth Orbit (LEO) satellite. Circling the earth only a few hundred miles up, LEO satellites have to keep moving to avoid falling into the atmosphere, so a given satellite passes over a stationary caller rather quickly. A call must be handed off from one satellite to the next to keep the session alive. This provides the advantage of global roaming for voice and data calls. LEO satellites fly low enough that dramatic improvements can be made on the earthbound transceivers: They can be low-powered, compact, and affordable devices.

8.2.7   Free-space laser

A wholly different category of transmission is based on laser-optic transmission operating in the near-infrared region of the light spectrum. Utilizing coherent laser light, these wireless line-of-sight links are used in campus environments and urban areas where the installation of cable is impractical and the performance of leased lines is too slow. Unlike microwave transmission, laser transmission does not require an FCC license, and data traveling by laser beam cannot be intercepted.

Free-space laser transmission can be a useful method for private network users to bypass the local exchange carrier for certain applications, such as point-to-point LAN interconnection. The lasers at each location are aligned with a simple bar graph and tone lock procedure. Monitors can be attached to the laser units to provide operational status, such as signal strength, and to implement local and remote loop-back diagnostics.

Some service providers now offer 10-Mbps transparent LAN services, which offer twice the bandwidth for the same price as a conventional land-line service.

The reason why free-space laser has not been used very often for business applications is that transmission is affected by atmospheric conditions, such as dense fog. The tiny water droplets diffuse the light and can weaken signals to the point that a connection is lost. But makers of these laser systems can overcome this problem to some degree by boosting the power of the laser so it cuts through the fog. In addition, service providers can engineer their networks so the individual links between sites are short enough so even the densest fog will not block the beam.

8.3   Analog Services and Lines

Analog switched services and private lines have been phased out on the long-distance portion of the public network in North America in favor of more reliable, higher bandwidth digital services and lines. But they are still common in the local-loop portion of the public network in North America and throughout the public network in most countries. Analog switched services and four-wire private lines offer a fairly narrow channel of not more than 4 kHz, which is good for voice calls and low-speed data transmission via modems at up to 56 kbps. This limited bandwidth—plus the disruptive effects of noise and other line impairments—makes analog services and lines unsuitable for supporting high-speed data.

The data applications of analog dial-up services include Internet access, remote LAN access, facsimile, remote diagnostics, and temporary backup to low-speed digital circuits. Dial-up service can be very economical for low-speed data applications that are accessed infrequently. Analog private lines are used when the applications are run frequently or continuously as in telemetry, remote banking via automated teller machines, and a variety of government and commercial information services available via kiosks. In multidrop configurations, analog lines can be an efficient and economical way to continuously poll systems to collect accumulated data and monitor performance status.

8.4 Digital Subscriber Line

Digital Subscriber Line (DSL) is a category of local-loop technology that turns an existing analog line, normally used for Plain Old Telephone Service (POTS), into a high-speed digital line over which a variety of broadband data services can be offered, including Internet access, corporate connectivity via VPNs, and extra voice channels.

The electronics at both ends of the local loop compensate for impairments that would normally impede high-speed data transmission over ordinary twisted-pair copper lines.

Provisioning issues.   In leveraging existing copper local loops, DSL obviates the need for a huge capital investment among service providers to bring fiber to the customer premises in order to offer broadband services. Instead, idle twisted pairs to the customer premises can be provisioned to support high-speed data services. Users are added by simply installing DSL access products at the customer premises and connecting the DSL line to the appropriate voice or data network switch via a DSL concentrator at the Central Office (CO) or a Serving Wire Center (SWC) where the data and voice are split out for distribution to the appropriate network (see Fig. 8.11).

There are about a dozen DSL technologies currently available—each optimized for a given level of performance relative to the distance of the customer premises from the CO or SWC. The farther away the customer is from the CO or SWC, the lower the speed of the DSL in both the upstream (toward the network) and downstream (toward the user) directions. The

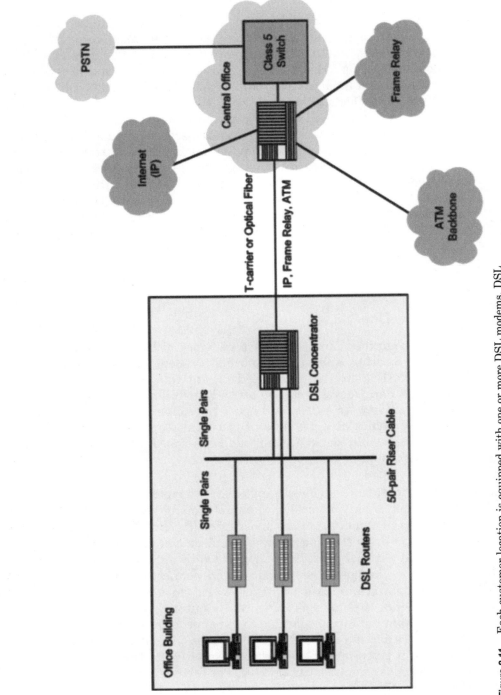

Figure 8.11 Each customer location is equipped with one or more DSL modems. DSL lines from multiple customers are concentrated at a DSL Access Multiplexer (DSLAM) that can be positioned in a building's equipment room or at the central office where voice and data are split out for delivery to appropriate networks.

closer the customer location is to the CO or SWC, the greater the speed in both directions.

When it comes to price, DSL is far more economical than other digital technologies, such as T1. Like T1, DSL is priced at a flat rate per month with unlimited hours of access. While T1 access circuits cost anywhere from $150 to $700 per month or more, depending on the carrier and location, DSL can cost much less for comparable bandwidth—as low as $29.95 a month for 256 kbps to $250 a month for 7 Mbps.

Security.   While DSL is more economical than T1 and offers more bandwidth, it is more vulnerable in terms of security. Although the access line is certainly more secure than cable because it is dedicated rather than shared, the security problem begins on the other side of the DSL Access Multiplexer (DSLAM) where traffic from multiple copper loops is aggregated for transport over a high-capacity fiber link to the Internet. Since the DSL connection is always on, it is possible for hackers on the Internet to find their way into the DSLAM and, from there, to individual computers on the other side. Since the DSLAM is not equipped with firewall capabilities, the customer must provide security.

Several vendors have addressed the need for security in their DSL products with built-in firewall capabilities. The units come with preconfigured firewalls to disallow certain types of traffic originating from the Internet by filtering packets on a per-connection profile basis for source and/or destination address, service, and protocol. Up to 255 rules are available in up to eight filter sets. The routers also support secure VPN access to corporate intranets and extranets via tunneling protocols and provide encryption for added protection. Security is further enhanced with Network Address Translation (NAT), which hides all IP addresses on the LAN behind a single statically or dynamically assigned IP address on the Internet.

For businesses with more stringent communications requirements, DSL may not be adequate. Not only is the amount of bandwidth affected by distance but the service does not provide consistent availability and reliability, and is not secure enough for the transmission of mission-critical information. To overcome these issues, network managers should consider T-carrier facilities.

8.5   T-carrier Facilities

T-carrier is a type of digital transmission that uses multiplexing techniques to accommodate many channels on the same physical line, whether it is a wire, coaxial cable, or microwave radio transmission system. This is accomplished by allocating each channel to a different time segment, as in the case of TDM.

T-carrier supports the widely available service (or private line) known as T1, which is a four-wire connection offering 24 bidirectional channels of 64 kbps each over 1.536 Mbps of bandwidth. With 8 kbps thrown in for link management functions, the aggregate bandwidth is 1.544 Mbps, which is known as Digital Signal Level 1 (DS1). In Europe, the United Kingdom, and other countries that adhere to standards issued by the ITU, the analog of T1 is E1, which accommodates 32 channels of 64 kbps each over 2.048 Mbps of aggregate bandwidth.

Depending on the carrier and service region, the economics of T1 are such that only five to eight voice-grade private lines justify the cost of a T1 line. Added benefits come in the form of more reliable, error-free transmission and capacity for growth. A single T1 facility can handle fixed- or variable-rate data. In addition, voice, data, video, and image transmissions can be integrated over a single facility for transport efficiency and economy. Businesses can realize additional savings through the use of T1 compression techniques such as Adaptive Differential Pulse Code Modulation (ADPCM), which increases the number of voice channels from 24 to 48. Proprietary voice compression schemes are also available to bring voice down to 2.4 kbps, providing even more cost savings.

T-carrier is also used to support Fractional T1 (FT1), which offers bandwidth increments of 64, 128, 256, 384, 512, and 768 kbps to support applications such as Internet access. T3 is an extension of T1 that is equivalent to 28 T1 lines, operating at the DS3 rate of 44.736 Mbps. With T3, users gain the bandwidth to support increased voice traffic and broadband applications, including multimegabit Internet access. In most cases, it only takes 8 to 12 T1 lines to cost-justify a T3 line.

8.6 Centrex

For companies that do not want to invest in their own communications infrastructure and get bogged down with management complexities, Centrex can provide a viable alternative to obtaining telephone services and advanced calling features. This service is intended as an alternative to buying a PBX—in fact, Centrex can be considered a remote PBX. Not only does Centrex free up a company's scarce capital for other purchases, it puts responsibility for maintenance and management on the telephone company. Today's Centrex services boast 100 percent feature parity with the most advanced digital PBXs currently available—including support for wireless communications and LANs.

The switches that implement Centrex services use computer-controlled time-division switching and have distributed architectures consisting of a host module and multiple microprocessor-controlled switching modules. These switches can directly interface with T-carrier systems to provide 24 digitized voice channels over twisted-pair wire at 1.544 Mbps. These capabilities enable Centrex offerings to include interfaces to the Digital Access and Cross-Connect System (DACS), thereby providing gateways to a variety of services over the public telephone network.

Centrex services offer a wealth of call processing and management features. Many Centrex offices routinely offer automatic route selection, local area networking, facilities management and control, message center services, and voice mail capabilities. In addition to basic rate Integrated Services Digital Network (ISDN), Centrex exchanges are in the process of being upgraded to provide primary rate ISDN service.

Centrex offers a variety of options to help communications managers monitor usage and control costs. Among these cost-management features is Station Message Detail Recording to Premises (SMDR-P), which transmits call records directly to the customer premises from the central office. SMDR-P arrangements provide virtually immediate access to call record data.

Centrex also provides on-line management features. With an on-premises terminal and an interactive software program, users can control the numbers, features, services, and billing codes assigned to each line within their systems. Not only can users review the status of their current Centrex configurations; they can also plan ahead to meet future communications demands by determining what changes need to be made and controlling the date of implementation. All such changes are input to a central management system on the carrier's network; individual Centrex exchanges poll the system daily for any customer changes and automatically update the telephone company's internal records.

Wireless communications is supported by some Centrex systems. The service enables employees in a building to send and receive telephone calls on low-power wireless telephones while moving around their offices. The system is based on a network of small, low-power transmission cells distributed strategically around the building and linked to a local exchange carrier by a central controller. This controller coordinates the handoff of calls from one cell to the next as the user moves about the workplace.

Despite the many advantages of subscribing to Centrex services, it does have its share of liabilities. In addition to being dependent on the local phone company for service and support, the contracts usually run from 5 to 10 years—the longer the commitment, the better the pricing. However, this can lock you, as a network manager, out of any new technologies that may come along and prevent your company from taking advantage of efficiencies and economies they may offer.

8.7   ISDN

Since 1996, ISDN has become increasingly popular for accessing the Web and improving the response time for navigating, viewing, and downloading multimedia content. Other applications of ISDN include videoconferencing, the delivery of multimedia training sessions, and the temporary rerouting of traffic around failed leased lines or frame relay networks and the handling of peak traffic loads (see Fig. 8.12). ISDN can also play a key role in various call center, Computer-Telephony Integration (CTI), telecommuting, and remote access (i.e., remote control and remote node) applications.

ISDN is a circuit-switched digital service that comes in two varieties. The Basic Rate Interface (BRI) provides two bearer channels of 64 kbps each, plus a 16-kbps signaling channel. The Primary Rate Interface (PRI) provides 23 bearer channels of 64 kbps each, plus a 64-kbps signaling channel. Any combination of voice and data can be carried over the B channels.

ISDN PRI was designed to be compatible with existing digital transmission infrastructures, specifically T1 in North America and E1 in Europe (2.048 Mbps). In fact, ISDN PRI is a digital service that rides over a T1/E1 facility. Because ISDN can evenly reduce both T1 and E1 into 64-kbps increments, the 64-kbps channels became the worldwide standard. The use of 64-kbps channels also allows users to migrate more easily from private T1/E1 networks to ISDN and build hybrid networks consisting of both public and private facilities. Table 8.5 compares ISDN PRI with T1/E1.

In both BRI and PRI, ISDN's separate D channel is used for signaling. As such, it has access to the control functions of the various digital switches on the network. It interfaces with Signaling System 7 (SS7) to provide message exchange between the user's equipment and the network to set up, modify, and clear the B channels. Via SS7, the D channel also gathers information about other devices on the network, such as whether they are idle, busy, or off. In being able to check ahead to see if calls can be completed, network bandwidth can be conserved. If the called party is busy, for example, the network can be notified before resources are committed.

The D channel's task is carried out very quickly, so it remains unused most of the time. For this reason, PRI users can assign the 64-kbps D channel to perform the signaling function for as many as eight PRI lines. However, this is rarely done because if the PRI line with the D channel goes out of service, the other PRI lines that depend on it for signaling also go out of service.

For BRI users, whenever the D channel is not being used for signaling, it can be used as a bearer channel (if the carrier offers this capability as a service) for point-of-sale applications such as Automated Teller Machines (ATMs), lottery terminals, and cash registers (see Fig. 8.13). Some carriers offer BRI users the option of using idle D channel bandwidth for always-on email.

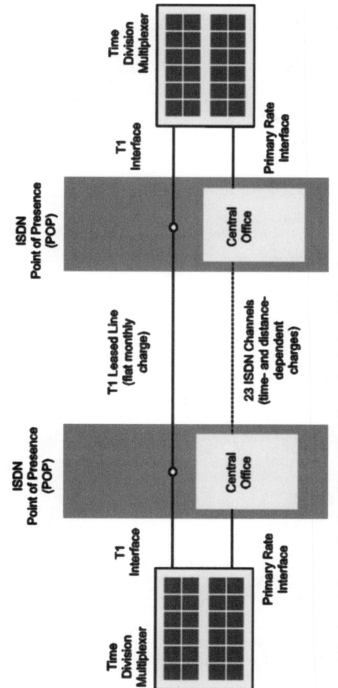

Figure 8.12 Among other applications, ISDN PRI can be used to back up T1 leased lines in case of failure, provide an additional source of temporary bandwidth to handle peak traffic loads, or support special applications such as videoconferencing on an as-needed basis. In addition, ISDN BRI can be used to back up frame relay Permanent Virtual Circuits (PVCs).

TABLE 8.5   A Comparison of ISDN PRI with T1/E1

ISDN PRI	T1/E1
Digital service	Digital facility
Circuit-switched	Dedicated
Any-to-any connectivity	Point-to-point connectivity
Shared bandwidth	Assigned bandwidth
Dial-up	Always on
Out-of-band signaling (D channel)	In-band signaling (bit-robbed)
Call handling features	No call handling features
Call-by-call channel assignment	Fixed channel assignments
Efficient bandwidth usage	Inefficient bandwidth usage
Optimized for voice, Computer-Telephony Integration (CTI) in call centers, videoconferencing	Optimized for data, compressed voice, PBX trunks

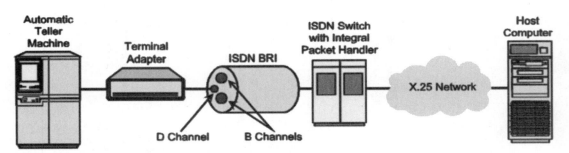

Figure 8.13   When ISDN's D channel is not used for signaling, it can be used for low-speed applications, such as electronic mail, while the B channels carry other voice and data calls.

With regard to ISDN PRI, there are two higher-speed transport channels called H channels. The H0 channel operates at 384 kbps, while the H11 operates at 1.536 Mbps. These channels are used to carry multiplexed data, data and voice, or video at higher rates than that provided by the 64-kbps B channel. The H channels also are ideally suited for backing up FT1 and T1 leased lines.

Multirate ISDN lets users select appropriate increments of switched digital bandwidth on a per-call basis. Speeds, in increments of 64 kbps, are available up to 1.536 Mbps. Multirate ISDN is used mostly for multimedia applications such as collaborative computing and videoconferencing, where the number of channels may vary with each session.

Over the years, ISDN has been touted as a breakthrough in the evolution of worldwide telecommunications networks—the single most important technological achievement since the advent of the telephone network itself in the nineteenth century. Others disagree and note that technologies such as IP, frame relay, and ATM have overtaken ISDN to the point of making it virtually obsolete. There are still problems with ISDN, however, due to inconsistencies in carrier implementation. For example, Global Crossing does not support 64-kbps "clear channel," which is required for video applications over ISDN, whereas other carriers do support clear channel and can pass video over ISDN with no problem. Global Crossing supports only 56 kbps, which is not suited for video applications. The debate over ISDN's relevance to today's telecommunications needs must be put into the context of specific applications. Just like any other service, ISDN will be adequate to serve the needs of some users, but not others.

8.8   Packet Data Services

Packet-based services are popular because they eliminate the need for more expensive private lines. The greater the distance between corporate locations, the greater the cost savings will be. Likewise, the more corporate locations there are, the greater the cost savings over private lines. The three packet services in vogue today are IP, frame relay, and ATM switching.

8.8.1   Internet Protocol

Transmission Control Protocol/Internet Protocol (TCP/IP) is a suite of networking protocols that is valued for its ability to interconnect diverse computing platforms very economically. Although TCP/IP has been in use for over 30 years, it continues to be enhanced with the addition of new protocols that extend its functionality to include support of voice, multimedia applications, and security. As a result, TCP/IP offers a mature, dependable environment for corporate users who need a common denominator for their diverse and sprawling networks.

Transmission Control Protocol.   TCP forwards data delivered by IP to the appropriate process at the receiving host. Among other things, TCP defines the procedures for breaking up the data stream into packets and reassembling them in the proper order to reconstruct the original data stream at the receiving end. Since the packets typically take different routes to their destination, they arrive at different times and out of sequence. All packets are temporarily stored until the missing packets arrive so they can be put in the correct order. If a packet arrives damaged, it is simply discarded and another one is resent. To accomplish these and other tasks, TCP breaks the messages or data stream down into a manageable size and adds a header to form a packet.

Internet Protocol.   The Internet is composed of a series of autonomous systems, or subnetworks, each of which is locally administered and managed. These subnets may consist of Ethernet LANs, ISDN, frame relay networks, and ATM networks, over which IP runs. IP delivers data between these different networks through routers that process packets from one Autonomous System (AS) to another.

Each node in the AS has a unique IP address. The IP adds its own header and checksum for error detection to make sure the data is properly routed. This process is aided by the presence of routing update messages that keep the address tables in each router current. Several different types of update messages are used, depending on the collection of subnets involved in a management domain. The routing tables list the various nodes on the subnets as well as the paths between the nodes. If the data packet is too large for the destination node to accept, it will be segmented into smaller packets.

Network performance depends on the resources available at the various hosts and routers—transmission bandwidth, buffer memory, and processor speed—and how efficiently these resources are managed. Although each type of resource is manageable, there are always tradeoffs between cost and performance.

Virtual Private Networks.   Traditionally, VPNs had been voice-oriented, allowing companies to interconnect their PBXs over a carrier's WAN at less cost than private lines. In the early 1990s, frame relay and ATM technologies were introduced to extend the VPN concept to data communications. In early 1997, a new trend emerged in which private data is routed between corporate locations worldwide over carrier-provided IP networks more economically than frame relay or ATM.

With a VPN, the carrier is responsible for network management, security, and Quality-of-Service (QoS) issues. In many cases, service-level guarantees are available from the carrier, which provides companies with credits for poor performance. Although frame relay and ATM continue to be used for VPNs—supporting voice in packet form as well as data—IP is now the most popular type of VPN. Sometimes a mix of services can be used to implement VPNs. For example, companies can use their current frame relay service for remote access, eliminating the need for additional VPN access circuits, routers, and modem pools for IP traffic.

IP-based VPNs are an increasingly popular option for securely interconnecting corporate locations over the Internet, including branch offices and telecommuters. They also can be used for e-commerce and making enterprise applications available to customers and strategic partners worldwide. Basically, this type of data service lets business users carve out their own

IP-based WANs within the carrier's high-speed Internet backbone. Security functions are performed on IP packets, which are then encapsulated, or tunneled, inside other IP packets for routing across the Internet. By drawing on the economies of transmission and switching that the larger Internet provides, VPNs offer substantial cost savings for companies of any size.

A tunnel is an end-to-end connection similar to a Virtual Circuit (VC) used to provision frame relay and ATM services. Since a tunnel goes only to a designated destination, it provides the ability to securely transport one or more protocols. The tunnel can handle both IP and non-IP traffic, which can be assigned various priorities to provide QoS. By themselves, however, many of the tunneling protocols do not provide native support for data security. To enhance data security, software-controlled encryption, authentication, and integrity functions can be applied to the packets for transport through the tunnel.

The Internet Engineering Task Force's (IETF) IPSec is one tunneling protocol that does provide for packet-by-packet authentication, integrity, and encryption. Authentication positively identifies the sender. Integrity guarantees that no third party has tampered with the packet stream. Encryption allows only the intended receiver to read the data. These security functions must be applied to every IP packet because Layer 3 protocols such as IP are stateless; that is, there is no way to be sure whether a packet is really associated with a particular connection. Higher-layer protocols such as TCP are stateful, but their connection-tracing mechanisms can be easily duplicated or "spoofed" by knowledgeable hackers. The security mechanisms associated with IPSec take care of these problems.

Carriers are beginning to offer classes of service for their VPN offerings. AT&T, for example, supports three classes of service for its private IP VPNs—high priority, low priority, and best effort. Network managers can assign these classes of service in any configuration to address the specific bandwidth demands of their network traffic.

Classes of service allow businesses to differentiate one application running on their network from another. The capability helps network managers use their network bandwidth more efficiently by letting them map applications into "classes," assign a priority to each class, and based on the priority, treat the classes differently.

For example, Voice over Internet Protocol (VoIP) and videoconferencing applications have stricter performance requirements than Internet browsing traffic. When both VoIP and Internet browsing traffic are on the same network, classes of service allow network managers to assign a higher priority to the VoIP traffic. That way, voice traffic is always serviced first so the voice quality can be maintained.

IP VPNs are available as fully managed, co-managed, or customer-managed services. They provide the means for companies to extend the reach of their networks globally. Selected segments of the VPN can be securely opened to business partners, suppliers, and clients, depending on unique needs. Customers of a managed VPN service pay a single, all-inclusive service price, which includes network access, CPE, software and hardware maintenance, VPN installation project management support, and 24×7 proactive management and monitoring. Local-loop access charges are additional.

In its early years of development and implementation, TCP/IP was considered of interest only to research institutions, academia, and defense contractors. Today, corporations have embraced TCP/IP as a platform that can meet their needs for multivendor, multinetwork connectivity. Because it was developed in large part with government funding, TCP/IP code is in the public domain; this availability has encouraged its use by thousands of vendors worldwide, who apply it to support nearly all types of computers and network devices. Because of its flexibility, comprehensiveness, and nonproprietary nature, TCP/IP is of considerable interest to organizations seeking an economical solution to a variety of communications and internetworking needs.

8.8.2 Frame relay

Before frame relay, the dominant packet technology was X.25, which was designed to operate over poor-quality voice-grade lines. Its store-and-forward method of transmission acted to correct transmission errors that occurred from node to node, making it a very reliable way to send data. However, the increasing use of digital lines made error correction within the network unnecessary because they were much more reliable than analog lines.

The result was that frame relay does not have the processing-intensive functions of X.25, including error correction, which is moved to the devices connected to the network. Frame relay still provides some error-detection capabilities, but bad packets are simply dropped—not retransmitted from the previous network node, as in X.25. It is the responsibility of the source or destination devices to correct such errors by requesting retransmissions of the missing packets. The most compelling advantages of a carrier-provided frame relay service include:

- *Improved throughput / low delay.* Frame relay service uses high-quality digital circuits end to end, making it possible to eliminate the multiple levels of error checking and error control. The result is higher throughput and fewer delays compared to legacy packet-switched networks like X.25.

- *Any-to-any connectivity.* Any node connected to the frame relay service can communicate with any other node via predefined Permanent Virtual Circuits (PVCs) or dynamically via Switched Virtual Circuits (SVCs).

- *No long-distance charges.* Since frame relay is offered as a service over a shared network, the need for a highly meshed private line network is eliminated for substantial cost savings. There are no distance-sensitive charges with frame relay, as there are with private lines.

- *Oversubscription.* Multiple PVCs can share one access link, even exceeding the port speed of the frame relay switch. In oversubscribing the port, multiple users can access the frame relay network—but not all at the same time—eliminating the cost of multiple private-line circuits and their associated CPEs for further cost savings.

- *Higher speeds.* Whereas X.25 tops out at 56 kbps, frame relay service supports transmission speeds up to 44.736 Mbps. If the frame relay switches in the network support Frame Relay Forum Implementation Agreement 14 (FRF 14), speeds at the OC-3 rate of 155 Mbps and the OC-12 rate of 622 Mbps over fiber backbones are possible.

- *Simplified network management.* Customers have fewer circuits and less equipment to monitor. In addition, the carrier provides proactive monitoring and network maintenance 24 hours a day.

- *Inter-carrier connectivity.* Frame relay service is compatible between the networks of various carriers, through Network-to-Network Interfaces (NNIs), enabling data to reach locations not served by the primary service provider.

- *Customer-controlled network management.* Allows customers to obtain network management information via in-band SNMP queries and pings launched from their own network management stations.

- *Performance reports.* Enables customers to manage their frame relay service to maximum advantage. Available network reports, accessible on the carrier's secure Web site include those for utilization, errors, health, trending, and exceptions.

- *Service level guarantees.* Frame relay service providers offer customers SLAs that specify availability as a percentage of uptime, round-trip delay expressed in milliseconds, and throughput in terms of the Committed Information Rate (CIR). If the carrier cannot meet the SLA, it credits the customer's invoice accordingly.

The virtual circuits have a CIR, which is the minimum amount of bandwidth the carrier agrees to provide for each virtual circuit. If some users are not accessing the frame relay network at any given time, extra bandwidth becomes available to users who are on-line. The CIR of their virtual circuit can burst up to the full port speed. As other users come on-line, however, the virtual circuits that were bursting beyond their CIR must back down to the assigned CIRs.

In the frame relay network, congestion can be avoided through control mechanisms that provide Backward Explicit Congestion Notification (BECN) and Forward Explicit Congestion Notification (FECN), which are depicted in Fig. 8.14.

BECN is indicated by a bit set in the data frame by the network to notify the user's equipment that congestion avoidance procedures should be initiated for traffic in the opposite direction of the received frame. FECN is indicated by a bit set in the data frame by the network to notify the user that congestion avoidance procedures should be initiated for traffic in the direction of the received frame. Upon receiving either indication, the endpoint (i.e., bridge, router, or other internetworking device) takes appropriate action to ease congestion.

The response to congestion notification depends on the protocols and flow control mechanism employed by the endpoint. The BECN bit would typically be used by protocols capable of controlling traffic flow at the source. The FECN bit would typically be used by protocols implementing flow control at the destination.

Upon receipt of a frame with the BECN bit set, the endpoint must reduce its offered rate to the CIR for that frame relay connection. If consecutive data frames are received with the BECN bit set, the endpoint

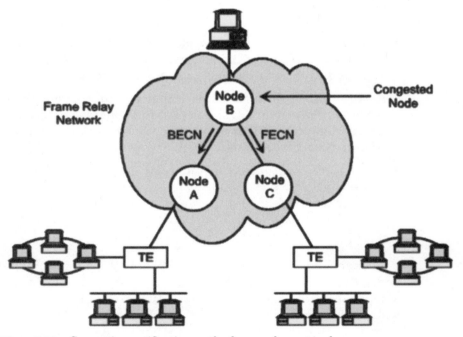

Figure 8.14   Congestion notification on the frame relay network.

must reduce its rate to the next "step" rate below the current offered rate. The step rates are 0.675, 0.50, and 0.25 times the current rate. After the endpoint has reduced its offered rate in response to receipt of BECN, it may increase its rate by a factor of 0.125 times the current rate after receiving two consecutive frames with the BECN bit clear.

If the endpoint does not respond to the congestion notification, or the user's data flow into the network is not significantly reduced as a result of the response to the congestion notification, or an endpoint is experiencing a problem that exacerbates the congestion problem, the network switches collaborate in implementing congestion recovery procedures. These procedures include discarding frames, in which case the end-to-end protocols employed by the endpoints are responsible for detecting and requesting the retransmission of missing frames.

Frame discard can be done on a priority basis; that is, a decision is made on whether certain frames should be discarded in preference to other frames in a congestion situation based on predetermined criteria. Frames are discarded based on their "discard eligibility" setting of 1 or 0, as specified in the data frame. A setting of 1 indicates that the frame should be discarded during congestion, while a setting of 0 indicates that the frame should not be discarded unless there are no alternatives.

The discard eligibility may be determined in several ways. The user can declare whether the frames are eligible for discard by setting the discard eligibility bit in the data frame to 1. Or, the network access interface may be configured to set the discard eligibility bit to 1 when the user's data has exceeded the CIR; in which case, the data is considered excess and subject to discard. For users who subscribe to CIR=0, which moves data through the frame relay network on a best-efforts basis subject to bandwidth availability, all traffic is discard eligible.

Frame relay charges differ by carrier and may differ further by configuration. Accordingly, frame relay service charges may include:

- Port charge for access to the nearest frame relay switch, which is applied to every user location attached to the frame relay network.

- Local-loop charge, which is the monthly cost of the facility providing access to the frame relay network. This charge may not apply if the customer's building is directly connected to the carrier's metro fiber ring, in which case the customer is charged only a one-time setup fee.

- Charges for the PVCs and SVCs, which are determined according to the CIR assigned to each virtual circuit.

- Burst capability, usually determined by the burst excess size. Most carriers do not specifically charge customers for bursting beyond the CIR.

- Customer premises equipment, which includes the frame relay or inter-networking access equipment optionally leased from the service provider and bundled into the cost of the service.

- IntraLATA/interLATA service. Usually, there is one price for "local" frame relay service and another price for "national" frame relay service. Neither is distance sensitive, however.

The X.25 protocol overcame the limitations of analog lines, but did so with a significant performance penalty, owing mainly to its extensive error-checking and flow control capabilities. In being able to do without these and other functions, frame relay offers higher throughput, fewer delays, and more efficient utilization of the available bandwidth—and at less cost than private lines.

8.8.3   Asynchronous Transfer Mode

ATM is a switching method for supporting voice, data, image, and video applications. Such networks may be accessed through a variety of standard interfaces, including T-carrier and fiber or via services such as IP and frame relay. The benefits of ATM over IP and frame relay include increased bandwidth, lower latency, and more efficient use of the available bandwidth through QoS enforcement.

Providing instantaneous bandwidth allocation and the ability to relay high-volume bursts of data, ATM uses fixed-sized cells of 53 bytes: 48 bytes for data transport and 5 bytes for overhead. The cell size has been fixed to simplify the switching of large volumes of data by hardware-based routing mechanisms, which results in extremely high transmission speeds.

Unlike legacy X.25 networks, where the switches operate in store-and-forward fashion to read the address information and implement other functions such as error correction, ATM uses virtual channel identifiers to route a cell through the ATM network to its destination. All cells enter the switching matrix in synchrony. In the event that two packets converge at the same output line, one of them is buffered. It is through buffer and queue management that the cells are properly ordered for ATM-level processing.

ATM also offers a consolidation solution for any company that maintains separate networks for voice, video, and data. The reason for separate networks is to provide appropriate bandwidth and preserve performance standards for the different applications. But ATM can eliminate the need for separate networks, providing a unified platform for multiservice networking that meets the bandwidth and QoS needs of all applications. Although the start-up cost for ATM is high, the economics of network consolidation mean that companies do not have to wait very long to realize return on their investment.

Quality of Service. ATM serves a broad range of applications very efficiently by allowing an appropriate QoS to be specified for each application. Various categories have been developed to help characterize network traffic, each of which has its own QoS requirements. These categories and QoS requirements are summarized in Table 8.6.

CBR is intended for applications where the PVC requires special network timing requirements (i.e., strict PVC cell loss, cell delay, and cell delay variation performance). For example, CBR would be used for applications requiring circuit emulation (i.e., a continuously operating logical channel) at transmission speeds comparable to DS1 and DS3. Such appli-

TABLE 8.6   ATM Quality-of-Service (QoS) Categories

Category	Application	Quality-of-Service Requirements			
		Bandwidth guarantee	Delay variation guarantee	Throughput guarantee	Congestion feedback
Constant Bit Rate (CBR)	Provides a fixed virtual circuit for applications that require a steady supply of bandwidth, such as voice,video, and multimedia traffic.	Yes	Yes	Yes	No
Varible Bit Rate (VBR)	Provides enough bandwidth for bursty traffic such as transaction processing and LAN nterconnection, as long as rates do not exceed a specified average.	Yes	Yes	Yes	No
Available Bit Rate (ABR)	Makes use of available bandwidth and minimizes data loss through congestion notification. Applications include email and file transfers.	Yes	No	Yes	No
Unspecified Bit Rate (UBR)	Makes use of any available bandwidth for routine communications between computers, but does not guarantee when or if data will arrive at its destination	No	No	No	No

cations would include private-line–like service or voice-type service where delays in transmission cannot be tolerated.

Variable Bit Rate-real time (VBR-rt) is intended for applications where the PVC requires low cell delay variation. For example, VBR-rt would be used for applications such as variable bit rate video compression and packet voice and video, which are somewhat tolerant of delay. Variable Bit Rate-non real time (VBR-nrt) is intended for applications where the PVC can tolerate larger cell delay variations than VBR-rt. For example, VBR-nrt would be used for applications such as data file transfers.

Available Bit Rate (ABR) is offered as a low-cost method of transporting applications traffic that can tolerate delay variations. The first application that offers traffic to the network gets to use the available bandwidth. Other applications that attempt to offer traffic to the network must wait until the bandwidth becomes free. If congestion builds up in the network, ABR traffic is held back to help relieve the congestion condition.

Unspecified Bit Rate (UBR) handles traffic on a best-effort basis with no guarantee of delivery. Newsgroup updates, network-monitoring messages, and file transfers are examples of nonessential traffic that are highly tolerant of delay and cell loss. If congestion starts building in the network, the ATM cells for such traffic are the first to be discarded to relieve the congestion.

Operation.   QoS enables an ATM switch to admit a CBR voice connection, while protecting a VBR connection for a transaction processing application, for example, while allowing an ABR or UBR data transfer to proceed over the same network. Each virtual circuit will have its own QoS contract, which is established at the time of connection setup at the User to Network Interface (UNI). The network will not allow any new QoS contracts to be established if they will adversely affect its ability to meet existing contracts. In such cases, the application will not be able to get on the network until the network is fully capable of meeting the new contract.

When the QoS is negotiated with the network, there are performance guarantees that go along with it: maximum cell rate, available cell rate, cell transfer delay, and cell loss ratio. The network reserves the resources needed to meet the performance guarantees, and the user is required to honor the contract by not exceeding the negotiated parameters. Several methods are available to enforce the contract. Among them are traffic policing and traffic shaping.

Traffic policing is a management function performed by switches or routers on the ATM network. To police traffic, the switches or routers use a buffering technique referred to as a "leaky bucket." This technique entails traffic flowing (leaking) out of the buffer (bucket) at a constant rate (the negotiated rate), regardless of how fast it flows into the buffer. If the traffic flows into the buffer too fast, the cells will be allowed on to the network only if enough capacity is available. If there is not enough capacity, the cells are discarded and must be retransmitted by the sending device.

Traffic shaping is a management function performed at the UNI of the ATM network. It ensures that traffic matches the contract negotiated between the user and network during connection setup. Traffic shaping helps guard against cell loss in the network. If too many cells are sent at once, cell discards can result, which will disrupt time-sensitive applications. Because traffic shaping regulates the data transfer rate by evenly spacing the cells, discards are prevented.

Inverse multiplexing over ATM.   Today even midsize companies with multiple traffic types and three or more distributed locations can benefit from ATM's sustained throughput, low latency, and adept traffic handling via appropriate QoS mechanisms. The availability of ATM routers with inverse multiplexing capabilities and $N \times T1$ access arrangements makes ATM suitable for mainstream use, particularly for companies who appreciate the benefits of ATM, but have been locked out of the service owing to its high cost of implementation.

In the past, T3 links were the minimum bandwidth required to access ATM networks, making the cost prohibitive for the vast majority of companies. Inverse Multiplexing over ATM (IMA) solves the bandwidth gap problem. With IMA, companies can aggregate multiple DS1 circuits to achieve just the right amount of bandwidth they need for their applications and pay for only that amount on an $N \times T1$ basis. The advantage of IMA is that such companies can scale up to the bandwidth they need, starting with a single T1, and then add links as more bandwidth is justified.

For example, when the bandwidth of four T1s is bonded by the IMA device the virtual connection through the service provider's network is provisioned at 6 Mbps. When the bandwidth of eight T1s is bonded by the IMA device, the virtual connection through the service provider's network is provisioned at 12 Mbps. Regardless of the number of T1 access links in place, the IMA device bonds them together, combining the bandwidth into a fatter logical pipe that can support mixed-media applications running over interconnected LANs (see Fig. 8.15).

8.9   Bundled Services

A relatively new trend in telecommunications services today is the bundling of multiple voice, data, and Internet offerings into attractively priced packages that meet the differing needs of small and midsize companies. The

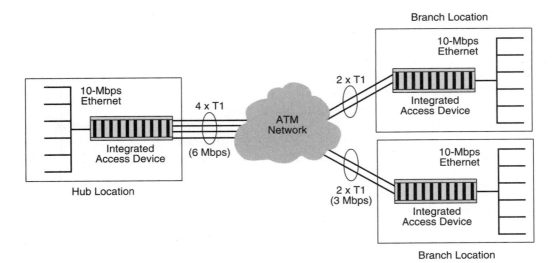

Figure 8.15   Inverse Multiplexing over ATM (IMA) allows the use of bonded multiple T1 access lines into and out of the service provider's ATM network, rather than forcing companies to use more expensive T3 access lines at each location. This makes it cost-effective for midsize companies to take advantage of ATM services to support a variety of applications.

strategy behind these bundles is to make it easier for companies to select services that best meet their needs, based on the number of lines or trunks in use, the number of employees in the organization, and bandwidth requirements for Internet access. Moving from one bundle to another as needs change can be done at an incremental difference in the monthly charge.

The value proposition of the bundling strategy is to offer customers multiple services over an integrated access connection at a lower price than services purchased individually or from multiple carriers. There is also the convenience of a single, consolidated invoice for all of the services and one carrier to deal with when making changes. In addition, flat rate pricing of the bundles provides businesses with a predictable monthly charge, which can make the communications budget easier to manage.

Depending on the carrier, there may be added incentives to go with their bundled services, such as ISDN PRI at no extra charge, free equipment, or email and Web hosting at no additional charge, which can help get e-commerce applications up and running quickly. There is wide divergence in the number of email addresses and the amount of Web space offered as part of the service bundles. Some service providers, on the other hand, do not include email or Web hosting in their bundles.

Bundled services represent an excellent option for smaller companies that do not have dedicated telecom or IT staff to manage the individual components of a voice communications system or data network, much less

Internet access, or the ability to deal with the various services and applications that a company runs over the Internet. The catch to bundled services is that the carriers do not permit them to be deconstructed; in other words, a company cannot pick out what it needs in a bundle and pay only for what it selects at the discounted rate. When it comes to bundles, it is an all-or-nothing proposition. So bundle selection takes some thought, taking into consideration current and near-term needs.

8.9.1 Decision criteria

Generally, the more bundles a service provider offers, the better fit there will be to a company's needs and the less risk there will be in overpaying for unneeded components. For this to happen, however, the service provider must offer a good range of bundles to choose from. A service provider that offers 40 bundles will have a better chance of meeting a company's needs more precisely than a service provider that offers only 5 or 15 bundles. Additionally, the more bundles a provider offers, the easier and less costly it will be to upgrade or change bundles in the future, since the incremental increase in the monthly charge will be much smaller with the carrier that offers more service bundles.

Although most carriers have similar features available, some carriers are more flexible than others when it comes to pricing for such features. With the voice component, for example, several base features may be included in the bundle price and others may be added at extra charge, including Direct Inward Dial (DID) numbers. If a company depends heavily on voice calls for much of its business, the availability and pricing of advanced calling features would be an important consideration.

With each bundle, the voice component typically comes with a predefined call allowance for domestic long-distance minutes. Call minutes beyond the allowance for that bundle incur a separate per-minute rate that is tacked on to the monthly bill. With some providers, the charges for calls that exceed the allowance are premium priced at 6 cents or more, versus call charges of 4.9 cents or lower per minute for "unbundled" calls. To choose the most appropriate bundle, therefore, requires that you, as a network manager, know at the start what the monthly LD call volume is for your company. You should know if the call volume has any seasonal fluctuations—up or down—that could affect the decision of which bundle or which carrier to choose. Generally, if the included LD minutes are not sufficient to cover at least 80 percent of a company's regular LD usage, the cost and rate structures for those additional minutes become more important.

Another important factor to consider from a long-distance perspective is the types of calls that may be made with the included minutes. For example, many bundled service providers today will include interstate

1+ calls, in-state 1+ calls, as well as toll-free calls (800, 888, 877, or 866). However, some do not include all types of calls, or they level a surcharge on in-state or toll-free calls. Due to varying regulations within states that affect the cost of in-state long-distance calls, some carriers might also charge more for in-state long distance, or place a limit on how many in-state long-distance calls may be made. Again, you will need to understand your own usage patterns in order to estimate how well a particular bundle or carrier will meet your company's needs.

With regard to the Internet component of the bundle, it is worth checking to ensure that the service provider has a Tier-1 Internet network. This is a network that is connected to all the major Internet peering points with redundant routes and optimized to move traffic from source to destination with the least amount of delay and the fewest number of hops. This type of network is also managed on a 24 $\times$ 7 basis to ensure maximum network availability and reliability. In contrast, Tier-2 or Tier-3 providers have regional backbones or no backbone at all, and therefore have little to no control over their customers' traffic or the network performance.

Without a Tier-1 service provider, a company risks poor application response time and more exposure to service outages of indeterminate length, which can negate the advantages of going with the service bundle. To ensure that a carrier's Internet access will properly support mission-critical applications, performance guarantees from the carrier should contractually support any speed, reliability, and availability claims that are made.

Sometimes service providers offer network-based security, Web site filtering, and anti-virus scanning for the Internet component. While these services protect against many common nuisances, they do not defend a company from the more severe threats emanating from the Internet—intrusions from hackers—which can result in the loss, damage, or theft of sensitive information. A hardware firewall at the customer site will provide this level of protection, but most of the time a system must be purchased from a separate vendor and qualified staff must be hired to manage it, which can be cost-prohibitive for smaller companies with already meager resources.

Internet security is extremely important today because of the potential cost of damage to companies that fail to take precautionary measures. Having the same service provider manage the access link, the router, and the firewall relieves the smaller company from this resource-intensive burden, while still allowing it to reap the advantages of bundled services.

Finally, the capability of a carrier to provide "one-stop shopping" should include Customer Premises Equipment (CPE) that limits the amount of work a customer has to do to use the service. In today's telecommunications environment, any bundled or T1 service from a reputable carrier should include an Internet router, preconfigured for service from the carrier. Some carriers will provide CPE that the customer has to manage after service is connected, while some carriers will continue to own and manage the CPE themselves. In either case, the company will have to manage its own LAN and workstations, but can connect the LAN directly to the router provided by the carrier. Both alternatives have advantages and disadvantages for you, as a network manager, but generally, the more the carrier is doing at no additional charge, the better.

8.9.2   Bundle examples

Let us say that a company needs 10 digital PBX trunks (two-way), 10 digital PBX DID trunks, 100 DID numbers, 3 business exchange lines, and a 384-kbps Internet access connection with only a 12-month commitment. The total monthly recurring charge could vary between $1000 and $1450 with a one-time installation charge of $250, depending on the provider and the city, as shown in Table 8.7.

If 384 kbps of bandwidth for Internet access is not the right fit, some service providers allow customers to upgrade to 512 kbps, 640 kbps, or 768 kbps. Most bundles come with a minimum of 128K or 256K bandwidth, with upgrades at an additional charge. The upgrade pricing deserves close scrutiny, since many providers charge premium prices for upgraded bandwidth.

Another way to save money on a service bundle is to go with a two- or three-year term. The service provider is more inclined to waive the one-time setup charge on a multiyear commitment, but will usually not waive this charge on a one-year commitment. Of course, some service providers will not negotiate on the installation charge at all. With terms longer than three years, you might risk being locked out of better pricing or improved technology that comes along in the interim.

TABLE 8.7   Example of Components and Costs of a Bundled Service

Services/components	Monthly recurring charges
23 total trunks and lines	$1170
DID trunk termination	$50
DID numbers	$15
384-kbps Internet bandwidth	$49
5750 long-distance minutes	Included
Total	**$1284**
One-time installation	$250

In the current economic climate, companies of all types and sizes are looking for ways to improve profitability by trimming the cost of operations. Among the first targets of cost cutting is telecommunications service, because cutbacks in this area can have an immediate and substantial impact. For example, a company with a 20 percent net profit margin has to sell $1000 more each month to have the same impact as a $200 cost savings. If the cost savings is significant enough, this tactic not only shores up the bottom line but it also has the effect of maintaining a company's competitive advantage, enabling it to ride out periods of low customer demand and prolonged volatility in capital markets. Bundled services can help companies meet the goal of cost savings and offer them a predictable monthly outlay, but without the need to forgo essential communications.

8.10   Conclusion

Communications managers are responsible for providing their organizations with equipment, media, and services that are adequate for their current and future needs. The sound combination of these elements can result in efficiencies and cost savings that have a significant impact on the competitiveness and long-term survivability of an enterprise.

With so many communications alternatives, the risk of error is great. Choosing the wrong service or product can result in severe penalties, including missed corporate objectives, damage to the company's competitive position, wasted capital resources, and poor return on investment. To minimize risk, as a network manager, you must continuously review the choices available, assess their cost-benefits trade-offs, and determine how some of the alternatives can best be combined to achieve a balance between efficiency and economy to sustain your company's competitive advantage.

Systems Integration

9.1 Introduction

Corporations large and small in just about every segment of the economy invest hundreds of billions of dollars annually in the technologies to build, upgrade, or expand their information systems. This includes all purchases of hardware, software, services, and support. There is ample reason for such activity. Corporate executives understand the role of information technologies in keeping their firms competitive. The quality of a company's systems holds the keys to cost control, serving customers better, increasing market share, and pursuing new business opportunities successfully. Clearly, Information Technology (IT) is being treated as a strategic resource rather than mere overhead expense, as in days past. Putting together IT systems that do all this, however, is no small undertaking.

Needs must be assessed, budgets hammered out, consultants hired, plans drawn, technologies reviewed, Requests For Proposals (RFPs) written, vendors evaluated, feasibilities studied, equipment installed, lines leased, and users trained. And that's only the beginning. More often than not, it is discovered that a variety of products from different manufacturers are required to meet organizational objectives.

Not only must diverse systems be seamlessly interconnected on the same network but users must also be provided with transparent access to every other element on the network. But rarely do in-house telecom and IT staff have the depth and breadth of expertise to accomplish this formidable task alone. Enter the systems integrator.

9.2   Systems Integration Defined

A *systems integrator* brings objectivity to the task of tying together disparate products and systems to form a seamless, unified network. This entails reconciling physical connections and overcoming problems related to incompatible protocols. The systems integrator uses its hardware and software expertise to customize the necessary interfaces. The objective is to provide compatibility and interoperability among different vendors' products at a price the customer can afford.

The systems integrator not only provides the integration of various vendors' hardware and software but can also tie in additional features and services offered through the public switched network. To do this, the systems integrator draws on its experience in information systems, telephony, and data communications. Added value is provided through strong project management skills and accumulated experience with customer requirements in a variety of operating environments. To do all this, the systems integrator must have in place a stable infrastructure capable of handling a high degree of ambiguity and complexity, as well as any technical challenge that may stand in the way of the integration effort. Accordingly, the staffs of some systems integrators may number several thousand people, many of whom work on-site at customer locations for the duration of the project.

With regard to networks, there are a number of discrete services that are provided by systems integration firms, including:

- *Design and development.* Network design, facilities engineering, equipment installation and customization, and acceptance testing

- *Consulting.* Needs analysis, business planning, systems-network architecture, technology assessment, feasibility studies, RFP development, vendor evaluation and product selection, quality assurance, security auditing, disaster recovery planning, and project management

- *Systems implementation.* Procurement, documentation, configuration management, contract management, and program management

- *Facilities management.* Operations, technical support, hot-line services, move and change management, and trouble ticket administration

- *Network management.* Network optimization, remote monitoring and diagnostics, network restoral, technician dispatch, and carrier and vendor relations

9.3   Role of Business Process Reengineering

Business process reengineering is a front-end activity that precedes systems integration. The end result of a reengineering design usually involves changing the computer environment, which, in turn, determines the kind of systems to buy and the degree of customization required. Of course, the type of integration project will determine the choice of systems integrator.

When organizations begin planning for new systems or technology migrations, they often find themselves in a dilemma: Should they first reorganize and redesign the various business processes, or should they incrementally add new systems to contain costs and minimize disruption to business operations? To illustrate how business process reengineering works, consider the case of a company that wants to automate workflow to enhance productivity, trim overhead costs, and improve customer response.

One approach to this dilemma is to perform a thorough workflow analysis. The place to begin is on a quantifiable piece of the business process. A work group or department should be selected that can benefit most from the introduction of automated workflow processing. Such operations as accounting, order fulfillment, claims processing, or publishing are ideal candidates and ones that could be used for evaluating future improvements or expansion.

A systems integrator that specializes in document management and imaging systems can help management make confident and informed decisions. In the evaluation and selection process, the systems integrator should demonstrate the ability to:

- Incorporate workflow and business reengineering techniques

- Provide line-of-business expertise

- Reduce risk through comprehensive analysis and reporting procedures

- Implement changes with minimal disruption to the company's daily operations

- Offer a suite of services to help meet the organization's total integration requirements

9.3.1   Predecision services

A systems integrator should offer predecision services that will help the IT organization determine where to apply the automated workflow and document imaging technology and project potential return on investment prior to buying a complete system. There are several components to this type of service: a focus review, requirements definition, and strategic impact analysis.

The *focus review* addresses an organization's strategic business objectives, identifying potential targets for improvement and providing a high-level cost-benefit analysis. The objective of the review is to develop a preliminary plan for an integrated solution within the existing work environment. The focus review also helps management zero in on which work group or department would gain the most benefit from automated workflow and imaging technology, based on such parameters as volumes of paper processed, content, frequency of access, user acceptance, and financial considerations.

Once the decision has been made to automate and image-enable a specific business process, the systems integrator should provide a *requirements definition*. This service builds on the preliminary input from the focus review to further define, analyze, and document the specific needs of the target work group or department. It lays the groundwork for all subsequent design and implementation activities. This service identifies the proposed project's major inputs, outputs, and volumes; defines the current workflow processes, identifying those that should be automated and those that should not; and defines all hardware, software, and future services that will be required for implementation.

With the current manual processing methods identified and understood, the systems integrator can then recommend appropriate steps to eliminate inefficient or outmoded practices to streamline operations. This analysis ensures that the work group or department's total requirements have been explored and understood prior to the company's commitment to purchase new systems. In addition, the requirements definition describes the types of third-party services that are required, such as consultants or document conversion services, which will be brought in under the direction of the integrator's project management team. The requirements definition also provides an initial installation schedule and cost estimates for implementing the solution.

The completion of these services results in a detailed design specification, which describes the actual solution system and how it will be implemented. The information included in this document includes the system configuration, workflow, end-user training, test plans, system acceptance scripts, and complete project schedules and timetables.

The *strategic impact analysis* takes these processes a step further, evaluating the potential effect of the proposed solution on the entire enterprise. The resulting report recommends an appropriate architecture, method of enterprise integration, and a specific implementation approach.

Business process reengineering is a methodology that involves identifying and applying appropriate technology to solving specific business problems, as opposed to simply looking at the technology in terms of features and functionality. When evaluating the cost-benefits of automating workflow and image-enabling various applications, this methodology includes conducting a detailed workflow analysis that starts with interviews with users in each department where the technology will reside. Users are questioned about what they do on the job and how information flows in and out of their environment. This gives evaluators a feel for what kind of supportive information technologies users need to do their jobs.

The workflow analysis may reveal the need for complementary solutions, such as Electronic Data Interchange (EDI), fax-on-demand, email, or

Internet connections to facilitate document distribution. There may be times when a combination of technologies will provide the solution to a workflow problem. It is even possible that an analysis done ostensibly to evaluate the need for automated workflow and document imaging might result in other useful recommendations. For example, it may be found that old procedures are still in place that have not been updated to reflect changes in organizational structure or in business practices.

9.3.2 Preinstallation services

Once management has signed off on the proposal, the systems integrator uses the information gathered in the predecision evaluation process to provide preinstallation services to ensure the completion of a detailed design and communications plan. This starts with the creation of a detailed design document, which defines all hardware and software up front so that the actual installation will be performed with minimal business interruption in the shortest possible time frame.

In the case of automating workflow, this document includes a complete description of document routing and the database structure, user forms, process automation, and the configuration of optical and magnetic tape storage. Eventually, this document will also be used to facilitate the implementation of the system acceptance test plan, which enables the customer to verify that all requirements are being addressed prior to system installation and cutover.

Next, the organization's workflow and image communications requirements are evaluated and verified prior to installation. This evaluation includes Local Area Network (LAN) requirements analysis, planning, design, installation, implementation, and support services. To realize the efficiencies and economies of image-enabled workflow systems, it is often required that imaged documents be accessible not only to local users over the LAN but also to remote users over the Wide Area Network (WAN) or telecommuters over a Virtual Private Network (VPN). This requires an evaluation of the impact of imaging on existing bandwidth.* In defining such requirements at the outset, unnecessary start-up delays can be eliminated.

9.3.3 Installation and implementation services

The systems integrator should provide site analysis and planning, so equipment and software can be installed immediately upon arrival with minimal impact on the workforce and current business operations. At this

---

* This is an important, but often overlooked, consideration. A typical bitmapped image is 15 to 20 times the size of the same information stored in a native text format. Implementing an imaged-enabled application over WAN links can cause serious bottlenecks unless more bandwidth is made available to handle the increased load.

phase of the project's life cycle, the systems integrator installs the hardware and software components and brings the system to an operational state. Afterward, the integrator initiates a verification process to confirm that these components are running properly. Next, all of the components of the system should be documented. This includes the software for Optical Character Recognition (OCR), image capture, workflow automation, and various Application Programming Interfaces (APIs) that facilitate the development of image applications from standard programming languages.

During the installation phase, the hardware and software elements are bridged. Then, using the results obtained during the design phase, the entire system is tested using the system acceptance scripts. Upon acceptance by Information System (IS) management, the system is put into service in the production environment.

9.3.4   Postinstallation services

Typically, image-enabled workflow systems do not operate in closed environments. The goal of the integrator's postinstallation services is to integrate the technology into mainstream business processes and to address conversion and support issues. Accordingly, the integrator should provide the technical experts who will integrate the imaging system with non-imaging applications such as facsimile, email, and Structured Query Language (SQL) databases that reside on mainframes or LAN servers. This allows current investments to be leveraged, while effectively integrating the various workflow and imaging technologies into an enterprisewide solution.

Document conversion from paper to digital can be an expensive and time-consuming undertaking when an organization makes the transition to automated solutions. This task is usually farmed out to a third-party service firm which, if contracted through the systems integrator, is managed as one of the project deliverables.

As part of postinstallation services, the systems integrator should provide for the ongoing support of the installed systems. This can include any number of support services such as on-site hardware repair service; overnight shipping of replacement components and modular subsystems; and access to technical staff via a toll-free number, a dial-up Bulletin Board System (BBS), or a Web site on the Internet.

9.3.5   Life cycle services

The systems integrator should provide management services spanning the entire project life cycle, including training. Typically, an integrator offers specialized courses that are conducted at regional training centers or on location at customer sites. In the case of image-enabled workflow automation, several training paths should be available, including courses for system administrators, managers, workflow analysts, and workstation operators. The training should range from basic imaging principles to overall system functions.

Once the new system is up and running and has been accepted by the IS manager, the integrator's project management team is no longer required. However, technical assistance retainer plans are usually available that include the on-site services of the project manager on either an ongoing or periodic basis.

9.4   Types of Integrators

In the United States and Canada alone, there are about 5000 companies that claim to do systems integration. Understanding the market players is important in the selection process because each has a different set of skills, experiences, and biases that can impact the integration solution. For example, the integrators who are subsidiaries of major manufacturers may promote solutions that favor the hardware and software products of their parent companies. Integrators that are subsidiaries of local or interexchange carriers may push solutions that rely on the new services the carrier wants to promote. All players claim to maintain an unbiased role in the process. In truth, every player brings its own biases.

Today's systems integrators include traditional firms, such as Electronic Data Systems (EDS), which rose to prominence by servicing the data-center environment; computer system vendors that specialize in their own product lines; and value-added resellers, with roots in the Personal Computer (PC) or desktop environment. Value Added Resellers (VARs) and distributors are also involved in systems integration. Systems integrators typically fall into the following categories:

- Data communications service providers

- Consultants

- IT shops

- Traditional service firms

- Equipment vendors, VARs, and distributors

Each type of firm has specific strengths and weaknesses, which should be taken into consideration when choosing a systems integrator. The wrong choice can delay implementation of new systems, inflate project costs, or even result in the wrong solution being implemented. It is therefore advisable to choose an integrator whose products and services are particularly pivotal to the business process under scrutiny. For example, if the integration project is such that the computer requirements are extremely well defined and no significant changes are expected in the applications, but entirely new carrier services might be involved to network the systems regionally, nationwide, or globally, it would be a mistake to select a computer vendor as the integrator. In this case the entity with the critical experience and expertise is going to be the data communications service provider.

9.4.1   Data communications service providers

Carriers, both local and long distance, have been in the integration services business for many years. As part of their data service offerings, they routinely assist companies in assessing needs, designing networks to interconnect systems from different manufacturers, selecting and installing Customer Premises Equipment (CPE), and managing the access and transport connections. Now that advanced services like Internet Protocol (IP)-based virtual private networks, frame relay, Gigabit Ethernet, Fibre Channel, and ATM are widely available, carriers are finding a lucrative market for integration services. These new customers can no longer cope with the mounting number of product and service choices, to say nothing of configuration complexity and ongoing management.

The large carriers and service providers have business units dedicated to systems integration services. AT&T, for example, touts its many years of experience in performing such services, both through its Federal Systems Group and through individual projects with large commercial customers. The Regional Bell Operating Companies (RBOCs), too, claim that systems integration is something they have been doing for many years. They and their competitors leverage their specialized skills, knowledge, and experience into full-fledged systems integration efforts.

In general, carriers are indeed strong in voice switching and transport. In fact, AT&T and the major telephone companies have a long history of success in joining together disparate private networks to the public network. But systems integration calls for expertise in many other areas. AT&T, for example, has not been able to successfully penetrate the computer market or the market for network management systems, even through acquisitions.* This indicates that it may be deficient in the data center and desktop environments. Although AT&T is staffed with LAN experts, it continues to approach LANs as merely an extension of the WAN. For many companies, this may be an overly simplistic approach, if

---

* Even through acquisitions, there is no guarantee of success. Having failed in its effort to build its own line of computers, AT&T bought NCR Corporation in 1990 to establish its presence in the personal computer market and expand its StarLAN product line. Having failed in that endeavor, AT&T divested its Global Information Solutions, Inc., unit, formerly known as NCR, and exited the computer business.

only because integrating LANs and legacy systems over the WAN requires detailed expertise in managing different LAN protocols.

A wide-area internet that integrates several different types of LANs and legacy systems requires that attention be given to a plethora of different physical interfaces, protocols, frame sizes, data transmission rates, quality of service issues, and security. Hardware and software configuration expertise is also required to perform any necessary customization. Knowledge of the various network operating systems is also essential. Complicating this task of network integration is that many of the tools that are available in the legacy environment are different from the ones available in the distributed environment.

Likewise, the RBOCs traditionally have not offered information systems expertise to their customers, which is required to merge business applications with the multivendor hardware architectures that reside at most customer sites. In fairness, much of this deficiency can be attributed to regulatory constraints. However, the RBOCs have always been criticized for their lack of strong project management skills and in-depth understanding of customer requirements beyond the Private Branch Exchange (PBX). In recognition of these shortcomings, the RBOCs have had to engage in teaming arrangements with computer makers, acquire service firms with a national presence, and engage in aggressive personnel recruiting efforts to fill gaps in their technical expertise.

The RBOCs typically work with their customers' designated long-distance carriers to put together the right blend of products, access options, lines, services, and features that will meet specific needs. Some offer remote network management services for LANs and WANs to customers within their service areas. This includes monitoring alarms and events, conducting diagnostic tests, troubleshooting problems, managing CPE, and dispatching repair technicians. Others work with customers to put together application-specific networks such as those for videoconferencing, telemedicine, distance learning, teleworking, and supercomputing.

9.4.2 Consultants

Communications consultants comprise another category of systems integrator. Over the years, the "Big Five" management consulting and accounting firms have expanded the scope of their activities to include communications consulting, planning, and integration. They bring a broad range of technical expertise and the intimate knowledge of client operations to integration projects.

The services offered by these firms include systems planning and design to implementation and management. Some firms specialize in helping organizations deal with their people, processes, and tools within the context of building heterogeneous client-server environments.

Although such firms tout their objectivity, this could be a two-edged sword. Some clients do not appreciate the way that the management and accounting side of the house keeps uncovering problems that require the services of the communications consulting group and vice versa. When this happens continually, the claim of objectivity goes right out the window. Although the extent of such practices differs from firm to firm, the customer really has no way of predicting such behavior and, when it is discovered, it is often after the fact. A check of references may alert the customer to potential problems, but only if the firm's behavior was clearly aggressive. Often, the offending firm maneuvers its clients so deftly that they believe they have made such decisions of their own volition. Although structural separations between business units can minimize this problem, there is really no way to verify adherence to such arrangements.

Another area of concern is that such firms maintain partnering relationships with hardware vendors, which may seem to compromise objectivity. When clients want advice on the best direction to move in, or the best carrier service or hardware to use, objectivity is what they are looking for and that is what most firms provide. Other times, clients want the consulting firm to propose a solution and be the turnkey provider. In this case, partnerships with vendors and carriers allow the integrator to move quickly on behalf of its clients.

This category of integrator also includes contract programmers and independent consulting firms. The limitation of the contract programmers is in the narrowness of their activities, which may include applications software, operating systems, media conversions, and communications protocols between PCs and mainframes. This type of firm plays a valuable role in systems integration, but usually as a subcontractor to the primary contractor, who typically assumes full financial responsibility and risk management for the entire project.

Independent consulting firms who are engaged in systems integration stress their objectivity. In not being associated with any vendor or service provider, they claim to have their customers' best interests in mind when evaluating hardware, software, and services.

While objectivity is indeed a valuable asset during the vendor selection process, systems integrators are often called in after the selection of hardware and software has been made. At that point, the objectivity of the independent consulting firm is a moot issue. Other criteria for choosing the independent consulting firm must be evaluated, such as its financial resources, areas of technical expertise, organizational stability, and project management skills. It is advisable to steer clear of consultants who stress their track record of "vendor bashing" or their reputation for "cutting carriers down to size." This kind of talk reveals poor interpersonal communications skills, which can unnecessarily prolong the systems integration project and drive up costs.

What most companies really need is a highly skilled communicator who is focused on the project at hand instead of on his or her own ego needs. Effective performance in systems integration hinges on the ability to interface well with multiple vendors, carriers, consultants, and subcontractors, as well as in-house staff. The consultant should come across as someone who will negotiate in good faith with all parties to implement the total integration plan.

Most independent consulting firms have very limited financial and organizational resources to draw on. This restricts the size and complexity of systems integration projects they can handle. It is worth discovering what that threshold is—preferably not from personal experience.

One way that integrators can broaden their expertise is by teaming up with those who do have the expertise. However, it is incumbent upon the buyer to ascertain the nature of such relationships. For example, complementary relationships should already be in place, and not be something that was hastily thrown together just to obtain a specific contract. Both parties should have evaluated each other's strengths and weaknesses well before entering into any formal or informal arrangement. Ideally, they should be able to demonstrate the integrity of the relationship by providing a list of projects that they have completed together. A check of references may reveal flaws in the relationship that bear further investigation

9.4.3 Information Technology shops

Sometimes a company's in-house IT staff may become expert at systems integration, as when a bank moves from an outmoded banking system to a best-of-breed mix of equipment from different vendors. Upon discovering that other banks have similar needs, the in-house staff may spin off into a separate profit center by offering systems integration services to other financial institutions. Such firms constitute a third category of systems integrator.

Hiring such firms may prove worthwhile if the organization's systems integration needs are confined to the data center. When it comes to wide area networking, however, relying on a firm whose experience is grounded solely in information systems may prove inadequate. The technologies that differentiate data centers and LANs from wide area networking are so dissimilar that specialized expertise is required for each. Even providing the links between LANs and WANs through such devices as routers, switches, and gateways requires more expertise than is usually found among data center professionals.

9.4.4 Traditional service firms

Closely related to the in-house IT shops that have gone commercial are the traditional service firms like TRW and EDS. These types of firms are grounded in information systems and, until the 1990s, did most of their business with the federal government. Now that more of the demand for such services is increasingly coming from the commercial sector, these firms are competing aggressively for corporate accounts.

TRW, for example, will operate, manage, and maintain any or all of a company's IT and telecom operations for a flat monthly fee. According to TRW, almost half of today's IT projects lack a clear understanding of Return-On Investment (ROI) and 40 percent are over budget, late, or fail to deliver altogether. TRW addresses this state of affairs with its concept of Information Partnering, which works to align IT operations with a company's strategic business goals, allowing it to focus on core business activities.

TRW's Information Partnering strategic outsourcing services consist of four tightly integrated service offerings. Enterprise Management Support aligns key business processes with business goals and objectives, integrating them with the optimum IT systems and applications. Infrastructure Services optimizes all IT assets for higher workforce productivity while aligning them with business goals and objectives. Collaborative Solutions optimizes the extended enterprise by enabling key suppliers, customers, and partners to involve themselves earlier and more directly in the product-service development cycle. Critical Asset and Information Protection provides enterprise security architecture to protect intellectual property, IT infrastructure, facilities, and people against intrusion and attack.

Whenever these integration firms encounter problems they cannot solve alone, they have the strategic relationships already in place with which to draw upon the appropriate expertise. This expertise comes from numerous subcontractors and vendors, who are often listed as the co-bidders when they pursue large contracts.

Using such firms has its share of risks. Some of these firms use systems integration as a cover to sell products, which run the gamut from management services to software and processing services. Some of these firms are so big that smaller customers may not get proper attention. Others are too busy chasing highly lucrative contracts to support their heavy infrastructures. Smaller projects may end up being perceived as quite trivial in the overall scheme of things—hardly worth the effort to complete on time and within budget.

9.4.5   Equipment vendors, VARs, distributors

Another category of systems integrator is the hardware vendor, of which there are several types: computer and telecommunications equipment manufacturers, data communications firms, and interconnect companies. Each seeks to leverage its existing software and turnkey systems expertise into integration operations. Recognizing customer concerns about objectivity, these firms usually handle systems integration through a separate division or subsidiary, which is chartered to engage in such activities apart from the firm's product sales and marketing efforts.

Computer manufacturers have strong expertise and integration experience with information systems, but are usually quite deficient when it comes to telecommunications. Interconnect companies are very good at marrying LANs with such devices as routers and gateways, but they too are deficient with respect to telecommunications. Although telecommunications equipment manufacturers—such as those who make PBXs—are strong in telecom, they are less proficient in LANs and other forms of data communications.

By treating voice as just another form of data—by digitizing, compressing, and managing it in combination with other types of traffic—the data communications firms can bridge the traditionally separate realms of IT and telecom. As such, they occupy a strategic position in the industry that is becoming more apparent as users seek to build hybrid networks consisting of both public and private elements. An independent data communications firm that is not aligned with either carriers or computer manufacturers has a lot to offer in bridging the two environments without users having to worry about the hidden agendas of either computer makers or carriers.

Skeptics argue that this type of hardware vendor is merely posing as a systems integrator, claiming that such operations are really reconnaissance missions designed to assess new business opportunities, steal accounts from other vendors, and promote proprietary solutions—all at the expense of customer needs.

Such firms have come to recognize, however, that they can be more successful if their products can link to the rest of the universe, rather than to only a small portion of it. In fact, their involvement in international standards setting and their experience in product design make some hardware vendors a viable choice for systems integrator. Of course, customers must delve into the qualifications and specific areas of expertise claimed by vendors who want to be their systems integrator. Vendors who are looking to gain a toehold in the systems integration market must be able to reach outside their own product lines and support open systems architectures to be successful in this endeavor.

As for the charge that vendors are using systems integration to steal accounts from other vendors, that is largely the result of one vendor outperforming another vendor, in which case the customer always ends up the winner. There's nothing much to complain about in that regard, unless of course you happen to be the vendor who lost the account.

Distributors, unlike other types of systems integrators, such as VARs, are not very good at doing customized tasks in twos or threes. Instead they excel at repetitive, high-volume tasks such as software loading and burn-in and testing. Loading UNIX shells, NetWare, and Windows is now a fairly standard chore for major distributors filling orders from systems integrators, whereas loading the applications are usually the responsibility of the VAR or primary systems integrator. The increased use of Windows and other graphical user environments on networks translates into greatly increased burn-in and testing of printers, memory upgrades, and interface cards.

Although distributors are expanding assembly and test centers, forming new divisions, and merging with other distributors, their real value is in making good partners for larger systems integrators, and not in dealing with the customer directly in such areas as business process reengineering, network design, and providing ongoing support.

9.5   Evaluation Criteria

Whether the choice for a systems integrator is a service provider, consulting firm, or hardware vendor, it should be an informed choice. Companies that make an informed choice can avoid two basic kinds of systems integration disasters: hiring a large integrator to minimize risk, then not paying sufficient attention to the project; or hiring a small, specialized integrator to get a focused approach, but finding out that it lacks the resources, staff, and skills to do the job properly. What follows are some tips for evaluating the candidates.

Basically, companies should be looking for an integration firm with the broadest amount of experience in a variety of technical fields, applications, and operating environments. Even if some of these points are remote considerations, they should be evaluated with long-term requirements in mind. After all, once a systems integrator is selected and considerable time and effort have been invested in developing the relationship, the client organization does not want to expend additional resources on nurturing new relationships with other systems integrators every time it has another requirement.

9.5.1   Technological leadership

To establish vendor or carrier claims of technological leadership and superiority, delve into their claims by finding out which unique features or capabilities their Research and Development (R&D) group is responsible for and why they are better than anything else on the market. Does the company hold any patents or software copyrights? Does it license proprietary technology to other manufacturers? Does it supply key subsystems for any products already on the market? Does it design and build key componentry based on Large Scale Integration (LSI) or Very Large Scale Integration (VLSI) to give its products price and/or performance advantages? If the answer to these questions is "no," find out what the vendor means when it describes its products or services as advanced, state-of-the-art, leading-edge, innovative, or unique. Many times, these words and phrases are just marketing jargon designed to attract interest.

After establishing the definitions of terms—and quite possibly finding out that the vendor cannot live up to them—inquire about the performance record of products or service offerings already deployed in networks similar to yours. Validating the answers by checking with references will help determine product or service quality, levels of customer satisfaction, and responsiveness to changing customer requirements. All of these are key ingredients that determine technological leadership.

9.5.2   Digital transport systems

Look for demonstrated expertise in network design, engineering, and implementation of large digital transport systems for a variety of rigorous applications, including those in retail, financial services, manufacturing, and government environments.

Evidence of such experience should include contract awards and successfully completed projects that involve installing and implementing multinode digital backbone networks. Ask for references that use similar hardware and software products planned for use on your network. When calling references, inquire about the firm's ability to meet work schedules, deal with multiple vendors simultaneously, get along with staff members, and stay within budget.

Look at the integration firm's experience in designing and installing hybrid networks consisting of such diverse elements as multiplexers, routers, gateways, and switches. Also look for experience in melding public and private facilities and services, including the Internet. With these building blocks, companies may mix and match several architectures instead of feeling constrained with a single networkwide architecture and, in the process, add elements of precision and control to network operations, while effectively positioning the company for other technology transitions. These building blocks can facilitate network expansion to include satellite and microwave, if necessary, as well as fiber-optic links. Even if these concerns are remote, companies should choose a systems integrator that will have the experience to meet future requirements.

Look at the experience of the firm in integrating LANs to WANs, using such devices as gateways, bridges, and routers. Experience with these network elements helps to establish the breadth and depth of a firm's knowledge of communications protocols, management systems, and specialized interfaces. Find out what experience the firm has in integrating networks that span international locations.

9.5.3　Network design

Evaluate the firm's network design tools. Be sure that they are capable of taking into consideration such factors as line topology, traffic load, facility costs, equipment types, communications protocols, hubbing arrangements, and the differing performance parameters of both voice and data. The tools should have the capability of pointing out errors in standards compliance and permit thorough testing of the design under various simulated loads.

These factors must be considered simultaneously with such variables as switch performance, which includes queuing, blocking, and reliability. Make sure that the design tool takes into account the type of traffic that the network must support: voice, data, image, or full motion video—or any mix of these. This is important because most of the current computer tools that are available address only a few aspects of the design problem and assume that the other aspects are fixed, essentially treating them as foregone conclusions.

Keep in mind that the information requirements of many computerized network design tools border on the onerous. A user might have to know the transaction message sizes of major applications, for example, and how often the applications are run during the peak busy hour at each location. Equipment and facility performance parameters such as delays must be provided. Currently experienced response time is often needed to calibrate and validate an existing baseline configuration model before a network with desired changes can be modeled with any degree of confidence.

9.5.4　Microwave and fiber-optic interfacing

The increasing sophistication of networks requires that the systems integrator have expertise in interfacing voice and data digital networks with microwave and fiber-optic systems. In the case of a hardware vendor, this can be determined by finding out how its products are being deployed by its customers.

For example, the deployment of the vendor's diagnostic modems on an oil company's microwave network connecting numerous oil platforms in the Gulf of Mexico to network control facilities in Texas might provide ample demonstration of success in integrating network elements to microwave. In this case, find out from this reference how much customization was required to tie the modems into the microwave network and determine the customer's level of satisfaction with the results.

To determine the vendor's experience in interfacing to fiber-optic transmission systems, find out if its T1 products are ever equipped with fiber termination cards to provide customers with the means to feed high-capacity fiber links. Follow up with references to determine their satisfaction with the results in terms of transmission performance, project completion, and final cost.

In the absence of direct experience in these areas, check into the firm's strategic alliances. Some firms have fairly mature alliance programs with other hardware vendors that provide critical components of the network along with the required technical expertise.

9.5.5   Network management

Network management systems unify computer and communications resources and transform them into strategic assets with which to improve a company's competitive position and long-term survivability. Selecting a systems integrator with experience in designing and implementing network management systems may be critical to the success of the integration plan.

Some data communications firms have demonstrated their expertise in network management by integrating host-based management systems with proprietary management systems that also tie into other element management systems and higher-level open systems platforms. A vendor's expertise in developing the links between network management systems gives them a unique perspective with which to provide objective advice concerning all aspects of network operations.

9.5.6   Project management

A systems integrator must absolutely understand the client's business needs and not simply approach everything as a technical problem. Look for a systems integrator that hires project managers who are specialists in core businesses, such as banking, retailing, manufacturing, or publishing. If the systems integrator does not understand a company's business, chances are the solution will not fully address critical needs. This does not mean that systems integrators know everything; in fact, it is unlikely that any one firm will have all of the expertise needed to solve a particular client's problems. Therefore, it is important to look at the systems integrator's vertical market partners for their specific core business expertise. The partner may be a VAR and, in some cases, may even be a competitor who possesses the required granularity of knowledge. To help differentiate among systems integrators, look at each company's background, skill set, and personnel, and whether it has handled similar projects.

The project management team should have extensive experience in all facets of data and voice communications, including system design, product development, integration, installation, and problem resolution. Another critical area is experience in move, add, and change management, which may be an ongoing activity, depending on the size and complexity of a client's network.

Find out if a dedicated project manager will be assigned to oversee all aspects of the integration project as well as its implementation. The project should be open to customization to meet your specific requirements. The resume of the assigned project manager should be furnished for evaluation along with an organizational chart showing lines of responsibility. This information may be used to support the decision-making process related to the selection of the systems integrator. It should not be changed later without prior notification and approval.

The project management team should be well versed in the use of computerized planning tools to coordinate and track your systems implementation plan. PERT, GANTT, and CPM should be among the tools used to develop and review implementation progress.

Once the equipment is installed and integrated into an entire system, the project management team should develop and oversee customized acceptance testing with client staff. At the same time, the project management team should prescribe appropriate levels of training for your staff, and be prepared to implement training if required.

The project management team should be able to assist in identifying specific inventory control and internal billing requirements, and then to recommend, develop, or customize software to meet those needs. At the same time, make sure the integrator's staff fully understands the project's objectives. There must be a clear definition of these objectives and full disclosure of any known constraints such as deadlines, budget, internal resources, and preferred brand choices.

Along with a careful assessment of integrators' project management skills, determine if they rely on any rapid implementation methodology and assess their ability to work with other companies or to work off-site. This can best be done by checking references. This is not because integrators might misrepresent what they do, but this assessment will ensure a thorough understanding of how well the integrator handles such things as project management, cost constraints, and meeting time deadlines. Other reasons for personally checking the references provided by the systems integrator is to learn from the trials and tribulations of others, determine how well the integrator worked in that company's environment, and assess the strengths and weaknesses of the contractor.

In any integration project it is advisable to stay involved and keep IT staff involved. Although the integrator will assume responsibility for the project's outcome, it is often a good idea to assign pieces of the project or oversight responsibilities to various in-house staff. In addition, regular review meetings should be conducted to maintain coordination between the systems integration team and in-house IT staff. By the time the systems integrator leaves, in-house staff should be well versed in the technical details needed to keep things running smoothly.

9.5.7   Facilities engineering

To successfully integrate diverse products from many manufacturers, the systems integrator should be staffed with professionals who understand all aspects of facilities engineering. This includes planning, scheduling, and coordinating the preparation of the site, as well as managing the total implementation plan. These activities ensure that the site meets all environmental, space, power, and security requirements before any equipment arrives.

The systems integrator should be able to configure, stage, integrate, and test entire systems before delivery to the installation site, especially if it is a third-party collocation site operated by a carrier or service provider. This proactive project planning approach defines activities and eliminates potential problem areas early in the systems integration process—when they are easier and less costly to correct.

9.5.8   Carrier services and pricing

The systems integrator should track all issues that may affect its customers and its ability to meet customer needs, including inter- and intraLATA service offerings and pricing. In staying abreast of service offerings and pricing, the systems integrator can show customers how to arrange their networks or redeploy network elements to take better advantage of opportunities that can translate into substantial cost savings. For example, instead of paying for a point-to-point T3 (45-Mbps) line to connect high-speed LANs across town, the integrator should be aware that a metropolitan Ethernet service offering 50 Mbps of bandwidth may cost considerably less.

9.5.9   Telco practices and procedures

Because today's networks typically traverse the serving areas of many telephone companies, the systems integrator must be completely familiar with their differing operations practices and procedures. This includes administrative as well as technical practices and procedures governing the installation of various devices to the local loop and inside plant.

To do this, the systems integrator should track and study the appropriate technical advisories and technical requirements publications, including those governing New Equipment Building Specifications (NEBS), which cover heat dissipation, power consumption, relay rack mounting, and alarm arrangements in central offices. The systems integrator should also be familiar with the Common Language Equipment Identifier (CLEI) coding scheme used by telephone companies.

Ideally, the project management team should include a former central office engineer, or an experienced technician with a telephony background—someone who knows the language of telephone companies and is familiar with their practices and procedures, as well as organizational structure.

9.5.10   Contemporary and future technologies

There is nothing more frustrating than dealing with professionals who do not keep abreast of new developments in technology. Such individuals are quite limited in what they can offer in the way of systems integration services. Make sure that the systems integrator has a formidable knowledge base on the transmission requirements to support a number of key applications such as videoconferencing, email, electronic document interchange, and remote database access. Make sure the systems integrator's expertise is not limited to standard terrestrial copper, but includes microwave, satellite, and fiber as well—if only to consider the alternatives.

It is also important to communicate to the systems integrator that the systems they install today must be able to seamlessly handle more sophisticated and more powerful applications in the future. The systems integrator must be able to leverage current systems, while providing the means to improve productivity, functionality, and ease of use as the company's needs change.

9.5.11   Project pricing considerations

Other factors that should be weighed when selecting among systems integrators are how they price a project-rate structure, maintain and support systems, and guarantee their work. One rule of thumb is to ask the integrator for a breakout of the estimated time to be spent on the project. While the hourly rate for integrator staff may be several times that of full-time employees, this cost may be offset by the urgency of the organization's needs. Do not be fooled into thinking that finding and hiring people at $35 to $60 an hour is better than paying $75 to $200 an hour for an integration project. If there is a tight deadline, the only way to meet it is to hire a ready-made project team from an integrator. To determine the commitment of a systems integrator, find out if there is a money-back policy if the results do not meet expectations, or if the company can withdraw from the project at any stage of implementation without penalty.

Service fees and product costs should be detailed separately. Ask in the proposal that the products be priced separately from the service. Sometimes a company might be better off getting the hardware and software from its own sources. It is really up to the customer to decide whether it wants to look at the project as the complete solution provided by integrators—hardware, software, and services. If it is a complicated project, there may be logistics to worry about. This means that having to configure and get delivery of the right items and get them at the right time can risk holding up the whole project schedule.

9.6   Alternative Arrangements

An alternative to hiring a systems integrator is to share integration responsibilities with a contractor. This can save as much as 30 percent of integration costs. However, projects that are comanaged can be hazardous because they tend to leave too many gray areas regarding expectations, responsibility, and accountability. Other common pitfalls include ill-defined projects, poor communication between the integrator and client, failure to define when the project ends, and weak project sponsorship on the user side. As in a standard outsourcing arrangement, it pays to put the requirements and expectations in writing. The contract must define all deliverables and include milestones to benchmark and evaluate the contractor's performance. Most important, there must be a statement that clearly indicates when the project is over.

Of course, there is always the option of relying on in-house staff for implementing the integration project. The homegrown approach is more apt to succeed when the project is relatively simple and confined. However, the more diversity there is in an organization's information systems and LANs, the more difficult it is to make things work together. Minimizing diversity may not be an option for many companies, especially if the business has been affected by mergers and acquisitions. Because these events are always a possibility, the integration effort should include hardware, software, and development tools that adhere to open system standards whenever possible. This will facilitate future integration efforts.

Assuming these hurdles can be surmounted, the homegrown approach can only be successful if the company is willing to take the time to develop the necessary internal expertise. That way, outside expertise can be relied on only on an exception basis. The advantage of using internally developed expertise is that existing staff members are closer to the end users and closer to the technologies they are using. Sometimes, outsiders may have a more difficult time getting end users involved in defining integration requirements. The integrator's interpersonal skills will determine its success in this endeavor.

9.7   Conclusion

Companies look to integrators for business solutions that improve their competitiveness in particular markets. While some companies have the expertise required to design and install complex systems and connect them over WANs, others are turning to systems and network integrators to oversee the process. Whether the choice is a carrier, computer company, traditional systems house, management consulting firm, or interconnect vendor, it should be an informed choice. This entails finding out the potential advantages and disadvantages of dealing with each type of firm.

The evaluation of various integration firms should reveal a well-organized and staffed infrastructure that is enthusiastic about helping clients reach their business objectives. This includes having the methodologies already in place, the planning tools already available, and the required expertise already on staff. Beyond that, the candidate must be able to show that its resources have been successfully deployed in previous projects of similar nature and scope.

Potential integration firms should be carefully screened for their service orientation. Their proposals should reflect an understanding of the client's industry, competitive situation, corporate culture, information systems environment, and networking needs. Network managers play an important role in helping their company determine how helpful the potential integration firm will be. Will it meet its unique business requirements? Will it provide additional resources to draw upon? Can the integrator be trusted to act in the customer's best interests as a partner?

In evaluating various firms for systems integration services, do not overlook the possibility of using a combination of consultants, hardware providers, contract programmers, and in-house IT or telecom staff with strong project management skills. With the involvement of in-house staff, control over the project can be exercised while the specialized expertise of other participants can be drawn upon. This arrangement also eliminates the possibility that any external participant will fulfill its hidden agenda of developing account control, possibly at the expense of meeting the client's integration needs in the most efficient and economical way possible.

At the end of the project, you, as the network manager, should do an analysis of what was accomplished, and use this opportunity to improve the knowledge and skills of in-house staff. Many integrators bundle knowledge transfer into their services to get clients up to speed quickly as the project winds down. It is up to the customer to take advantage of it.

Help Desk Operations

10.1 Introduction

Today's distributed computing environment, characterized by desktop processing, resource sharing via Local Area Networks (LANs), and global interconnectivity via Wide Area Networks (WANs), requires resources that will satisfy the growing requirement for end-user assistance. The reason is as simple as it is compelling: Ignoring pleas for help not only results in poor returns on technology investments but it can also result in lost staff productivity, which can impede organizational responses to customer needs and competitive pressures.

One way to efficiently and economically service the needs of a growing population of computer and communications users is to set up a help facility that is staffed and equipped to handle a wide variety of end-user problems. This facility is often called the *help desk,* a concept that originated decades ago in the mainframe environment.

Briefly, the help desk acts a central clearinghouse for support issues. Trained staff field problems from end users and attempt to solve them over the phone at a specially equipped workstation before calling in support contractors, carriers, or in-house technicians. Experienced help desk operators can answer up to 80 percent of all calls without having to pass them to another authority. If the problem cannot be solved over the phone, the operator dispatches a technician and monitors progress to a satisfactory conclusion before closing out the transaction.

The in-house help desk can prevent new computer users, who may not have basic computer skills, from damaging files and possibly tying up network resources. Aside from handling trouble calls from users, help desks can play a key role in supporting such services as order and delivery tracking; moves, adds, and changes; and vendor performance monitoring.

Establishing a help desk to coordinate the resolution of computer and communications system problems offers a number of benefits that are realized daily. Users have a single telephone number to remember; support personnel are assured of an orderly, controlled flow of tasks and assignments; and management is provided with an effective means of tracking problems and solutions. Finally, the help desk provides users with a safety net. Knowing that someone is available to solve their problems, or even to help them find their way through unfriendly documentation, gives users confidence so they are more willing to learn new applications.

10.2   Help Desk Functions

Help desk functions can range from simply resolving problems to overseeing an entire Personal Computer (PC) population, data processing system, and network, including inventory and maintenance procedures. Many products are modular, allowing these and other functions to be phased in enterprisewide. In addition, some products let the user customize applications and redesign any screen. Some products have a dynamic indexing feature that tracks lookups and recommends reindexing on frequently used fields. In most cases, forms are designed to be simple so that entry-level and part-time support staff may require little or no training. Many products are Windows-based and are therefore capable of displaying multiple information sources simultaneously. Each window is dynamically resizable, so the integrity of the form is maintained, with all fields and text visible.

While specific support goals vary according to organizational needs, the following general capabilities are provided by most help desk software:

- Call management
- Problem logging and prioritization
- Trouble ticket processing and tracking
- Reference database
- Problem tracking and escalation
- Maintenance history
- Trend analysis
- Management reports

Basic help desk software displays incoming call information at the help desk console. Information may include the caller's ID, equipment ID, time and date of the call, network connection information, number of recent calls, the caller's department, and the caller's location.

Each problem is logged at the time it is called in. Problems can be prioritized according to severity levels. The help desk employs suitable resolution procedures to address each level, from catastrophic to routine. Routine calls are resolved immediately, and all appropriate information regarding the problem solution—including help desk operator, time, and date—is documented. As many as 80 percent of all problems reported to the help desk may fall into the routine category.

Catastrophic problems require more intensive help desk diagnosis and greater technical knowledge to begin the tracking procedures and identify the problem. This may include determining required skill levels and referring the problem to an expert who typically has more experience and access to sophisticated diagnostic tools. Relevant details regarding user or equipment identity, expected resolution time, and type of problem are recorded and tracked.

10.3   Types of Help Desks

There are several types of help desks. The most common type of help desk provides network, system, and application support for internal corporate users. This type of help desk system is used to support telecom and data communications users. In recent years, the role of the help desk has expanded, creating the need for different types of help desk products. These products share the same basic core technology as the traditional help desk, but specialize in the support of particular constituencies.

- *Telecom support.* Telecom support includes the elements of a traditional help desk and is often augmented with integrated voice response systems with a fax-back capability to deliver solutions to problems. This enables the help desk to extend support to a 24×7 schedule. These capabilities are important because 90 percent of the time end users are just asking questions about how to implement certain calling features. An automated response capability can greatly relieve the work burden on help desk staff, which would then be freed up to focus on real problems.

- *Customer service center.* This type of help desk service is designed to support the special needs of customers. It includes call and customer tracking, problem resolution, and workflow management. Call management can be achieved with an Automated Call Distributor (ACD) integrated with the help desk system.

- *Quality control facility.* This help desk is aimed at companies with extensive quality assurance, engineering, and release management organizations. The software tracks and manages product change requests, defects, test cases, and corrective actions. Some products even support the international ISO 9000 quality assurance standards.

- *Sales support.* This help desk automates the complete sales and marketing process. It tracks leads and sales opportunities and sends fulfillments. It can be used to coordinate resources throughout the sales cycle and develop marketing campaigns.

These different support systems can even be integrated into an enterprisewide support system. In this case, a replication system is used to synchronize data across distributed support organizations. This allows multiple organizations and locations to share a consistent and complete view of corporatewide support information, while minimizing the administrative overhead normally associated with distributed systems.

10.4   Help Desk Installation

To ensure maximum effectiveness, network managers should plan the help desk function concurrently with the network design. In this way, the help desk's projected role and user service objectives will reflect overall network goals. A well-documented network plan defines the types of service(s) the network will provide, projected traffic volumes and utilization, and end-user service-level commitments. It will also define the means of achieving the necessary results. Using this information, network managers and planners can establish service-quality standards. This standards process can be further refined to create a help desk plan that:

- *Refines the network's purpose.* Help desk services should reflect and enhance the network's reliability and availability.

- *Expands the network's functions.* Specific network functions and services for end users depend on the domain served by that network. User demands on the help desk reflect the functions it performs using the network. The help desk's role and services should reflect those demands.

- *Tracks network-generated reports.* The help desk should assist users in generating appropriate reports through a variety of means including email and facsimile.

- *Accounts for user types.* Help desk staffing will depend on the number of network users, their applications, and how often they use specific system functions.

10.5   Infrastructure Requirements

There are a number of infrastructure requirements that also deserve attention to ensure the success of the help desk. For example, an easy-to-remember phone number should be selected for the help desk, and it should be listed in the corporate phone directory. Companies with multiple sites spanning different service areas should consider an 800 number for their centralized help operations.

Help desk phones should have labor-saving features such as last number redial, call waiting, conferencing, and message waiting. The operators should be equipped with cordless phones and pagers to give them mobility. And, depending on the call volume and size of the help staff, an Automated Call Distributor (ACD) could prove useful in implementing a menu system, whereby callers select appropriate expertise based on the nature of their problem.

Other options that facilitate problem solving include email, facsimile, and voice mail, which can overcome differences in time zones. Bulletin Board Systems (BBSs) allow users to look up common problems for a recommended solution before calling the help desk operator. Electronic conferencing allows users and help desk staff to collaborate on a problem and converse in real-time via messaging. Voice response systems can be used to provide answers to routine questions, thus freeing up help desk operators to work on more complex problems, or provide basic assistance when the help desk is shut down.

The help desk computer should have special software that tracks the details of calls (call volume, duration of calls, time to answer calls) so that performance statistics can be compiled for analysis with the goal of improving the support operation. There should also be a database where trouble histories can be maintained to expedite future problem solving, to justify hardware and software purchases, and track product failure rates.

10.6   Operation

When a call comes in to a typical help desk, an operator logs the name of the caller and enters the kind of equipment being used and the nature of the problem in the appropriate fields of a call registration screen. This information is automatically logged with a date-time stamp and stored in a database. The database is used by the operator to keep a problem and resolution history. When the same user calls the help desk operator again, profile information on the caller is displayed, facilitating problem resolution.

If the problem is unknown to the operator, a database search is done via a keyword or topic. Some databases are based on expert systems, in which case problem resolution is automated by either decision-tree logic or rules-based technology. An alternative database search-and-retrieval method is case-based reasoning, which takes a "by example" approach to problem solving. This entails help desk operators and higher-level technicians entering actual problems and solutions into the database using free-form English text. This approach eliminates the need for time-consuming and expensive programming typical of expert systems.

If the problem cannot be resolved immediately by the help desk operator or through a database lookup, a trouble ticket is issued and the problem is handed off to a higher level of expertise, such as a technician or network manager. Problem-resolution status is monitored by special tracking software, which issues alerts at specified time intervals until the trouble ticket is closed.

Help desk personnel can solve most problems over the phone with the aid of databases. However, another tool that they can draw on is remote control software that allows them to view the computer screens of callers to determine the source of a problem and take control of their machines to provide a solution. Figure 10.1 illustrates the relationship of remote control software to other response mechanisms and the relationship of these response mechanisms to other help desk functions.

10.6.1   Remote control software

Any user who has ever placed a call to in-house support personnel knows how frustrating it can be to explain a problem over the telephone. Novice users have a particularly hard time determining what information—and how much—to provide. On the other end of the phone line, technical support professionals are often equally handicapped. Unable to see what's happening at users' terminals, they struggle to solve problems blindly.

These problems can be overcome with software products that provide help desk operators with the ability to access remote computer systems, allowing support staff to monitor users' terminals as if they were there in person. If necessary, the support person can even guide the user through the problem by entering appropriate keyboard input. This provides an effective vehicle for quick and efficient troubleshooting and makes it possible for organizations to centralize their end-user support functions. This means technical support people no longer have to run from location to location to fix problems.

Remote control software is most useful to help desk staff that supports LAN users at the applications level. As noted, as many as 80 percent of all trouble calls are applications- rather than hardware-related. Because the help desk operator can solve so many problems remotely, technicians do not have to waste time looking for nonexistent hardware problems.

There are about a dozen remote control application packages currently available that provide bidirectional remote support over a LAN. These products differ in capabilities, features, and pricing. Generally, they offer a set of remote support and diagnostic tools that enable help desk operators and other support people to use their time more productively.

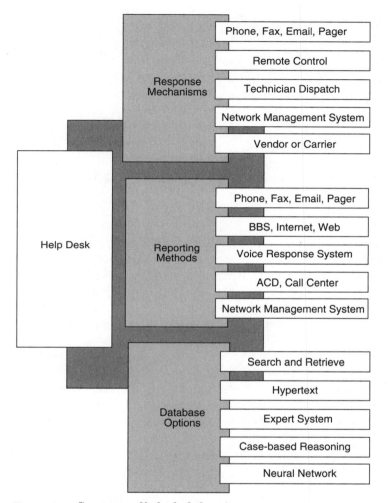

Figure 10.1   Summary of help desk functions.

There are a number of capabilities that are especially useful to help desk operators, such as screen echo. This allows a help desk operator to initiate a session to view the user's workstation screen. By viewing the end-user's screen or witnessing keystrokes as they occur, the help desk operator can often determine the exact cause of the problem.

Another useful capability is remote run, which allows the help desk operator to take over a workstation and operate it remotely from his or her own keyboard, thus locking out the user. This capability is often used when problems cannot be detected simply by watching remote video activity. The help desk operator can even join applications in mid-session.

With an integral messaging capability, help desk operators can compose, send, save, and recall window-type messages from a message library and direct them to any or all workstations. This feature permits important status messages, for example, to be sent quickly to individual users as required, or general-interest messages to be broadcast to all workstations in the local or target cluster, or to workstations at remote nodes.

Help desk operators can communicate interactively with select workstation users in conversation mode, allowing both ends of the connection to converse via keystrokes displayed in separate message windows. The conversation mode is toggled on and off by the help desk operator, who can continue the questioning or instructions to the user as appropriate until the problem is resolved.

Often it is necessary to verify the installation of certain system services at a remote site, such as a print spooler. A partition status function permits the help desk operator to review the memory allocation of the target workstation, without disrupting the user. In addition, this function allows the help desk operator to confirm operating system levels installed, services currently installed in memory, and other vital information to ensure the proper operation of various services.

A reboot workstation function allows the help desk operator to reboot local cluster workstations or remote workstations, and resume the testing of target locations once network communications are restored.

If a workstation user is going to be working on a sensitive task such as payroll processing, he or she can toggle off the remote control software functions to prevent help desk access to that application. Upon completion of the task, the remote control software functions can be toggled back on.

10.6.2   Security

In the hands of unauthorized users, such remote control functions can be misused to wreak havoc on a LAN, as well as invade the privacy of individual workstation operators. To protect the privacy of users and the integrity of each user's data, most remote control software products include security mechanisms.

First, only those capabilities required by the help desk operator can be made available to that operator. Also, individual workstation users can selectively enable or disable any or all target functions that help desk operators can perform on their workstations. This can also be done on a clusterwide basis.

Second, when a help desk operator dials into a remote workstation, cluster, or LAN, the proper node-level password must be entered to gain access.

Third, workstation users are alerted by an audio and visual notification whenever the help desk operator initiates a remote control session. The workstation user can retrieve identification of the node location and, with a single keystroke combination, the Identification (ID) of the user initiating the remote session. If there is no match to an authorized help desk operator, the workstation user can terminate the session with a single keystroke combination.

10.7   Staffing

When staffing the help desk, emphasis should be placed on staff with strong interpersonal communications skills. Technical training can always be used to upgrade competence, but people skills can take a lifetime to develop. As the primary point of contact between the users and the rest of the organization, the support staff must have well-developed people skills so that users feel their problems are being given the attention required. The staff must also be self-motivated and self-directed, since the workload is not set by a schedule but rather by the ringing of the phone or the receipt of email messages.

In addition to being self-directed and possessing strong interpersonal communications skills, the support staff should have intensive, in-depth technical backgrounds. After all, users require that the support person be able to effectively handle their problems, and most hardware and software products have their little quirks and nuances that become understandable only through experience. Although this kind of information can be shared among support people, the need for formal cross-training may be warranted as organizations increasingly move toward multivendor and multi-platform environments.

This brings up the need for another helpful quality: resourcefulness. No matter how hard somebody works at it, no individual on the help desk staff is going to have the answer to every question. This is where a requirement for creativity and networking skills comes into play. Good help desk people get to know who the experts are in the company—who to call when they get stumped. It might be someone else on the information-center staff, the technical-support department of a software company, or even a power user in another department. Knowing where to get the answer may be more important than knowing the answer itself.

Where are such people found? Chances are the right people to staff a help desk are working at the company now. Look for a reasonably mature, energetic person with fair-to-good technical skills. The technical skills do not have to be highly developed at first, but the person should be a fast learner and exhibit a genuine interest in helping others. Since there is always something new to learn, it helps to have someone who is "ego-involved" with his or her job—a person whose value system is such that he or she is incapable of yielding to failure or becoming easily discouraged by the lack of immediate results.

If help desk operators must be recruited from outside the company, you, as a network manager, must pay special attention to the interview process. An effective interview technique is to hold a group session of potential candidates. The interviewer should throw out some technical problems for discussion that are likely to be addressed by the help desk. The interviewer asks everyone to participate and discuss how they would solve the problems, even commenting on the ideas of the other participants. During the session, the interviewer should note the following attitude characteristics of each candidate:

- Willingness to participate in the discussion
- Appearance of negativity displayed to the ideas of others
- Ability to offer helpful comments
- Ability to listen without interrupting others
- Overall attention span

To evaluate a candidate's effectiveness in communicating with a non-technical person on the other end of the help line, have the candidate take a novice computer user through the process of resolving a problem over the phone. Then ask the user to evaluate the experience.

If there is still any doubt about an individual's attitude or technical qualifications, that person should not be hired as a help desk operator. If the candidate pool does not permit such selectivity, a probation period should be used to determine if that person is capable of handling the workload effectively in this potentially fast-paced work environment.

There are several benefits that can accrue to the company as a result of having a well-qualified support group. First, there is the immediate payback on capital investments in computer and communications technologies, since they will be used to optimal advantage with a help desk in place. Second, productivity will also be improved, since outages by users and interruptions of other telecom and IT staff will be minimized. Third, with increasing reliance on information systems and data networks to support mission-critical applications, their problem-free availability to users can improve corporate success.

Assuming the availability of a strong support team, the next challenge for you, as a network manager, will be in motivating high performance. Since staffing a help desk may be viewed as a dead-end job, providing proper incentives is another key in building an effective support group. Once the right people have been hired and trained, the organization is vulnerable to another problem—how to keep them. Turnover is quite high among service and support personnel, usually because these specialized individuals do not have a career path. With nowhere else to go within the organization, they are usually on the lookout for greener pastures elsewhere. This means the organization must be prepared to continually reinvest resources in recruitment and training to fill in any gaps in expertise.

Staff continuity can be a key factor in the success of the help desk. In the effort to keep qualified service and support staff, there are several incentives that may be tried. For example, after a specified time with the company, staff members may qualify for full reimbursement of college tuition, including textbooks and lab supplies. A home notebook computer may be provided with a dial-in capability for remote diagnostics, in turn permitting flexible working hours. In some cases, tax advantages from a home office constitute another potential incentive.

An annual three-day leave to attend a career-related seminar or trade show with all expenses paid might prove to be another viable incentive, as would subscriptions to various technical journals and book clubs. An advisory role in formal corporate working groups would heighten the help desk operator's visibility within the organization and provide a fair amount of ego gratification as well. If the individual has strong instructional skills as well as interpersonal communications skills, these could open up a role in corporate new-hire orientation or in-house training programs.

Any one or a combination of these incentives is more economical than continually replenishing help desk personnel. If these suggestions are not very appealing, it is worth the effort to ascertain what each staff member values most, so that an effective incentive program can be tailored to each person's career needs.

It is not always easy to determine whether a candidate is suitable for the task until his or her work is evaluated under actual stressful conditions. It may be necessary to temporarily place candidates in the help desk environment for evaluation before permanent assignment, reinforcing the notion of career progression.

10.8   Staff Responsibilities

To achieve maximum effectiveness, it is essential to define help desk staff responsibilities, which typically include the following:

- *Being available during the designated time.* Help desk hours vary by network size and coverage, frequency of use, type of business, and its geographical locations. Help desk personnel should be informed of the hours during which the help desk is open and always be available during that time.

- *Answering and recording user calls.* Although other forms of communications should be available, the primary means of communications between users and help desk staff is through the telephone. Help desk staff should be trained in proper telephone manners.

- *Defining and categorizing user problems.* Help desk personnel should be trained in effective questioning skills to accurately categorize and isolate users' problems. Effective problem identification is critical to the help desk's success.

- *Maintaining problem records.* All problem calls should be categorized, recorded, and assigned a trouble ticket number for tracking purposes.

- *Guiding end-user problem solving.* Help desk staff should have the ability to guide users step-by-step in solving problems. The staff should expect users to make mistakes and experience difficulty in resolving problems and should help them extricate themselves from unexpected situations.

- *Updating problem reference lists.* Help desk staff should maintain contact lists of power users, professionals, and vendors who can help correct problems that the help desk cannot immediately solve. Network managers should create procedures to be used when determining which source to call, how to explain the problem, and how to assign and track the responsibility if necessary.

- *Closing a trouble ticket.* When the help desk has solved a problem, the trouble ticket is closed. Help desk staff should call the originator to confirm problem resolution.

Help desk staff should have narrowly defined responsibilities, and these responsibilities should be made known to all department managers and network end users. The following are some of the responsibilities that the help desk should *not* be responsible for:

- *Service and maintenance.* It is not uncommon for users to expect the help desk staff to actually perform maintenance and other service duties. Users as well as the help desk staff should know that maintenance and service are performed by the designated technical staff and not by the help desk.

- *Training.* The help desk is not a full-time training department. While duties performed by the help desk often support user training, users and help desk staff must clearly understand that the facility's primary responsibility is to help users solve their problems. End-user training is the responsibility of individual department managers using internal or external formal training procedures. Help desks should perform remedial and product-specific training as an operational adjunct service, not as the primary training path. In organizations that lack a formal training staff, help desk organizations schedule complete training sessions as systems and networks evolve.

- *Visits to the end user.* Help desk staff should not leave the help desk to gather information directly from users. When visits to user locations are necessary, the person who is best capable of addressing the problem—technician, installer, or trainer—should be dispatched. Today's network support tools include email and remote control support, which expand productivity and minimize the need for personal visits.

10.9 Help Desk Tools

One of the help desk's biggest challenges is determining the network's current status when a user calls with a performance-related problem. Matching the caller's reported symptom(s) with possible causes and interpreting the result are difficult, demanding specific technical problem determination and resolution skills and effective support systems. These tools are absolutely necessary to meet help desk responsibilities in an effective and professional manner and must contain the following capabilities:

- Problem logging
- Problem identification
- Problem isolation
- Problem correlation
- Problem resolution
- Problem record closing

Effective help desk problem tracking and management requires special support tools—not adaptations—that automate help desk functions. Available products and add-ons are broadly classified into the following types:

- *Problem category displays.* These include displays (terminals, workstations, etc.) and dynamic network maps that display data containing current network and system problem determination information. Artificial Intelligence (AI) applications that offer device-specific, multiple-choice scripts that assist help desk staffers to determine problem categories and possible solutions also fall into this group.

- *Dynamic, real-time summary presentation interfaces.* These include terminals or overhead displays that project meaningful information about current system status into critical business areas, reducing redundant problem calls. These devices include information panels that continually track network performance levels and can be customized to actively forewarn both help desk personnel and selected end users about impending network problems.

- *Automation aids that simplify routine help desk tasks.* These tools log routine information about network devices and nodes into help desk automation tools. Information gathered includes terminal and/or caller ID, network location and address, the time/day/date of the call, most recent call from the same source, and other relevant information.

- *Automated inventory collection.* These tools scan servers and workstations to generate hardware and software inventories without having to visit each machine. This information can assist in problem resolution, pinpointing the need for such things as more Random Access Memory (RAM) or disk space, or the need to upgrade software or install a patch.

- *Multisession windowing utilities.* These facilities let help desk personnel navigate from session to session without exiting one session and logging on to another. These navigation tools enable help desk staff to duplicate the caller's on-screen procedure and mimic the session parameters in an effort to pinpoint the cause of the error.

- *Remote control software.* These facilities allow help desk staff to access a user's workstation and actually take control of it to identify and correct problems.

- *Statistical evaluation programs.* These facilities generate various customized reports about help desk activities that allow multiple views of data, stimulate alternative thinking, and help uncover latent problems.

- *Forms-based problem reporting via email.* These tools allow users to report problems via email. An alternative to busy phone lines, help requests sent via email are automatically logged and responded to based on severity level. If appropriate, the response can be delivered via email as well.

- *Prepackaged content modules to speed development of knowledge databases.* These facilities address specific problem-solution domains such as operating systems and particular software applications, allowing overextended help desk staff to get a head start in compiling knowledge databases that aid in problem resolution.

- *Workflow engines for information distribution.* These facilities couple document creation, routing, and distribution processes to communicate support information to all help desk operators, customer support staff, field personnel, or remote self-support users, enabling all to benefit immediately from new information.

In addition to generating customized reports and logging a variety of information, many help desk systems automate such routine tasks as network inventory management, software distribution and installation, and maintenance scheduling.

10.10   Role of Expert Systems

Communications managers are starting to appreciate the synergy between help desk environments and expert support systems. With expert systems, organizations can distill the essence of what application and communication experts know, encode it, store it in a database, and make it accessible when needed. This way, whenever the resident expert is unavailable, that accumulated knowledge and expertise can still be applied, translating users' problem descriptions into solutions for others to use.

When an expert system is installed, experts representing all technical areas typically take turns working and expanding the help desk knowledge base, constructing an information repository or special database that can then be accessed and operated by nonexperts. Fields for problem resolution actions, physical and logical network objects, and technical topics let the expert pick out and enter appropriate solutions. Problem answers are captured in a transaction file.

At the end of each week, an expert edits the activity file and deletes or modifies any inappropriate or incorrect answers, then routes the usable information to the main database repository. Once this answer base is complete, other nonexpert help desk personnel can use it to guide callers through diagnostic sequences beginning with a simple description of the problem. Each problem-symptom set is progressively narrowed until a solution is reached.

Expert systems let help desk implementers automate several problem determination or procedural functions and interpret incomplete information. These knowledge databases contain the heuristic reasoning chains that experts often employ to solve problems. Unlike traditional databases, which store passive data and facts, knowledge bases and inference engines make decisions using active data and dynamic information under variable conditions. Automating the help desk function in this way vastly improves quality of service while dramatically decreasing service costs. In addition, it releases experts to work in other areas, provides consistent answers to questions, and improves the help desk staff's general credibility. Some expert systems use the database to generate graphics and textual reports on what types of hardware and software cause the most problems. The organization can then use this information to guide future purchase decisions or plan system modifications.

Knowledge-based expert systems and technologies encompass Intelligent Text Retrieval (ITR) and hypertext, case-based reasoning, rule-based reasoning, and neural networks. Each logs and records problem information and advises and assists help desk operators. These systems store complex data and relationships, allowing relatively unskilled help desk staff to solve problems far above their knowledge level.

Entering information manually to populate knowledge databases can be enormously time-consuming, particularly for an already overextended help desk staff. There is now a growing market for prepackaged information that can be readily integrated into help desk systems. The data generally is formatted as raw text, in a database structure customized for each specific help desk application. While these products will not resolve all problems submitted to the help desk, they can provide the basis for getting started quickly.

10.10.1   Intelligent Text Retrieval

Keyword search and text retrieval systems are extremely useful support tools for most help desks. Replacing many of the dry manuals with fast data retrieval methods—usually from Compact Disk-Read Only Memory (CD-ROM) storage—allows help desk operators to answer extremely specific end-user questions. Some ITRs even contain pictorial views of devices, which are linked to expanded descriptions and operating procedures to help operators gain additional insight into supporting and resolving problems.

10.10.2   Case Based Reasoning

Case Based Reasoning (CBR) makes use of a collection of previously solved problems and associated symptoms, indexed and stored by case in a relational database. CBR uses two basic elements: *cases,* often called the *knowledge base,* and an *inference engine* containing the logic to extract solutions. Users enter free-form problem descriptions and the CBR system searches its case-base to find nearly identical situations and the associated or recommended solutions. Using CBR systems is fairly easy, but designing and building them are difficult and tedious. Every known case and solution must be entered into the system before meaningful results can be expected. CBR gives help desk staff the ability necessary to manage complex knowledge tasks in order to provide effective results.

Every new case enhances the derived information in the knowledge base and increases the probability of correct diagnosis. One strong benefit associated with case systems is their simple way of adding additional cases and building and expanding existing information. Nonexperts can simply create new case scenarios as the networks and applications change; and every solved problem becomes another case, automatically extending the knowledge base.

10.10.3 Rule Based Expert Systems

Rule Based Expert Systems (RBES) are procedural systems that analyze both data and relationships between items. They store rules sets for accomplishing a certain task or procedure, for example, how to reset a system or evaluate particular alarms. These systems, like content-based reasoning systems, use inference engines—actually software logic—but in a different manner. RBES procedures are more similar to programming than writing procedure manuals. For this reason, an RBES is considered more difficult to implement. Updating the rules base is a completely manual process, creating another problem. These systems also require expert programming support as the knowledge base expands, making them less responsive to rapid system and network changes. This is a major limitation of the fast-paced environment of many help desks.

10.10.4 Neural networks

The newest technology applied to help desks is *neural networks,* also called *cognitive processing.* This technology combines mathematics, computer science, fuzzy logic, and neuroscience, as well as conventional text parsing, to integrate new information into a help system without specifically having to program it. Neural nets use pattern recognition logic to re-create the human deductive process using data element layers to emulate the brain's neural construction. Each element layer connects to higher elements using weighted values that indicate the connection strength. By numerically or logically summing these values through the neural network, the most likely solution is reached, often with reasonable alternative choices. While such products are relatively expensive and more difficult to manage, neural networks represent one of the most promising tools for help desk support environments.

10.11 Delivering Support via the Internet

The traditionally centralized help desk is giving way to a more distributed approach, in keeping with the trend toward increasingly downsized and decentralized corporate operations, with its emphasis on client-server and remote access technologies. Accordingly, alternative ways of delivering help desk support are commonly implemented. For example, many companies once provided bulletin boards that enabled remote users and customers to dial in to databases that provided answers to common problems. The big advantage of bulletin boards was that they were available 24 hours a day. These have been supplanted in recent years by Web-based systems on the Internet, which are easier to administer.

A user at a branch office that is not connected to the corporate backbone network, for example, can dial in to the company's Web server and fill out a standard help request form. At the push of a button, the completed form is sent via email to the help desk, where it is logged and responded to according to the reported severity of the problem.

Users can perform a range of actions such as authorizing a change request and adding new information to an existing trouble ticket. Help desk applications and data generate Web forms (schemas) and hyperlinks dynamically (see Fig. 10.2). This frees up help desk managers to focus on serving customers and improving business processes, instead of constantly maintaining static Web pages.

Help desk vendors, too, are recognizing the potential of economical and ubiquitous Internet connections for augmenting help desk operations and offer add-on modules that integrate the Web with their products. Such modules permit remote users to use browsers to submit a trouble ticket and periodically check its resolution status. Via the Web, users can also have access to the help desk database or third-party databases to find solutions to problems on their own.

Organizations can also use such software to provide customer support via the Web. Some products offer the means to create a password-protected home page, containing customizable incident logging forms with direct links to a customer support center. Customers can access the forms through the support center via the Web and log incidents directly. Aspects of the forms are customizable, including screen color, fonts, fields, and workflow rules. For companies that must provide customer support worldwide, some products allow the creation of different forms in different languages or the creation of a single multilingual form.

The Web is being used to deliver internal network support as well. Routers, switches, and remote-access servers can be configured remotely over the Web. Such network devices contain an integral ROM-based home page to display and configure network device settings, enabling users to view and interact with the devices using a Web browser. Using hypertext links for quick movement between management functions and on-line resources, a remote user can get immediate answers to setup or troubleshooting questions. Different views and access privileges can even be created that vary by user login and password.

Leveraging existing Internet connections and exploiting the forms-handling capabilities of the Web to extend help desk and customer support functions worldwide frees up staff from handling time-consuming telephone calls

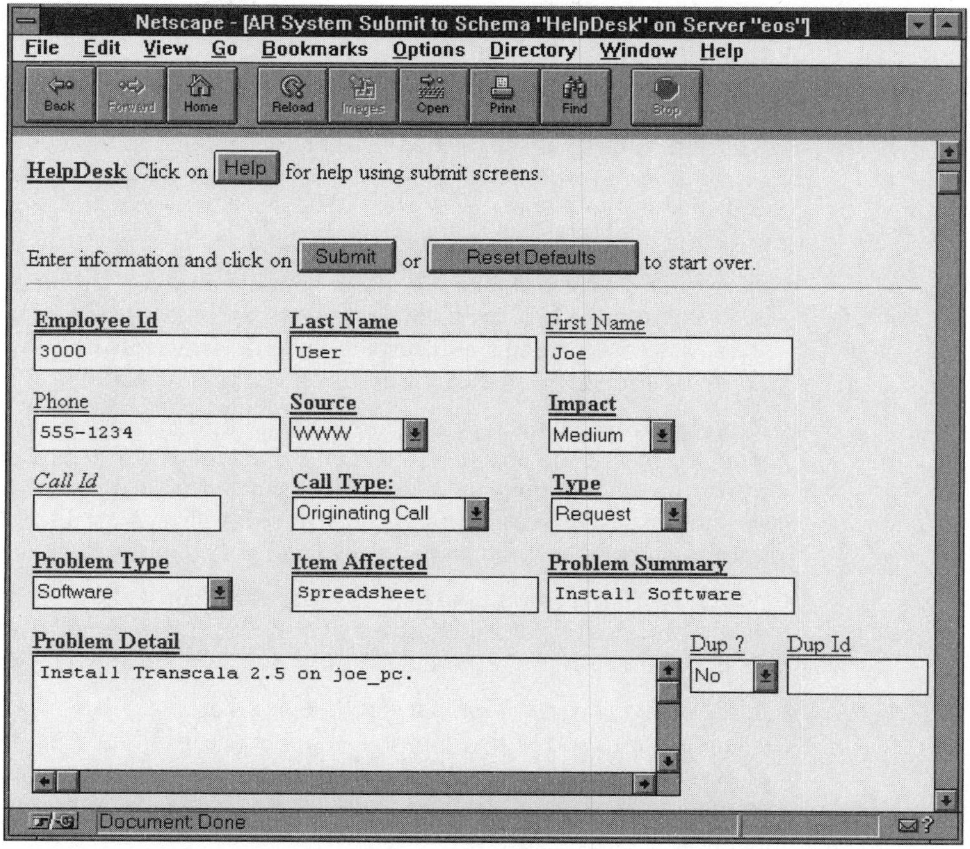

Figure 10.2 Remedy Corporation's Action Request System has extensions to the Web. The company's ARWeb client lets organizations create a Web-based help desk that can be accessed by anyone with a Web browser. In this case, a user can access a Web form for reporting a problem to the help desk.

and can shield them from some of the abrasiveness inherent in verbal exchanges. Delivering support electronically can also lower stress levels among support staff.

10.12   Outsourcing the Help Desk

Many computer vendors offer help desk services for the distributed computing environment. Such services go beyond the subscriber's network to monitor and control the individual endpoints—the workstations, servers, and hubs. Using remote control software on a dial-up or dedicated link, workstations and servers are monitored periodically for such things as errors, operational anomalies, and resource usage. Upon receiving reports of these or other events, the vendor's technical staff is supposed to respond within the contract-specified time frame to resolve the problem.

The specific services provided by the vendor may include one or more of the following:

- *Local event management.* This service accepts events created by application and systems software and filters these events according to user-defined criteria.

- *Integrated event monitoring.* This service enables the vendor's centralized technical staff to monitor events received from remote managed sites. Each technician can have a specific view into the event database, allowing for different functional areas of responsibility.

- *Remote access.* This service enables the vendor's technical and management center staff to view or control any remote workstation or server as well as any memory partition on a workstation or server. The target screen is viewed in a window on the technician's monitor and the commands work as if they were input locally.

- *Software distribution.* This service provides controlled electronic distribution and installation of system and application software at remote sites. License management also may be included in this type of service.

- *Asset management.* This service keeps a complete inventory of the subscriber's hardware and software. This is accomplished via periodic scans of workstation and server configuration files. Move, add, and change requests are tracked as well to keep the inventory current.

10.13 NMS-Integrated Help Desks

One of the newest trends is to integrate help desk and trouble-ticketing functions within a Network Management System (NMS). Such integration allows you, as a network manager, to coordinate all activities on the network, from problem detection to resolution, on a single computer screen.

When an alarm is generated by the NMS that can alert you to the status of any device on the network, a device identification code is passed to an Element Management System (EMS), which locates and displays the physical connectivity. After the information is collected by the NMS and EMS, the help desk software produces a trouble ticket listing the symptoms of the problem and its probable causes. Recommended actions are also detailed on the trouble ticket.

In addition to inventory information, a circuit trace can be added to the trouble ticket, which provides a map of all the connections to an ailing LAN segment. The trouble ticket also facilitates the logging of user complaints, automatically prioritizes complaints that have not been resolved, and keeps a history of recurrent problems.

This comprehensive management solution improves the information-gathering process used by network managers to make critical decisions, eliminating duplication of effort and allowing different components of the network to be managed as a single enterprise. The solution also allows for the integration of problem history, planning, and resource management functions using one platform. This, in turn, improves the user's ability to keep vital network devices efficiently up and running.

10.14 Conclusion

Today's help desks are designed to facilitate the reporting and resolution of end-user problems and, as such, assume a key role in keeping corporate computer systems and networks operating smoothly with minimal disruptions.

Although a help desk costs money, it can pay for itself in many ways that, unfortunately, can be hard to quantify. The fact is most companies have millions of dollars invested in computer and communications systems that support complex applications. They also have millions of dollars invested in people whose productivity depends on the proper functioning of these assets.

To ensure that both people and technology are utilized to optimal advantage, there must be an entity in place that is capable of solving the many and varied problems that inevitably arise. For most companies that entity is the help desk, which can ease the burden of managing information processing resources by eliminating many of the routine problems that occur on a daily basis.

Network Integration

11.1 Introduction

Distinct from systems integration, which focuses on getting different computer systems to communicate with each other at the applications level, *network integration* is concerned with getting diverse, far-flung Local Area Networks (LANs), client-server systems, and host-centric data centers interconnected over the Wide Area Network (WAN), using various carrier facilities and services. Typically, network integration requires that attention be given to a plethora of different physical interfaces in addition to protocols, frame sizes, data transmission rates, traffic types, quality of service issues, and security. Companies also face a bewildering array of carrier facilities and services to extend the reach of information systems globally over high-speed WANs. To get everything working properly requires careful hardware and service selection, software customization, and applications tweaking.

Data communications was once a relatively simple function, conducted in a host-centric environment, primarily under the auspices of IBM's Systems Network Architecture (SNA). After more than 25 years, SNA is still a stable and highly reliable architecture. It has undergone a few enhancements, including Advanced Peer-to-Peer Networking (APPN), a routing scheme for client-server computing in multiprotocol environments that links devices without requiring the use of a mainframe. Despite the widespread adoption of distributed computing, including client-server architectures, the mainframe is still valued for its ability to handle mission-critical applications. Additional advantages include accounting, security, and management tools, which enable the organization to closely monitor all aspects of performance and to contain costs. The sheer financial investment in legacy SNA hardware and software—estimated to exceed several trillion dollars worldwide—provides ample incentive to protect and leverage these assets within the distributed computing environment of LANs and WANs.

Convinced of the many benefits of distributed computing over LANs and WANs, companies in every segment of the economy are faced with the daunting task of integrating diverse systems, applications, and even networks over a smaller set of carrier facilities and services. As enterprise networks continue to grow and expand to include mobile professionals, telecommuters, small branch offices, and far-flung international locations, the need to interconnect LANs, information systems, and data centers becomes even more urgent. Among the key challenges facing network managers today are how to tie together incompatible LANs, meld legacy systems and LANs, and consolidate multiprotocol traffic over a single WAN backbone. Furthermore, all this must be done at minimum cost and without inflicting performance penalties on end users.

11.2   Internetworking Devices

Traditionally, three types of intelligent devices have been used for internetworking: bridges, routers, and gateways. Each operates at different layers of the Open Systems Interconnection (OSI) reference model, which dictates the level of functionality that each device provides (see Fig. 11.1).

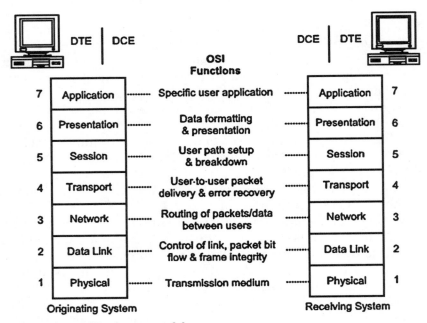

Figure 11.1   OSI reference model.

11.2.1   Bridges

At OSI Layer 2, *bridges* can be used to create an extended network that greatly expands the number of devices and services available to each user. For example, bridges provide connectivity to LANs at remote locations via Permanent Virtual Circuits (PVCs) provisioned with frame relay or Asynchronous Transfer Mode (ATM) services. LANs can also be extended to other locations through carrier-provided Ethernet services, which are now available in most major cities.

Bridges can also be used for segmenting large LANs into smaller subnets to improve performance, control access, and facilitate fault isolation. The bridge does this by monitoring all traffic on the subnets that it links. It reads both the source and destination addresses of all the packets sent through it. If the bridge encounters a source address that is not already contained in its address table, it assumes that a new device has been added to the local network. The bridge then adds the new address to its address table.

The bridge isolates traffic by examining the destination address of each packet. If the destination address matches any of the source addresses in its table, the packet is not allowed to pass over the bridge because the traffic is local. If the destination address does not match any of the source addresses in the table, the packet is allowed to pass on to the adjacent network. This process is repeated at each bridge on the internetwork until the packet eventually reaches its destination. Not only does this process prevent unnecessary traffic from leaking onto the internetwork, it acts as a simple security mechanism that can screen unauthorized packets from accessing other resources.

In examining all packets for their source and destination addresses, bridges build a table containing all local addresses. The table is updated as new packets are encountered and as addresses that have not been used for a specified period of time are deleted. This self-learning capability permits bridges to keep up with changes on the network without requiring that their tables be manually updated.

Bridges can also be used to interconnect LANs that use different media, such as twisted-pair, coaxial and fiber-optic cabling, and microwave. In office environments that use wireless communications technologies such as spread spectrum and infrared, bridges can function as an access point to wired LANs. On the WAN, bridges even switch traffic to a secondary port if the primary port fails. For example, a full-time wireless bridging system can establish a modem connection on the public network if the radio link is lost due to environmental interference (see Fig. 11.2).

The routing capabilities of bridges are generally very limited. For example, source routing is a bridging method originally developed by IBM for

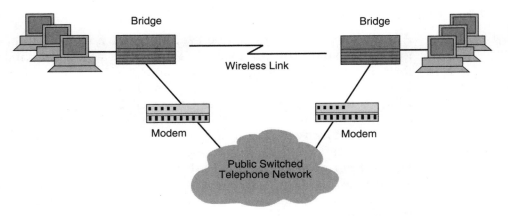

Figure 11.2 Full-time wireless bridging system with optional dial-up backup capability.

interconnecting its token ring networks. This method relies on information contained within the token to route information between LANs. Each bridge on the network receives a route-discovery packet from the source bridge. The devices append path information and return the packet to the source. From the appended information, the source device determines the most appropriate path for subsequent packets. However, source routing does not always choose the best routes through the WAN because line use varies at any given time. Furthermore, source routing has no effective way to route traffic around congested links or to redistribute traffic over multiple paths.

There are some techniques that have been developed over the years for overcoming this and other limitations of source routing, which are discussed later in this chapter.

11.2.2 Routers

Routers operate at the network layer of the OSI reference model (Layer 3). While bridge functionality is employed on frame relay, ATM, and Ethernet networks, router functionality comes into play on higher-level networks, such as Internet Protocol (IP). Routers are used to connect networks that use different protocols at both the data link and physical layers, while bridges are used to connect networks with the same protocols at the data link and physical layers. Today's routers include integral bridges, allowing the same device to handle both functions simultaneously.

The network layer provides the information required to switch and route data to its intended destination. This allows a router to offer more advanced and more complex services than a bridge. For example, a router actively selects the path between the source and destination nodes, basing its selection on such factors as distance (measured in terms of hop count), network congestion, and transit delay. The use of routing protocols permits a dynamic exchange of information between all routers on the network. Because the routing tables of each router can communicate with each other and share information about the network, they reflect the same network landscape.

There are two types of routing protocols: distance vector and link state. *Distance vector protocols* issue periodic broadcasts that propagate routing tables across the network. Such protocols, one of which is the Routing Information Protocol (RIP), are adequate for small, stable networks, but not for large, growing networks where the periodic broadcast of entire tables can consume an inordinate amount of network bandwidth. *Link state protocols,* such as Open Shortest Path First (OSPF), operate more efficiently. Instead of issuing periodic broadcasts of entire tables, they send routing information on a flash basis to reflect only the changes to network connections.

The decision to bridge or route is based on these and several other considerations. Table 11.1 summarizes the issues that merit consideration.

As noted, there are devices that combine the functionality of bridges and routers. These devices handle the packets in an appropriate fashion at the time they go out over the internetwork. For example, legacy protocols such as Digital Equipment Corporation's (DEC) Local Area Transport (LAT) and Local Area VAX Cluster (LAVC) and IBM's Network Basic Input/Output System (NetBIOS) protocols have no network layer and, consequently, no network address. Since these protocols cannot normally be routed, they must be bridged. But even these protocols can be routed with techniques that entail encapsulation into Transmission Control Protocol/Internet Protocol (TCP/IP). These techniques include IBM's Data Link Switching (DLSw) and the RFC 1490 standard, which are discussed later.

11.2.3   Gateways

Gateways are used to interconnect dissimilar networks or applications. Gateways operate at the highest layer of the OSI reference model—the application layer (see Fig. 11.3). A gateway consists of protocol conversion software that usually resides in a server, switch, mainframe, or front-end device. One application of gateways is to interconnect disparate networks or media by processing the various protocols used by each so that information from the sender is intelligible to the receiver, despite differences in network protocols or computing platforms.

For example, when an SNA gateway is used to connect asynchronous Personal Computers (PCs) to a synchronous IBM SNA mainframe, the gateway acts as both a conduit through which the computers communicate

TABLE 11.1   Relative Merits of Bridges and Routers

Properties	Bridges	Routers
Reliability	Operate at the data link or MAC layer. Protocols provide some error detection, but do not guarantee message delivery.	Operate at the network layer Some protocols guarantee message delivery.
Network availability	Most bridges are not tolerant of network failure: They cannot route around failed links or points of congestion. Such conditions not only diminish bridge performance, but can also result in lost messages.	Routers are highly tolerant of network failure. They operate within a highly meshed WAN that provides multiple paths between message source and destination.
Transit delay	Since bridges perform little processing, they generally introduce minimum transit delays But the arrival of packets to a bridge at a rate that exceeds its processing rate can congest links and/or bridges, which can result in packet loss.	Since routers perform more sophisticated network layer processing, they can introduce some transit delay. However, such delay can be offset by the availability of multiple routes through the internetwork.
Error detection	Bridges perform data link error checking.	Routers perform error checking at both the data link and network layers.
Frame size	Bridges operate best when source and destination networks support identical packet sizes.	Routers fragment large packets and perform reassembly to mediate network differences.
Security	Bridges provide rudimentary security, such as destination address checking, to keep unauthorized packets from entering an adjacent network.	Routers offer greater protection against unauthorized access to data and resources. In addition to packet filtering, access to the internetwork can be controlled by protocol filtering. Other firewall capabilities can be added to routers as well.
Cost	Bridges are less expensive than routers, but the savings may be offset by the inefficient use of available bandwidth.	Routers are more expensive than bridges. The cost differential is influenced by such factors as the number of ports and protocols supported, as well as advanced features such as quality of service mechanisms and security capabilities.

and as a translator between the various protocol layers. The translation process consumes considerable processing power, resulting in relatively slow throughput rates when compared with other interconnection methods—hundreds of packets per second for a gateway versus tens of thousands of packets per second for a bridge. Consequently, the gateway may constitute a potential bottleneck when utilized frequently, unless the network is optimized for that possibility.

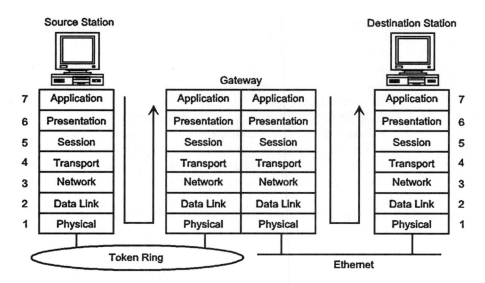

Figure 11.3   Gateway functionality in reference to the OSI model.

In addition to its translation capabilities, a gateway can check on the various protocols being used, ensuring that there is enough protocol processing power available for any given application. It also can ensure that the network links maintain a level of reliability for handling applications in conformance to user-defined error rate thresholds.

Gateways have a variety of applications. In addition to facilitating LAN workstation connections to various host environments, such as IBM SNA 3270 systems and IBM midrange systems, they facilitate connections to X.25 packet-switching networks. Other applications of gateways include the interconnection of various email systems, enabling mail to be exchanged between normally incompatible formats.

In some cases, gateways can be used to consolidate hardware and software. An SNA 3270 gateway shared among multiple networked PCs can be used in place of IBM's 3270 Information Display System or many individual 3270 emulation products. Although the IBM system is a standard means of achieving the PC-host connection, it is expensive when used to attach a large number of standalone PCs. The relatively high connection cost per PC discourages host access for occasional users and limits the central control of information.

If the PCs are on a LAN, however, one gateway can emulate a cluster controller and thereby provide all workstations with host access at a very low cost. Cluster controller emulators use an RS-232C or compatible serial interface to a host adapter or communications controller, such as an IBM 3720 or 3745. One gateway can emulate an entire cluster of terminals and support up to 254 simultaneous sessions

A relatively new type of gateway provides connections between the Internet and the Public Switched Telephone Network (PSTN), enabling users to place phone calls from their multimedia PCs or conventional telephones over the Internet or a carrier's managed IP data network and vice versa. This arrangement allows users to save on long-distance and international call charges.

The IP-PSTN gateways perform the translations between the two types of networks. When a standard voice call is received at a near-end gateway, the analog voice signal is digitized, compressed, and packetized for transmission over an IP network. At the far-end gateway the process is reversed. The packets are decompressed and returned to their original digital form for delivery to the nearest Class 5 central office and then to the local subscriber.

The gateways support one or more of the internationally recognized G.7xx voice codec specifications for toll-quality voice. The most commonly supported codec specifications are as follows:

- *G.711.* Describes the requirements for a codec using Pulse Code Modulation (PCM) of voice frequencies to achieve 64 kbps, providing toll-quality voice on managed IP networks with sufficient available bandwidth.

- *G.723.1.* Describes the requirements for a dual-rate speech codec for multimedia communications (e.g., videoconferencing) transmitting at 5.3 and 6.3 kbps. This codec provides near toll-quality voice on managed IP networks.*

- *G.729A.* Describes the requirements for a low-complexity codec that transmits digitized and compressed voice at 8 kbps. This codec provides toll-quality voice on managed IP networks.

The specific codec to be used is negotiated on a call-by-call basis between the gateways using the H.245 control protocol. Among other things, the H.245 protocol provides for capability exchange, enabling the gateways to implement the same codec at the time the call is placed. The gateways may be configured to implement a specific codec at the time the call is established, based on predefined criteria, such as the use of the:

- *G.711 only,* in which case the G.711 codec will be used for all calls.

- *G.729 (A) only,* in which case the G.729 (A) codec will be used for all calls.

- *Highest common bit rate codec,* in which case the codec that will provide the best voice quality is selected.

---

* The Mean Opinion Score (MOS) used to rate the quality of speech codecs measures toll-quality voice as having a top score of 4.0. With G.723.1, voice quality is rated at 3.98, which is only 2 percent less than that of an analog telephone.

- *Lowest common bit rate codec,* in which case the codec that will provide the lowest packet bandwidth requirement is selected.

This capability exchange feature provides carriers and Internet Service Providers (ISPs) with the flexibility to offer different quality voice services at different price points. It also allows corporate customers to specify a preferred proprietary codec to support voice or a voice-enabled application through an intranet or IP-based Virtual Private Network (VPN).

11.3   Methods of Integration

Methods used to integrate different types of LANs and meld legacy systems with LANs include translation, encapsulation, segmentation, emulation, and speed matching. An understanding of these methods is necessary for successfully implementing any network integration effort.

11.3.1   Translation

One way to integrate different types of LANs is through *translation.* Since the packet structures of Ethernet, token ring, and Fiber Distributed Data Interface (FDDI) LANs are fairly similar (see Fig. 11.4), differing only in terms of length, translation is a fairly straightforward process. A special kind of bridge, called a *translating bridge,* reads the data link layer destination addresses of all messages transmitted by Ethernet devices, for example. If the destination address does not match any of the source addresses in its table, the packet is allowed to pass on to an adjacent token ring.

Using a translating bridge for this purpose has a serious drawback: Such devices cannot fragment packets. A token ring or FDDI packet of 4500 bytes cannot be placed on an Ethernet LAN, which is limited to supporting packets of no longer than 1528 bytes, including overhead. To make this scheme work, the token ring or FDDI devices must be configured to transmit packets at the Ethernet packet length. However, it is not very practical to ratchet down the performance of high-speed LANs just to accommodate traffic to a slower LAN.

11.3.2   Encapsulation

Encapsulation is the process of putting one type of data frame into another type of data frame so that the appropriate receiving device will recognize it (see Fig. 11.5). It is a process that allows different devices using multiple protocols to share the same network. Encapsulation entails adding information to the beginning and end of the data unit to be transmitted. The added information is used to perform the following tasks:

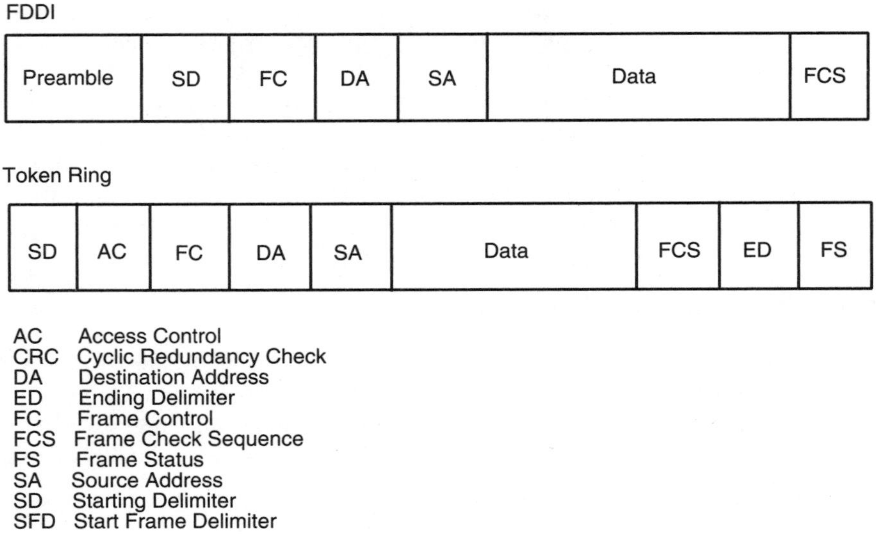

FDDI

Preamble	SD	FC	DA	SA	Data	FCS

Token Ring

SD	AC	FC	DA	SA	Data	FCS	ED	FS

AC   Access Control
CRC   Cyclic Redundancy Check
DA   Destination Address
ED   Ending Delimiter
FC   Frame Control
FCS   Frame Check Sequence
FS   Frame Status
SA   Source Address
SD   Starting Delimiter
SFD   Start Frame Delimiter

Figure 11.4   Comparison of packets by LAN type.

- Synchronize the receiving station with the signal

- Indicate the start and end of the frame

- Identify the addresses of sending and receiving stations

- Detect transmission errors

At the destination device, this envelope is stripped away and the original frame is delivered to the appropriate end user in the usual manner. There is significant overhead with this solution because one complete protocol runs inside another. But any increase in transport capacity requirements and associated costs is usually offset by savings from eliminating the need for separate networks.

11.3.3   Segmentation

Segmentation is the process of breaking down large packets into smaller packets for transmission over the network. At the receiving end, the smaller packets are reassembled. Segmentation and Reassembly (SAR) is used in ATM networks and is a function handled at the ATM Adaptation Layer (AAL). Among other things, this layer is responsible for segmenting the offered information into 53-byte cells and at the receiving end, reassembling it back into its native format.

Because ATM has the ability to accept any type of traffic and reduce it to a series of 53-byte cells for transmission through the network positions,

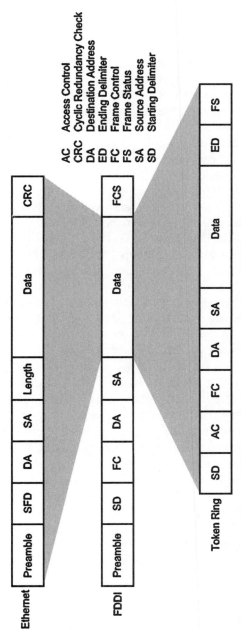

Figure 11.5 Encapsulation of Ethernet and token ring packets into an FDDI packet.

AC	Access Control
CRC	Cyclic Redundancy Check
DA	Destination Address
ED	Ending Delimiter
FC	Frame Control
FS	Frame Status
SA	Source Address
SD	Starting Delimiter

it is an integration solution for any company that maintains separate networks for voice, video, and data. The reason for separate networks is to provide appropriate bandwidth and preserve performance standards for the different applications. But ATM can eliminate the need for separate networks and provide a unified platform for multiservice networking that meets the bandwidth and Quality-of-Service (QoS) needs of all applications. Various categories of performance have been developed to help characterize network traffic, each of which has its own QoS requirements.

Constant Bit Rate (CBR), for example, is intended for applications where the virtual circuit requires special network timing requirements (e.g., strict cell loss, cell delay, and cell delay variation performance). CBR would be used for applications requiring circuit emulation (e.g., a continuously operating logical channel) at transmission speeds comparable to DS1 and DS3. Such applications would include PBX trunking where delays in voice transmission cannot be tolerated.

Variable Bit Rate-real time (VBR-rt) is intended for applications where the virtual channel requires low cell delay variation. For example, VBR-rt would be used for applications such as variable bit rate video compression, and packet voice and video, which are somewhat tolerant of delay. Variable Bit Rate-near real time (VBR-nrt) is intended for applications where the virtual circuit cannot tolerate larger cell delay variations than VBR-rt. For example, VBR-nrt would be used for applications such as data file transfers.

Available Bit Rate (ABR) is intended for routine applications and when the customer seeks a low-cost method of transporting bursty data for non-critical applications that can tolerate delay variations. The application that offers traffic to the network first gets to use the available bandwidth, while other applications must wait until the bandwidth becomes free. If congestion builds up in the network, ABR traffic is held back to help relieve the congestion condition.

Unspecified Bit Rate (UBR) is intended for nonessential traffic, such as news and network monitoring, which is delivered on a best-effort basis. Although the carrier will attempt to deliver all ATM cells received over the virtual channel, there is no guarantee of delivery for UBR traffic. If there is any network congestion, this may result in loss of ATM cells to relieve congestion in the network.

11.3.4   Emulation

Emulation has traditionally been associated with terminal emulation, a method of allowing PCs to mimic 3270 terminals for communication with hosts in the SNA environment (discussed in Sec. 11.4). There is also LAN Emulation (LANE), a more recent technology that is used to integrate Ethernet or token ring LANs into ATM networks. LANE provides the backbone infrastructure that allows existing Ethernet- and token ring–attached stations to keep their current application interfaces,while allowing high bandwidth servers and routers to be connected directly via ATM.

The addresses used in ATM networks are not the Media Access Control (MAC) addresses that underlie IEEE 802.3–compatible protocols. To allow ATM-to-LAN communication, MAC addresses must be mapped to ATM addresses. This is done by the LANE software. For the ATM-to-LAN link, the sending device must ask the LANE server for the address of the receiving node. The ATM sender then uses that address to contact the remote LAN device.

By hiding the use of ATM from applications, LAN emulation allows legacy applications to operate unchanged over high-speed ATM networks, permitting companies to preserve their existing investments. On the other hand, because these existing applications do not understand such things as class-of-service contracts and other stipulations of ATM, they will not gain all the benefits of ATM. Eventually, the use of LAN emulation will fade away, with the development of new "ATM-aware" applications. For now, LAN emulation is an important capability that can help justify the migration to ATM.

11.3.5   Speed matching

Integrating 10-Mbps Ethernet with 100-Mbps Fast Ethernet is much easier. This is because Fast Ethernet leaves intact legacy Ethernet's existing MAC layer, which uses Carrier-Sense Multiple Access with Collision Detection (CSMA/CD), and adds another layer to support 100-Mbps networking. This allows for a very economical migration path to higher bandwidth. Furthermore, 100-Mbps devices can share common cabling with 10-Mbps devices, and bridging the two involves a relatively simple speed-matching function.

Ethernet has proved to be highly scalable in terms of bandwidth. Gigabit Ethernet at 1000 Mbps is becoming popular and uses the same Category 5 cabling as 10/100 Mbps Ethernet, enabling organizations to preserve their original investment in existing cable infrastructure. To ease the integration of multispeed networks, vendors support 10/100 Mbps from the same network interface card, and some support 10/100/1000 Mbps from the same card.

11.4   Terminal Emulation

One of the simplest (and oldest) ways for PCs to access SNA and other host applications is through terminal emulation. For a PC to communicate with a proprietary host system in the data center, it must be made to do something it was not designed to do—emulate a terminal so that it can be recognized as such by the mainframe. In the IBM environment, 3270 terminal emulation is used, which permits synchronous data transfer between PCs and the mainframe. With 3270 terminal emulation, data is exchanged in a format that is readily acceptable to the host. Basic emulation software may come packaged with modem software, or a more feature-rich version may be purchased separately that operates over LAN, TCP/IP, and leased-line connections. The software can reside on a server or on each PC on the network.

A number of 3270 terminal emulation products have become available over the years, additionally providing 3278 or 3279 terminal emulation and supporting both direct and dial-up connections to 3174, 3274, and 3276 controllers without the requirement for additional mainframe software. In addition to allowing the user to save terminal screens, these emulation products allow the user to hot key between PC and terminal sessions. Windows and 3270 profiles give the existing keyboard dual functionality that reflects the two configurations.

There is still a lot of manual configuration associated with using terminal emulators. A big part of the job of such emulators is to convert between ASCII keystrokes (including escape sequences) and 3270 keys such as PA1 through PA3 and PF1 through PF24. The system administrator must create the mappings. To make a 3270 key map for the emulator, the administrator must know what the mapping in the protocol converter is, and then assign the ASCII values (characters or sequences) that correspond to each 3270 key to the desired PC key. It is the responsibility of each site administrator to document the key mappings used by its protocol converters. Once it is known how the ASCII values correspond to each 3270 key, then the key bindings can be created.

11.5   SNA-LAN Integration

Because legacy systems promise to be around for quite some time, companies face the challenge of routing SNA traffic while still deploying client-server protocols. A growing number of companies are also looking for ways to transport multiple LAN protocols over such packet-switched architectures as frame relay and ATM. The ultimate goal of SNA-LAN integration is to have a single network infrastructure that is shared by SNA and other LAN protocols.

There are several good reasons for trying to combine legacy SNA networks with multiprotocol LAN internetworks. The primary reason is cost. By combining networks, there is only one infrastructure to deal with rather than two, which translates into cost savings. Another reason is flexibility. With all traffic on one network, end users can have access to traditional SNA applications as well as newer LAN services, such as groupware and collaborative applications.

Users value LANs for their connectivity to a variety of corporate resources and WANs for their ability to interconnect distributed LANs. Multiprotocol routers make this fairly easy to do over high-speed leased lines and carrier services such as T1, metro Ethernet, frame relay, and ATM. But users also value SNA for its reliability in supporting mission-critical applications on the mainframe, many of which are not easily converted to client-server and other distributed environments. To avoid the expense of duplicate networks, users are looking at ways to integrate incompatible SNA and LAN architectures over the same facilities or services. Aside from cost savings and flexibility, the consolidation of such diverse resources over a single internetwork offers several other benefits, such as:

- Eliminating the need to provision, operate, and maintain duplicate networks: one for SNA, one for token ring, and another for non-IBM environments.

- Allowing slow leased-line SNA networks to take advantage of the higher speeds and reliability offered by LAN and/or WAN links.

- Consolidating diverse traffic types and minimizing potential points of network failure, providing a more resilient infrastructure.

11.5.1   Integration issues

While the objectives of SNA-LAN integration are fairly clear-cut, the differences between the protocols make the objectives difficult to achieve. Because SNA is a deterministic technology, its protocol suite requires entirely different routing methods—ones that free up bandwidth, eliminate session time-outs, and route over peer-to-peer LAN connections without imposing an undue performance penalty on users.

This immediately disqualifies Ethernet. Running SNA traffic over Ethernet is not advised because access is controlled with the contention-based procedure called CSMA/CD. This procedure requires that devices on the Ethernet compete for time on the network when they sense that the network has become idle. If two or more stations try to access the network at the same time, their data collide, forcing each station to back off and try again at staggered intervals. Thus, Ethernet's method of media access control poses an insurmountable obstacle for SNA traffic.

Token ring is a much better solution. In fact, SNA works well when Front-End Processors (FEPs) connect the host to terminal users over the token ring network. The token passing scheme is deterministic and, when time-slice parameters are properly configured, this can prevent terminal-to-host sessions from timing out. However, when traffic must traverse logically adjacent LANs via bridges and/or routers, there are several problems that must be addressed.

One problem with SNA is that it must be given priority treatment through the network to avoid session time-outs. If a return acknowledgment is not received in a timely fashion, the SNA protocols interpret this to mean that the connection has been dropped and the session must be terminated. Although capabilities can be added to routers to distinguish between SNA and non-time-sensitive traffic, another problem is that even when prioritization is available, often it is applied only on the first access port into the network and is not carried through on intermediate nodes throughout the network. (Some solutions to this problem are discussed in the next section.) The resulting delay can cause SNA sessions to time out. Finally, there is always the possibility that too much traffic will congest the link. These problems are prevented by locally spoofing the acknowledgment messages that normally originate from the remote node. Configuring the router for local acknowledgment also minimizes unnecessary traffic, which helps control congestion on the network.

11.5.2   Integration approaches

The predominant protocol used between IBM hosts and terminals in an SNA network is the Synchronous Data Link Control (SDLC) protocol. Encapsulating SDLC packets within TCP/IP or another routable protocol alleviates the fixed-path limitation of SNA. It does this by stuffing SDLC data packets into a TCP/IP envelope, which can then be routed across the internetwork (see Fig. 11.6). This routing method offers the advantages of adaptive and dynamic routing, including best-route selection, load sharing, and the use of redundant links, which are not available under SNA.

However, SDLC encapsulation within IP solves some problems while creating others. SNA was designed with the assumption that a reliable, connection-oriented data link exists beneath the upper layers of the protocol. The SNA data link control layer provides deterministic delivery of packets. Encapsulating SDLC packets within IP violates SNA's inherent design. In a large IP network with a potentially large end-to-end delay, the nondeterministic nature of IP can pose a serious delay problem.

Under SNA, the primary SDLC station (usually at the host end of the link) and the secondary SDLC station (usually at the establishment or cluster controller end) exchange polls and acknowledgments. The primary station maintains a timer and terminates the session if the secondary station does not respond to polls quickly enough. One of the problems with IP is that it cannot guarantee delivery of SDLC frames before the timer expires.

Another problem has to do with the size of the packets needed to handle simple functions, such as the exchange of frames to keep an SNA session alive when no data is being transmitted. These session keep-alive frames are 2 bytes; if encapsulated in a TCP/IP packet with a 20-byte TCP header and a

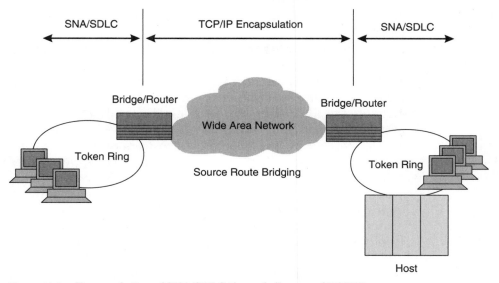

Figure 11.6 Encapsulation of SNA/SDLC through the use of TCP/IP

20-byte IP header, the 42-byte IP packets would continuously traverse the IP network even when the SNA end stations have nothing to send. This adds unnecessary traffic to the network and can cause congestion on the links.

One solution would be to overconfigure the network so that there is always enough spare bandwidth available to mitigate the possibility of congestion. But this inflates the cost of the network. As noted in Sec. 11.5.1, a better solution entails the use of routers that answer the polls locally, thus preventing unnecessary traffic from getting on the network in the first place. This local polling termination solution is based on Logical Link Control 2 (LLC2), which is a protocol developed by the Institute of Electrical and Electronic Engineers (IEEE) to facilitate SNA data transmission over token ring LANs. Previously, the communications controllers performed LLC2 polling of remote devices. Now, the LLC2 capability is built into bridges, routers, and standalone protocol converters.

With local termination, the router connected to a 37×5 communications controller, for example, acknowledges the polls from the 37×5 controller, while the router connected to the 3×74 establishment controller issues polls to the 3×74 controller. Local acknowledgment of the polls obviates the need for poll response data to be sent over wide area links and also helps keep SNA sessions active during congestion or after a failure while the router tries to find a new data path.

While this feature works well, it also makes network management more difficult. What was once a single session now becomes three sessions: SNA

LLC2 local acknowledgment from end node to router, a TCP/IP session between routers, and local acknowledgment from remote router to remote end node.

Another solution is to assign priorities to the traffic, so that delay-sensitive SNA/SDLC traffic can be sent out by the router before delay-tolerant traffic. Via the router's management system, priority may be assigned based on protocol, port, source, or destination address, or any bit pattern defined by a priority control filter. As many as a few dozen filters can be created for each protocol supported on each of the router's synchronous interfaces. For each protocol, some products also allow users to assign priority levels to each application—urgent, high, normal, and low. Low-priority traffic is queued, while high-priority traffic is forwarded immediately.

Still another priority control mechanism is bandwidth reservation, an approach that divides the available bandwidth among the various user protocols, such as TCP/IP, SNA/SDLC, or Internet Packet Exchange (IPX). Received frames are placed in their respective protocol-related queues and then passed to a serial link queue from which they are sent out over the serial link. Frames with a protocol assigned a higher percentage of bandwidth are passed to the serial link queue more often than those assigned a lesser amount of bandwidth. During low traffic periods, existing traffic is assigned the unused bandwidth by passing frames of the corresponding protocols more frequently over the serial link. In this way, bandwidth reservation ensures uninterrupted service during peak traffic periods.

11.5.3   Advanced solutions

While there are a number of vendor-specific solutions for integrating SNA with LANs and TCP/IP internets, vendors and users have coalesced around a few standard approaches, specifically:

- Advanced Peer-to-Peer Networking (APPN) or High Performance Routing (HPR)

- Encapsulation or Data Link Switching (DLSw)

- Frame relay Permanent Virtual Connections (PVCs) per RFC 1490

APPN extends SNA to PC LANs by letting midrange processors communicate on a peer-to-peer basis, while HPR streamlines SNA traffic so that routers can move the data around link failures or outages.

Generally, APPN is used when SNA traffic must be prioritized by class of service, which routes traffic directly to end nodes, or when SNA traffic must be routed peer to peer without going through a mainframe. HPR is used when traffic must be sent through the distributed network without disruptions. HPR provides link-utilization features that are important when moving to packet-switched LANs, such as ATM or frame relay, and provides congestion control for optimizing bandwidth. This method offers a faster routing path than APPN because it ensures that SNA traffic is prioritized and routed to all network nodes. HPR's performance gain over APPN comes from its end-to-end flow controls, which are an improvement over APPN's hop-by-hop flow controls.

Another SNA routing technique is DLSw, which is used in environments consisting of a large installed base of mainframes and TCP/IP backbones. DLSw assumes the characteristics of APPN and HPR routing and combines them with TCP/IP and other LAN protocols. DLSw encapsulates TCP/IP and supports SDLC and High-Level Data Link Control (HDLC) applications. It prevents session time-outs and protects SNA traffic from becoming susceptible to link failures during heavy congestion periods.

Like DLSw, frame relay uses encapsulation to transport all protocols, including SNA/APPN. It provides SNA-guaranteed bandwidth through frame relay's permanent virtual circuits and, compared to DLSw, uses very little overhead in the process.

Advanced Peer-to-Peer Networking.   APPN is IBM's enhanced SNA technology for linking devices without requiring the use of a mainframe. Specifically, it is IBM's proprietary SNA routing scheme for client-server computing in multiprotocol environments. As such, it is part of IBM's LU 6.2 architecture, also known as Advanced Program-to-Program Communications (APPC), which facilitates communications between programs running on different platforms.

APPN routes SNA traffic natively across PC LANs. Large IBM shops that must prioritize and route traffic in a peer-to-peer fashion are good candidates for APPN. HPR, used exclusively in the SNA environment, enhances SNA routing and adds to APPN by further prioritizing and routing SNA traffic around failed or congested links. It is used in situations where bandwidth is critical, especially in packet-switched networks.

Included in the APPN architecture are Automatic Network Routing (ANR) and Rapid Transport Protocol (RTP) features. These features route data around network failures and provide performance advantages, closing the gap with TCP/IP. ANR provides end-to-end routing over APPN networks, eliminating the intermediate routing functions of early APPN implementations, while RTP provides flow control and error recovery. To these features, HPR adds a very advanced feature called Adaptive Rate Based (ARB) congestion prevention.

ARB uses three inputs to determine the sending rate for data. As data is sent into the network, the rate at which it is sent is monitored. At the destination node, that rate is also monitored and reported back to the originating node. The third input is the allowed sending rate. Together, these inputs determine the optimal throughput rate, which minimizes the potential for packet discards to alleviate congestion.

By enabling peer-to-peer communications among all network devices, APPN helps SNA users connect to LAN networks and more effectively create and use client-server applications. APPN supports multiple protocols, including TCP/IP, and allows applications to be independent of the transport protocols that deliver them.

APPN's other benefits include allowing information routing without a host, tracking network topology, and simplifying network configuration and changes. For users still supporting 3270 applications, APPN can address dependent LU protocols as well as the newer LU 6.2 sessions, which protect a site's investment in applications relying on older LU protocols.

Data Link Switching.   DLSw was developed by IBM in 1992 as a way to let users transport SNA traffic over TCP/IP. Because DLSw entails a fair amount of processing, it imposes a considerable resource burden on networking nodes, typically routers. Nevertheless, DLSw eliminates session time-outs and offers predictable response time.

In DLSw, SNA frames are encapsulated within TCP/IP packets. And while encapsulation inherently involves some amount of duplicative network handling, the encapsulation aspect of DLSw is not too burdensome from a network-processing point of view. It is the additional, higher-level processing of DLSw that makes it so much more formidable.

For example, DLSw uses a form of spoofing to prevent host-issued SNA polls from continually being sent over the underlying TCP/IP network. A DLSw router at the host end intercepts and responds to these host polls, while another DLSw router at the remote end conducts its own polling exchange directly with a communicating SNA station.

Another task is maintaining SNA-session information across the TCP/IP router network. This involves additional processing on all the intermediate DLSw routers in the TCP/IP network—not just the periphery routers that directly interface to the SNA host and workstation.

Among the strengths of DLSw is that it works across any type of WAN and it can automatically locate destinations via MAC addresses. Its weaknesses are that it has very high overhead for the tasks it carries out, its dynamic alternate routing capability cannot cope with failed central-site bridge and/or routers, and it has no built-in traffic prioritization scheme. In addition, scalability issues may limit the number of TCP sessions that can be supported.

Frame relay.   RFC 1490 leaves the issue of rerouting in the event of path failure to the frame relay network. Frame relay networks tend to be more robust and less prone to errors than other networks. Not only do frame relay networks operate over high-quality digital facilities, most frame relay networks provide transparent alternate routing in the event of path failure. If this level of fault protection is not enough, RFC 1490–compliant routers can act as frame relay switches and complement the WAN's transparent rerouting. Not only do RFC 1490 solutions provide the high availability required by SNA-based mission-critical applications, RFC 1490 routers provide dial backup, eliminating resilience as a potential issue.

For most networking professionals responsible for IBM-centric networks, it is often difficult to choose between RFC 1490 and DLSw. If frame relay is to be used as the infrastructure for multiprotocol internetworks—which seems likely in North America because of increasingly attractive pricing—RFC 1490 can become the clear choice for such integration. It provides a highly optimized, cost-effective means of transporting SNA/APPN traffic and encapsulates other multiprotocol traffic across the WAN. Offering a native encapsulation technique for frame relay, RFC 1490's overhead for SNA/APPN data traffic is roughly 25 percent of that employed by DLSw. This low overhead can make a significant difference to application response times, overall network congestion, and network data usage charges.

The advantages of RFC 1490 do not obsolete DLSw. In fact, they each have a role to play, but in different networking scenarios. Frame relay environments, whether public or private, tend to be the domain of RFC 1490, whereas DLSw tends to be practical in IP-based backbones—such as WANs built around the use of the Point-to-Point Protocol (PPP) over leased lines—or backbones that use the X.25 protocol instead of frame relay. If the network is not going to be frame relay–centric, then DLSw will most likely be the best option.

DLSw should only be seriously considered if the network has some native TCP/IP traffic. Otherwise, TCP/IP and the administrative and bandwidth usage overhead—the routing table updates required by TCP/IP—will be introduced into the network just to support DLSw. Thus, if the traffic mix of a network consists of SNA/APPN with Internet Packet Exchange/Synchronous Packet Exchange (IPX/SPX) and NetBIOS, as is likely in many IBM-oriented networks, introducing TCP/IP in order to use DLSw is probably not a good idea. Even if there is native TCP/IP traffic flowing over a frame relay WAN, RFC 1490 is inevitably going to be a more efficient and streamlined solution, both short- and long-term, than DLSw.

11.6   Outsourcing Network Integration

For organizations that lack in-house expertise to handle network integration, migration, and management, outsourcing these responsibilities might be a viable option. There are outsourcing firms that specialize in migrating host-centric SNA environments to multiprotocol LANs and WANs, for example. Carriers, too, have come up with network management and migration services aimed at IBM-centric data centers moving from multidrop analog, and private line arrangements to more advanced transmission capabilities, such as frame relay.

Large carriers with extensive data service offerings and comprehensive service plans are capable of migrating companies from private leased-line services, for example, to frame relay and ATM services. Their managed SNA services typically include Frame Relay Access Devices (FRADs) that transport SNA traffic over frame relay permanent virtual circuits. Such services include installation, on-site maintenance, and around-the-clock network monitoring and management of both the FRADs and access lines. Customers also have the option of using a router to integrate LAN networks with non-LAN–based SNA devices to connect all network points as they migrate from host-based networks to distributed computing. Another option lets customers directly connect a frame-relay circuit to the front-end processor of a host computer using a standardized format that lets any kind of data, including SNA, be transferred over frame relay without the need for other external devices.

The carrier provides network analysis and design to identify the most appropriate data service for a company's needs, a service transition period, financial incentives, internetworking of different services in use at various company sites, and equipment and services to integrate existing network protocols with frame relay. The carrier may also offer various options for handling mixed services and technologies in the customer's networks, provide leasing options for CPE, and a payment plan that allows customers to pay a single charge to maintain old and new services during the transition period.

Such services are aimed at removing the financial obstacles to migrating private lines to frame relay and ATM. These services also make it easier to move from the host-centric to distributed computing environment.

11.7   Conclusion

Throughout their development histories, SNA networks and LANs have moved along separate paths. Now with trends as diverse as corporate downsizing, distributed computing, and participation in the competitive global economy, companies are virtually forced to consolidate the two types of networks. Given the vast installed base of SNA and LAN networks, the movement toward integration makes sense, especially with the advent of reliable integration techniques and services. In fact, the business case for interconnecting the two types of networks is so obvious that justifying it with cost savings, return on investment, operational benefits, and simplified administration may not even be necessary.

While some companies have the expertise required to design and install complex networks, others are turning to network integrators to oversee the process. The evaluation of various integration firms should reveal a well-organized and staffed infrastructure that is enthusiastic about helping to reach the customer's networking objectives. This includes having the methodologies already in place, the planning tools already available, and the required expertise already on staff. Beyond that, the integrator should be able to show, with references and contact information, that its resources have been successfully deployed in previous projects of a similar nature and scope.

Navigating Service Level Agreements

12.1 Introduction

The concept of the Service Level Agreement (SLA) dates back to the 1960s, when Information Technology (IT) departments used these agreements to measure technical services such as data center uptime. Today, SLAs are more elaborate in nature and scope. They can be applied to Web and e-commerce hosting, metropolitan area Ethernet, Virtual Private Network (VPN), Internet access, and network security as well as to the wide variety of other voice and data services delivered over T-carrier, optical carrier, or wireless facilities.

When Service Level Agreements are applied to the services an IT department provides to other business units within an organization, the result could be better management of capital resources (computer systems and their peripheral complement) and a higher level of contribution of the IT department to the success of the company based on the more effective use of data processing resources. At the same time, a successful implementation of an intracompany SLA can give network managers the documentation they need to justify staffing levels and budgets. It can also provide a set of objectives IT staff must work toward to earn raises and bonuses.

The Service Level Agreements offered by carriers and other service providers define guaranteed levels of network performance. AT&T and Sprint are among the largest carriers that offer guarantees on network performance that include metrics on packet loss and latency for their data services, which include Internet, frame relay, and Asynchronous Transfer Mode (ATM). Service Level Agreements are available for other services as well, from traditional T-carrier services to Internet Protocol (IP)–based VPNs, intranets, and multicompany extranets. Telecom service providers feel compelled to offer written performance guarantees to differentiate their offerings in the increasingly competitive market brought about by the Telecommunications Act of 1996.

Service Level Agreements are even becoming popular among Internet Service Providers (ISPs) as a means to lure business applications out of the corporate headquarters to outsourced Web hosting arrangements. To do this successfully, the ISP must operate a carrier-class data center, offer reliability guarantees, and have the technical expertise to fix any problem, day or night. The SLA may also include penalties for poor performance, such as credits against the monthly invoice if network uptime falls below a certain threshold. Some ISPs even guarantee levels of accessibility for their dial-up remote-access customers.

Whatever the specific service arrangement, an SLA is a contract between the provider and the customer that defines the services provided, the metrics associated with these services, acceptable and unacceptable performance levels, liabilities on the part of the provider, and what actions are to be taken in specific circumstances. Carriers and service providers write these documents in language so specific as to minimize exposure to liability. Usually the extent of liability for a service provider is limited to credits that are applied to the customer's next monthly invoice.

Service-level management, on the other hand, refers to the people, systems, and tools that allow the organization to monitor compliance with the SLA for each type of service. Sometimes the service provider will make the tools available to the customer in the form of reports that can be accessed on its Web site. Depending on the service, the customer may choose to implement third-party tools that give it independent verification of SLA compliance.

12.2   Enterprise SLAs

As noted, the concept of the SLA originated within IT departments. Long neglected as a result of the migration of Personal Computers (PCs) in the workplace, SLAs are starting to be rediscovered as a means of justifying the recentralization of computing resources into the data center or the expansion of an existing data center, especially when the IT department provides mission-critical services to other business units within the organization. Here, SLAs are gaining more attention because fundamental changes are under way in the way IT departments are run. Today, many IT departments are being operated as profit centers, and there is more awareness of customer satisfaction, particularly with application response time.

If the IT department is going to take control of computing resources away from another department, it must have the tools to prove it is still meeting that department's needs. Consequently, platform and third-party vendors are providing IT managers with more tools for SLA monitoring and enforcement. A communications server, for example, might come with the tools to monitor and automatically enforce SLAs. If a priority application is being denied the bandwidth called for in the SLA with the finance department, for example, the performance monitor alerts IT staff that the SLA has been breached. Meanwhile, the system automatically makes adjustments by choking off bandwidth for one or more other routine applications and assigning the bandwidth to the priority application. The performance monitor then issues an alert that the SLA has been restored.

Whereas the SLAs of telecom service providers and ISPs are fairly standard offerings that are promoted to attract new customers, an enterprise SLA must be customized for each business unit down to the application user group. The enterprise SLA is written by the IT department using baseline performance data gathered by various monitoring tools and information gathered from end users about their needs and expectations regarding levels of service.

An enterprise SLA can provide the IT department with a structured approach to fulfilling its responsibilities to the company. When properly implemented, it not only reduces end-user complaints but it can also shift the focus of IT from a reactive to a proactive approach to problem solving. When complaints arise, IT can go right to the processes that the complaints come from to determine whether there is an IT problem or a business process problem. This eliminates a lot of the finger-pointing that might otherwise go on within the organization.

The SLA also lets IT prioritize its tasks and helps IT see when service level targets are being missed so that it can plan to improve. At the same time, the enterprise SLA can be structured to penalize the IT department for poor performance. Bonuses, or the lack of bonuses, are among the most effective ways to encourage peak performance and response time to problems.

Not limited to information systems, enterprise SLAs can also be applied to private networks. Vendors offer tools that monitor network performance and alert administrators to performance violations and trends that may lead to violations. Tivoli, for example, offers a product called Network Connectivity and Performance Manager. It includes a module called SLA Administrator (see Figs. 12.1 and 12.2). According to the model used in SLA Administrator, an SLA violation occurs whenever an average metric value exceeds the value defined for that metric in the service level. An SLA trend is generated based on a statistical model composed of the historical average metric values of the SLA.

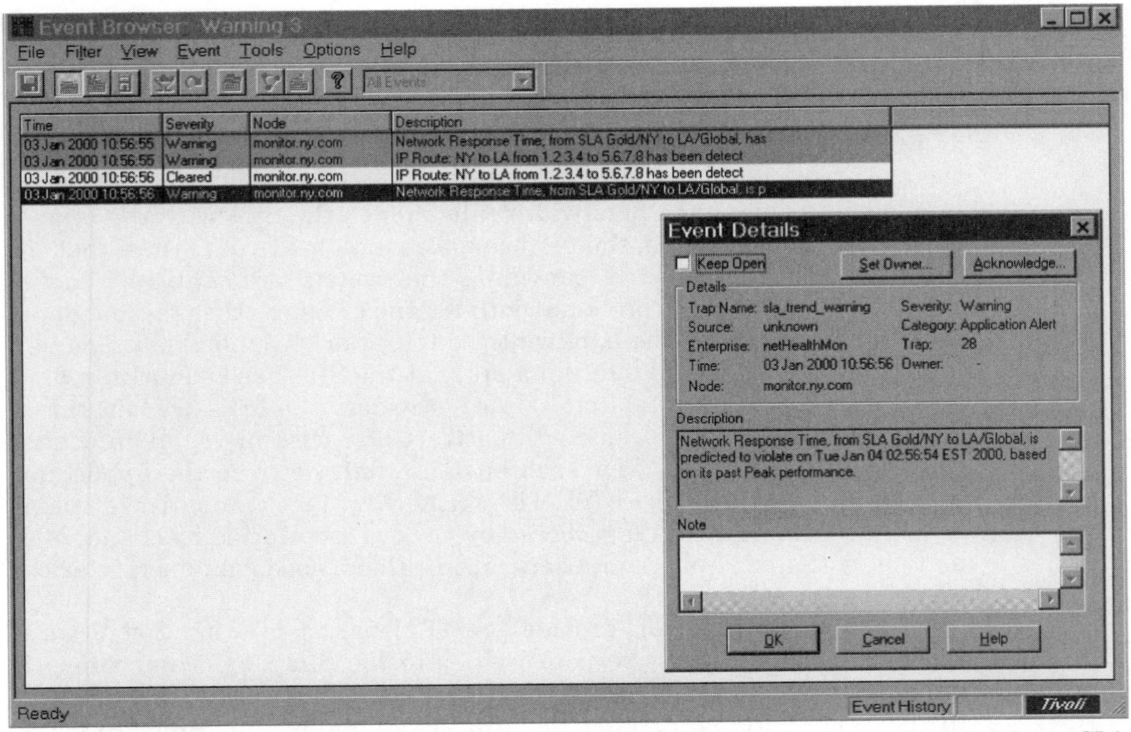

Figure 12.1 In this example, from Tivoli's SLA Administrator, the event details are displayed for an SLA trend-warning event that corresponds with a service named Gold, on the IP route from New York to Los Angeles, and schedule name Global. Other events are displayed for IP route up/down and violations.

Whenever a gradual transition to a metric value occurs and the transition occurs over data with low variance, the model will determine if the projected measured value will violate the service level metric over the next two agreement intervals. An SLA trend warning is then issued in the form of a Simple Network Management Protocol (SNMP) trap sent to the network management system. If the average metric values for an SLA that previously sent an SLA trend warning are transitioning significantly away from the service level value, then an SLA trend clear event is issued, also in the form of an SNMP trap.

12.2.1 Performance baselining

Performance baselining is a procedure for understanding the behavior of a properly functioning network so that deviations can help identify the cause of problems that may occur in the future. The only way to know a network's normal behavior is to analyze it while it is operating properly. Later,

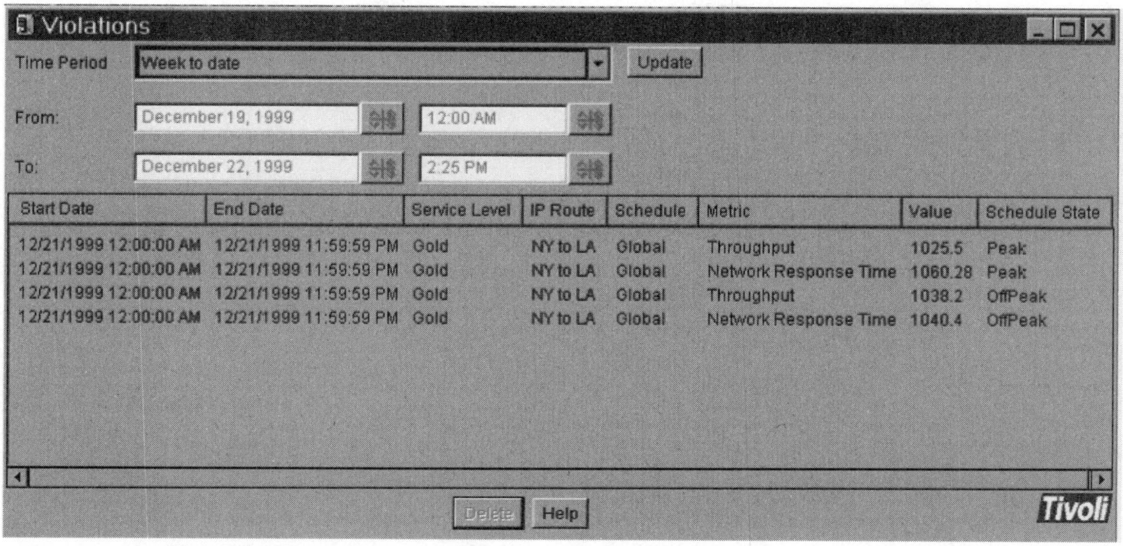
Violations
Time Period
Week to date
Update
From:
December 19, 1999
12:00 AM
To:
December 22, 1999
2:25 PM
Start Date
End Date
Service Level
IP Route
Schedule
Metric
Value
Schedule State
12/21/1999 12:00:00 AM 12/21/1999 11:59:59 PM Gold NY to LA Global Throughput 1025.5 Peak
12/21/1999 12:00:00 AM 12/21/1999 11:59:59 PM Gold NY to LA Global Network Response Time 1060.28 Peak
12/21/1999 12:00:00 AM 12/21/1999 11:59:59 PM Gold NY to LA Global Throughput 1038.2 OffPeak
12/21/1999 12:00:00 AM 12/21/1999 11:59:59 PM Gold NY to LA Global Network Response Time 1040.4 OffPeak
Delete
Help
Tivoli

Figure 12.2 An SLA violation contains the following information: the start and end dates of the evaluation period when the violation was detected, the service level name, IP route name, schedule name, metric name, the value that exceeded the metric value, and the schedule state (peak or off-peak).

technicians and network managers can compare data from the properly functioning network with data gathered after conditions have begun to deteriorate. This comparison often points to the right steps that lead to a corrective solution. The savings in the time it takes to resolve problems will contribute to increased network availability.

Information requirements. The first step in baselining performance is to gather appropriate information from a properly functioning network. Much of this information may already exist, and it is just a matter of finding it.

For example, many enterprise management systems have the capability of automatically discovering devices on the network and creating a topological map. This kind of information is necessary for knowing what components exist on the network and how they interact physically and logically. For Wide Area Networks (WANs), this means locations, descriptions, and cable plant maps for equipment such as routers, bridges, and network access devices [Local Area Network (LAN) to WAN].

In addition, information on transmission media, physical interfaces (T1/E1, V-series, etc.), line speeds, encoding methods and framing, and access points to service provider equipment should be assembled. Although it is not always practical to map individual workstations in the LAN portions of the WAN, or to know exactly what routing occurs in the WAN cloud, knowing the general topology of the WAN can be useful in tracking down problems later.

To fully understand how a network behaves, it is necessary to know what protocols are in use. Later, during the troubleshooting process, the presence of unexpected protocols may provide clues as to why network devices appear to be malfunctioning or why data transfer errors or failures are occurring.

Some network problems begin to occur after new devices or applications are installed. The addition of new devices, for example, can cause network problems that have a ripple effect throughout the network. A new end-user device with a duplicate IP address, for instance, could make it impossible for other network elements to communicate. Or a badly configured router added to the network could produce congestion and connection problems. Other problems occur when new data communications are enabled, or existing topologies and configurations are changed. A log of these activities can help pinpoint causes of network difficulty. In addition, previous network trouble—and its resolution—is sometimes recorded, and this, too, can lead to faster problem identification and resolution.

Often, previously gathered data can provide valuable context for newly created baselines. Previously assembled baselines may also contain event and error statistics and examples of decoded traffic based on network location or time of day. Usage trends and normal error levels over time may be found in a system's statistics log. These logs may have been gathered over long periods, yielding valuable information about the history of network performance. Many protocol analyzers let technicians specify the period over which this kind of data is logged, the interval between log entries, and the type of statistics to log. The log file can be exported to a spreadsheet or other application program for off-line analysis.

LAN traffic on the WAN. With knowledge of what kind of LAN traffic to expect on the WAN, technicians and network managers will have a better idea of the analysis that might have to be performed later. In addition, knowing how LAN frames might be handled at end stations can help troubleshooters make a distinction between WAN problems and end-station processing problems.

Knowing the WAN traffic type (voice, data, video, etc.) can also help troubleshooters estimate when network traffic is most likely to be heavy, what level of transmit errors can be tolerated, and whether it even makes sense to use a protocol analyzer. For example, an analyzer may incorrectly report errored frames and corrupt data when it is attempting to process voice or video traffic based on data communication protocols.

Peak usage. Knowing when large data transfers will occur—such as scheduled file backups between LANs connected across the WAN—can help you, as a network manager, predict and plan for network slowdowns and failures. It can also help technicians schedule repairs so that WAN performance is minimally impacted. Some of this information can be obtained from interviews with network administrators or key users. Other times, it must be gathered with network analysis tools. Getting comprehensive baseline data may entail gathering this information at regular intervals at numerous points throughout the network.

Baseline information is compared with current information to see network changes. For example, to isolate failing devices or connections, the number of errors most recently recorded is compared to the previous number of errors that occurred over a similar time interval. Current network problems can result from subtly changing conditions that are detected only after examining a series of baselines gathered over time. For example, congestion problems may become apparent only as new users are added to a particular part of the network. Examining historical trends can help isolate these situations.

Performance baselining provides a profile of normal network behavior, making it easier for technicians and network managers to identify deviations so appropriate corrective action can be taken. This "snapshot" of the current network can also be used as the input data for subsequent performance modeling. For example, network administrators and operations managers can use the baseline data to conduct what-if scenarios to assess the impact of proposed changes. A wide variety of changes can be evaluated, such as adding servers, network bandwidth, or application workloads and relocating user sites. During analysis, performance thresholds can be customized to highlight network conditions of interest. These capabilities enable users to plan and quantify the benefits of feature migrations and to make more accurate and cost-effective decisions regarding the location and timing of upgrades.

12.2.2   SLA development considerations

An enterprise Service Level Agreement describes what is important for end users to obtain from IT in terms of service. Typically, a separate document is created and maintained for each major application user group, since needs differ from one group to another. The metrics could be Online Transaction Processing (OLTP) response times, batch turnaround times for end-of-day reports, actual hours of system availability, and bandwidth availability. In essence, the SLA documents what a particular group of workers and managers need from IT to best fulfill their responsibilities to the company.

Each application can be categorized by the amount of resources necessary to support it, which is typically referred to as *workload characterization* and involves understanding how elements of the application use resources, both automated and manual. When this level of understanding is achieved, an information base is established that can be used to effectively manage IT resources and develop meaningful SLAs.

A well-managed data center, for example, would never allow an uncoordinated increase in the number of users of a particular application. If users are added without first understanding their impact on existing network and system resources, it is likely that the response time service level for that application will be exceeded. The extra resources consumed by the application to support the additional users could cause the service levels of other applications to be exceeded as well. In this situation, the IT department could easily find itself in the position of violating the SLA and, in the process, lose credibility among top management and the business units it is intended to serve.

Network managers usually have access to online software tools to monitor system performance and resource utilization, which will give you a general idea of how all systems on the network are behaving at any given moment. This information enables networks and systems to be managed effectively and provides the starting point for developing the SLA.

Careful examination of the data will provide a picture of how systems are being used and how a particular application user group perceives its level of service. Most tools allow this information to be displayed graphically over time and include an export capability to put the data into a form that can be manipulated further by a spreadsheet or statistical package. The following tips can help you get started with SLAs:

- *Set priorities.* Work with business units to set priorities. Instead of technology-oriented goals, focus on metrics that users can understand, such as application response time and system availability.

- *Set measurable goals.* Instead of buying new tools to obtain performance baselines, start with the performance metrics that are already in place, if possible, then plan ways to improve them.

- *Tie IT costs to SLAs.* Users who must pay for services tend to have more realistic expectations regarding what they really need and what they can expect to get. If there is no charge-back mechanism in place, the cost data can at least be shared with business unit managers for the purpose of cost containment.

- *Keep users informed.* Users are interested in how their experiences compare with those of other users. Provide comparison reports that show how each work group's IT services are performing. When they see that their service levels are not out of the ordinary, they may be less likely to complain.

- *Never consider the SLA finished.* An SLA should be considered a living document. It must be constantly reevaluated and refined. Hold regular meetings with business units to review possible changes in traffic, new applications, and potential server demand. Such reviews can be used to build a consensus on appropriate courses of action.

The SLA is not the objective, but merely the documentation of a continuing management effort. If end users are unhappy with performance, the focus should be on solving the problem, not revisiting the SLA process and possibly tearing it up as unworkable.

12.2.3 SLA components

When writing an SLA for internal use, it is recommended that information be gathered about current network performance levels and that the expectations of business units and end users be quantified. Once these are known, the IT department's next task is to write the SLA. At a minimum, it is recommended that the document should contain the following components:

- *Background.* This section should contain enough information to acquaint a nontechnical reader with the application, and to enable that person to understand current service levels and why they are important to the continued success of the business.

- *Parties.* This section should identify the parties to the agreement, including the responsible party within IT and the responsible party within the business unit and/or application user group.

- *Services.* This section should quantify the volume of the service to be provided by the IT department. The application user group should be able to specify the average and peak rates, and the time of day they occur. The user may be provided with incentives to receive better service or a reduced cost for service, if peak resource usage periods can be avoided.

- *Timeliness.* This section should provide a qualitative measure of most applications to let end users know how fast they can expect to get their work accomplished. For OLTP applications, for example, the measure might be stated as "95 percent of transactions processed within 2 seconds." For more batch-oriented applications, the measure might be stated as "reports to be delivered no later than 10:00 A.M. if input is available by 10:00 P.M. the previous evening."

- *Availability*. This section should describe when the service will be available to the end users. The end users must be able to specify when they expect the system to be available in order for them to achieve their specified levels of work. The IT department must be able to account for both planned and unplanned system unavailability and work these factors into an acceptable level of performance for end users.

- *Limitations*. This section should describe the limits of IT support during conditions of peak period demand, resource contention by other applications, and general overall application workload intensities. These limitations should be explicitly stated and agreed to by all parties to prevent finger-pointing and dissatisfied end users down the road.

- *Compensation*. This section should put teeth into the agreement for both parties using some form of compensation. Ideally, a charge-back system should be implemented in which end users are charged for the resources they consume to provide the service they expect. This gives business units the incentive to apply management methods that optimize costs and performance. If a charge-back system is impractical, the costs should still be identified and reported back to the business units to show where the IT resources are going. The frequency and format of this information should also be described in this section.

- *Measurement*. This section should describe the process by which actual service levels will be monitored and compared with the agreed-upon service levels, as well as the frequency of monitoring. This section should also include a brief description of the data collection and extrapolation processes, and how users are to report problems to IT.

- *Renegotiation*. This section should describe how and under what circumstances the SLA should be changed to reflect changes in the environment.

Once written, the SLA should be distributed to all parties for review. Questions and problems need to be resolved before the SLA can be put into effect. Once a consensus is achieved, all parties will have a stake in its successful implementation. When the SLA is ready for implementation, the IT department must implement procedures to determine if service levels are being met. Additionally, IT needs to be able to forecast when the service levels can no longer be met owing to growth or other external factors.

12.2.4 Service level enforcement

It is not enough to have a clearly defined SLA and voluntary compliance to guarantee the service levels of various applications; there must be enforcement mechanisms in place to ensure that the SLA is not violated.

Enforcement can be automated with bandwidth management solutions, which are especially useful for IP networks. As more companies move to an intranet model that relies on information sharing and Web navigation, it is becoming necessary to ensure high-quality network services for mission-critical business applications. Additionally, today's IP network applications are both bandwidth intensive and time sensitive. They often require support for voice, video, and multimedia, all of which eat up scarce bandwidth. Bandwidth management tools ensure that users and applications share this resource appropriately.

Allot Communications is one of dozens of vendors that offer a bandwidth management solution for IP nets. The company's NetEnforcer enables the allocation of high priority and service guarantees to mission-critical applications. It can ensure the performance of delay-sensitive applications and can limit bandwidth allocation and delay when application response time is critical. NetEnforcer enables the corporate network manager to control resources such as bandwidth, servers, applications, and users. It also monitors and accounts for traffic usage based on clients, servers, application, access time, and DiffServ (Differentiated Services) tagging.

NetEnforcer also enables restrictions to be placed on noncritical applications. This, in turn, can open up clogged network pipes. Email servers, for example, can be limited to a specified amount of bandwidth. Likewise, Web browsing can be controlled to an acceptable limit. These actions will cause a substantial improvement in critical application response time. Critical applications such as Citrix, SAP, Baan, Oracle, and PeopleSoft can be assigned appropriate bandwidth based on their business priority.

Orchestream offers Service Activator, a bandwidth management solution that can be sold to enterprises with the Quality of Service (QoS) module switched on, allowing companies to manage the classes and levels of bandwidth within their own networks. Service Activator offers policy control as the means of implementing QoS in conjunction with the installed base of routers. This allows network managers to set rules to control the users and applications that can access the different classes of service available on the network. Policies are based on rules that simply relate the ability of subjects (users and/or applications) to access network resources (data, systems, and other resources). The power of the policy-based approach is that it is possible to manage the access controls for thousands of individuals as easily as for a single user. The rules can be varied dynamically as business needs demand.

For example, enterprise administrators can partition the network into multiple performance classes tailored to the needs of specific applications. There can be a guaranteed minimum bandwidth service class for mission-critical applications, such as Citrix WinFrame, SAP R/3, or Oracle. There can be a low-latency performance class for delay-sensitive applications, such as Voice over Internet Protocol (VoIP) and video conferencing; and a standard class for applications that can tolerate some delay, such as email or intranet access. Under this prioritization scheme, the key traffic in the guaranteed minimum bandwidth class will get network service, even in times of congestion.

One of the oldest bandwidth management products is Packeteer's PacketShaper, which is a policy-based, application-adaptive bandwidth management hardware product that controls the use of WAN bandwidth and delivers end-to-end QoS on a per-application-flow basis. PacketShaper includes several mechanisms for bandwidth allocation, and these can be used individually or in combination to control the rate of traffic flow:

- *Transmission Control Protocol (TCP) rate control.* Improves the throughput for a limited-capacity link by letting network administrators go beyond simple traffic priorities to set kilobit-per-second partitions for each classified traffic flow.

- *Partitions.* Limit how much of the network can be used by a single traffic class. They function like frame relay Permanent Virtual Circuits (PVCs), which allow other applications to share the unused bandwidth when it becomes available.

- *Policy-based class of service.* Allows policies to be set to protect specified traffic, cap bandwidth-intensive traffic, and give priority service to latency-sensitive sessions. If a class of service cannot be satisfied owing to changing network conditions, this feature allows the network administrator to decide whether to deny access, squeeze in another user, or, for Web requests, redirect the request. The network administrator can choose different options for different traffic classes.

PacketShaper provides network administrators with a mechanism to see what effect constantly changing traffic patterns have on network performance. PacketShaper provides a broad picture of any segment anywhere on the network. This enables the network administrator to see if huge bursts of activity are occurring on a segment, for example, and what application is causing it, all in real time.

12.3 Service Provider SLAs

Telecommunications service providers, Internet service providers, and Application Service Providers (ASPs) are responding to competitive pressures by offering SLAs that guarantee high levels of network performance. Performance guarantees are important for weaning enterprise customers away from private networks that have traditionally supported mission-critical business applications. Performance guarantees also help to further differentiate the services of competitors and offer you, as a network manager, another tool with which to compare service providers.

12.3.1 Performance metrics

The SLAs from carriers and service providers define such performance metrics as latency, packet loss, and availability.

Latency. This metric refers to the round-trip delay between the time a request packet is sent to a destination and the time a response packet is received. Latency is reported in milliseconds (ms). Carriers constantly measure the latency (speed) of core areas of their network using data collected by pings via the Internet Control Message Protocol (ICMP). Data is collected from designated routers in key network hubs worldwide, usually in five-minute intervals. Monthly latency statistics are derived from averaging all samples from the previous month.

Targets vary by carrier, but 65 ms or less for regional round-trips within Europe and within North America (for U.S. and Canadian customers) is considered very good performance, while 85 ms or less for regional round-trips within Europe and within North America for non-U.S. and non-Canadian customers is considered very good performance, as is 120 ms or less for transatlantic round-trips between London and New York.

Packet loss. This is a measure of packets sent to a destination that do not elicit corresponding return packets. Packet loss is reported as a percentage of total packets sent. This should be a very low figure, such as 0.016. Instead of packet loss, some carriers use the metric packet delivery, which is also expressed as a percentage. For regional round-trip traffic within Europe and North America, for example, an SLA might promise a packet delivery rate of 99 percent or greater.

Availability. This metric is a measure of network uptime, or the time the network is actively handling customer traffic. Availability is reported as a percentage of the total uptime. For example, 99.999 percent uptime, often referred to as "five nines," is viewed as the highest availability rate that can realistically be achieved. It translates to less than five consecutive minutes of downtime per year. A guarantee of 99.99 percent—"four nines"—availability ensures no more than five consecutive minutes of outage per month. The guarantee of 99.9 percent —"three nines"—availability means that there will be no more than 45 consecutive minutes of outage per month.

Service Level Agreements usually spell out conditions that do not qualify as "unavailability," such as downtime resulting from network maintenance, circuits provided by other carriers, acts or omissions of the customer, or "acts of God," civil disorder, natural cataclysm, or other occurrences beyond the reasonable control of the service provider.

Some carriers and service providers include additional metrics in their SLAs, such as provisioning time and restoration time.

- *Provisioning time.* This metric refers to the agreed-upon due date of a circuit. If the due date is missed, the recurring charge for that circuit may be waived for one month, for example. This applies to ATM or frame relay PVCs as well.

- *Restoration time.* This metric refers to how quickly the carrier or service provider can return a circuit to full operation after a service-affecting outage. If a customer reports a frame relay service outage (even if the problem is with local access), for example, and it is not restored in four hours, as stipulated within the SLA, recurring charges for the affected ports and PVCs may be waived for one month.

12.3.2   Sample SLA

There are no industry standards for writing Service Level Agreements. Although there are some common elements, such as performance metrics associated with a particular type of service, carriers and service providers set their own rules for complying with the SLA and for issuing credits to the customer for performance violations. If the SLA is structured properly, the carrier or service provider should not lose very much money in credit allowances to its customers.

AT&T, for example, considers an interruption of its ACCUNET T1.5 Service as having occurred when there has been a loss of continuity, or when 300 or more seconds of transmission contain errors in a 15-minute period. This means, barring a cable cut, which results in loss of continuity, that the service must be shown to have errors for five minutes or more during a 15-minute interval before the customer can qualify for a credit under AT&T's Service Assurance Warranty. The customer's recurring charges for the channel and associated rate elements in the month that the interruption occurs are the basis for calculating the credit allowance for that month. Discounts due to pricing plans are applied prior to the application of credit allowances.

Under the Service Assurance Warranty, if more than one interruption is reported on an ACCUNET T1.5 Service (including a local channel) in a given month, each subsequent interruption is considered independently in calculating total credits for that channel on the following month's bill. The cumulative credit allowances, however, may not exceed 100 percent, per channel, in a given month and credit may not be carried over to subsequent months. Table 12.1 summarizes AT&T's credit allowance for interruptions of its ACCUNET T1.5 Service.

In the example of how credit allowances are applied, summarized in Table 12.2, it is assumed that an ACCUNET T1.5 channel is priced at $8550 per month.

TABLE 12.1   Summary of Credit Allowances for Interruptions of AT&T's ACCUNET T1.5 Service

Length of interruption	Credit per interruption, %
1 minute up to, but not including, 1 hour	5.0
1 hour up to, but not including, 2 hours	10.0
2 hours up to, but not including, 3 hours	15.0
3 hours up to, but not including, 4 hours	20.0
4 hours up to, but not including, 5 hours	25.0
5 hours up to, but not including, 6 hours	30.0
6 hours up to, but not including, 7 hours	35.0
7 hours up to, but not including, 8 hours	40.0
8 hours up to, but not including, 9 hours	45.0
Over 9 hours	50.0

SOURCE: AT&T Business Service Guide (Version 1.0), effective 7/31/01.

TABLE 12.2   The Application of Credit Allowances on an ACCUNET T1.5 Channel Priced at $8550 per Month

Channel	Duration	Credit allowance	Credit
Month 1			
Trouble #1	1 hour, 15 minutes	10%	$ 855.00
Trouble #2	2 hours, 59 minutes	+ 15%	+ $1,282.50
Total credit		25%	$2,137.50
		Following month's bill	$6,412.50
Month 2			
Trouble #1	35 minutes	5%	$ 427.50
Trouble #2	11 hours, 43 minutes	+ 50%	+ $4,275.00
Total credit		55%	$4,702.50
		Following month's bill	$3,847.50
Month 3			
Trouble #1	20 minutes	5%	$427.50
Total credit			
		Following month's bill	$8,122.50

SOURCE: AT&T Business Service Guide (Version 1.0), effective 7/31/01.

12.3.3 SLA tools

The increasing use of SLAs among carriers and service providers has resulted in a proliferation of performance measurement and reporting tools that provide users with the documentation they need to confront service providers when network performance falls below thresholds stated in the SLA. There are SLA reporting tools for such services as Digital Subscriber Line (DSL), IP, and frame relay as well as application-specific SLA reporting tools for e-commerce, Web hosting, managed firewall, VoIP, and enterprise applications leased from ASPs.

With regard to DSL, for example, service providers can generate a variety of SLA reports for both internal and external use. Performance management and capacity planning reports help service providers determine if their DSL equipment is optimally provisioned, and help them anticipate the need for more capacity. Service providers can also use the SLA reports to diagnose historical or chronic problems and alert their network operation centers to potential problems.

Service Level Agreement reports can also be generated for business customers, enabling them to see if a service provider is fulfilling its DSL performance commitments. The SLA reports for each business customer are separated and made available on a secure Web server (see Fig. 12.3). The ability to provide SLA reports not only reduces carriers' costs and increases revenues by enabling them to offer usage-based billing and reporting but it also enables them to move beyond offering best-effort Internet access services to offering enterprise-class services that are capable of supporting mission-critical applications. The ability to back up their SLA guarantees with performance-monitoring data enables service providers to offer tiered service offerings that differ by QoS.

12.4 Types of SLAs

There are SLAs for every type of voice or data service offered by carriers and service providers, only a few of which are discussed here because of space limitations.

12.4.1 Internet SLA

There are several metrics associated with Internet SLAs, the most important of which are those for availability, latency, and packet delivery. A good SLA will stipulate that the carrier's TCP/IP network will be available 100 percent of the time. If the carrier fails to meet this guarantee during any given calendar month, the customer's account will be credited according to the formula described in the SLA. The "network unavailability" metric

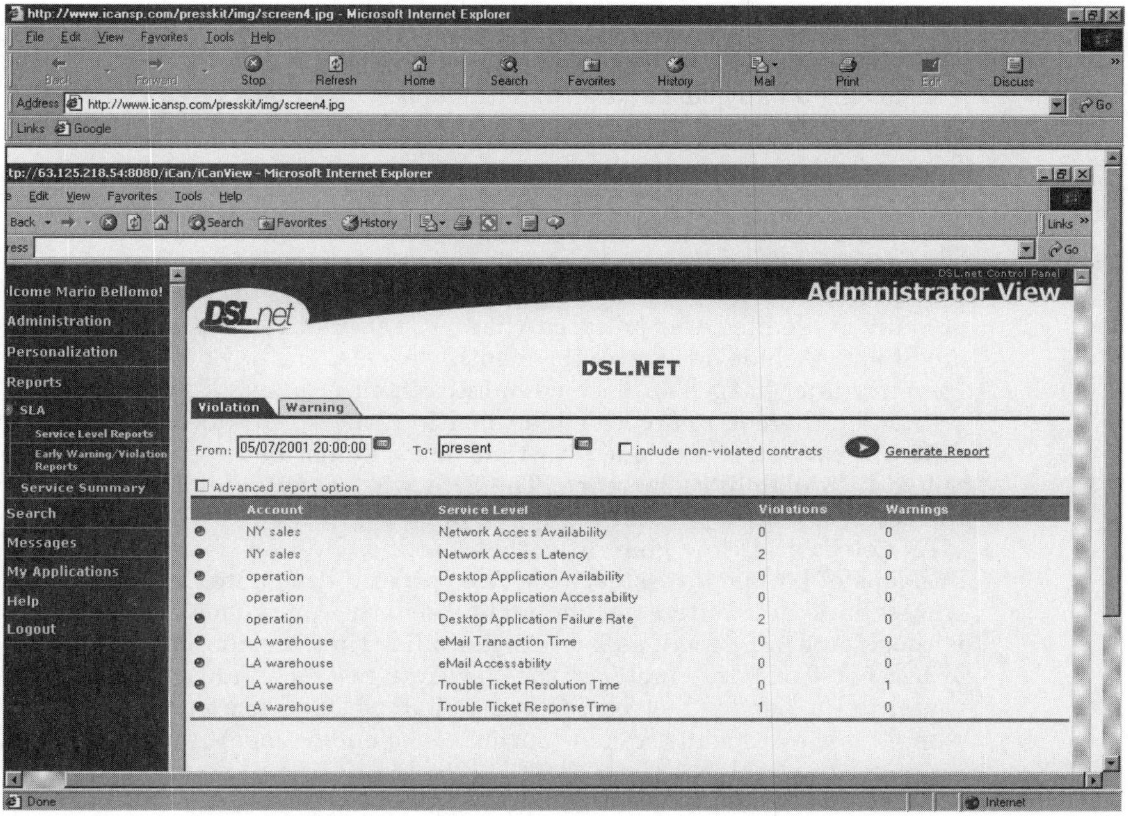

Figure 12.3 DSL.net's SLA violation report, available to customers on its secure Web server, provides advance warning indicators and violation reports from iCan Assure, enabling administrators to address issues early and make the proper adjustments.

consists of the number of minutes that the carrier's TCP/IP network, or the access circuit, was not available to the customer.

Some carriers want customers to report network unavailability and do so within a specified time period, such as within five days of the occurrence, to qualify for the credit. Other carriers are proactive and automatically credit the customer's account without requiring that the customer report the occurrence. There are usually numerous caveats that exempt the carrier from having to issue credits for network unavailability. Heading the list is unavailability resulting from

- Carrier maintenance

- Problems on telephone company circuits outside the contiguous United States
- Problems on any customer-ordered telephone company circuits
- Customer's applications, equipment, or facilities
- Acts or omissions of the customer
- Acts that are outside of the reasonable control of the carrier

If the carrier fails to meet the availability guarantee, the customer's account is credited the prorated charges. Depending on the carrier, the credit may include one day of the monthly fee and one day of the telephone company line charges for each cumulative hour of network unavailability, or fraction thereof, in any calendar month. If the access line comes from a different carrier, a problem with that line may not be covered under the network availability guarantee. The SLA will be very specific about how the credit is calculated and where it can be applied.

As noted, a latency guarantee that specifies average round-trip transmissions of 65 ms or less between the carrier's designated inter-regional transit backbone routers (i.e., hub routers) in the contiguous United States is considered very good, as is average round-trip transmissions of 120 ms or less between a hub router in the New York metropolitan area and a hub router in the London metropolitan area. Latency is measured by averaging sample measurements taken during a calendar month between hub routers.

If the carrier fails to meet any network latency guarantee in a calendar month, the customer's account is automatically credited for that month. Depending on the carrier, the credit may consist of prorated charges for one day of the monthly fee for the service. Credits will not be issued if the carrier's failure to meet the latency guarantee is attributable to causes beyond its reasonable control. These causes will be described in the SLA.

Carriers and service providers offer packet delivery guarantees on Internet services, which specify delivery rates of 99 percent or greater between hub routers in North America and, for international service, between hub routers in select cities in North America and select cities in Europe. Packet delivery is measured by averaging sample measurements taken during a calendar month between hub routers. Current and historical network performance statistics relating to the packet delivery guarantees may even be posted on the carrier or service provider's Web site, making comparison shopping easier for potential customers. Table 12.3 shows how a carrier might do this.

If the carrier or service provider fails to meet any packet delivery guarantee in a calendar month, the customer's account is credited for that

TABLE 12.3   Packet Delivery Statistics (in Percent) Issued by a Carrier and Posted on Its Public Web Site

	Transatlantic packet delivery	Europe packet delivery	North America packet delivery	Panama packet delivery
		2002		
Jan	99.958	99.862	99.933	99.947
		2001		
Dec	99.975	99.933	99.860	99.993
Nov	99.772	99.789	99.809	99.963
Oct	99.590	99.866	99.791	N/A
Sep	99.953	99.907	99.663	N/A
Aug	99.851	99.914	99.829	N/A
July	99.802	99.840	99.827	N/A
June	99.400	99.829	99.795	N/A
May	99.624	99.887	99.324	N/A
Apr	99.960	99.072	99.249	N/A
Mar	99.599	99.766	99.482	N/A
Feb	99.491	99.805	99.239	N/A
Jan	99.999	99.945	99.911	N/A
		2000		
Dec	99.880	99.766	99.957	N/A
Nov	99.936	99.915	99.942	N/A
Oct	99.875	99.901	99.867	N/A
Sep	99.994	99.489	99.926	N/A
Aug	99.950	99.972	99.942	N/A
Jul	99.945	99.978	99.961	N/A
Jun	99.700	99.956	99.906	N/A
May	99.896	99.992	99.988	N/A

month. Typically, the credit will include the prorated charges for one day of the monthly fee for the Internet service. No credits are issued if failure to meet a packet delivery guarantee is attributable to causes outside of the reasonable control of the carrier or service provider, as defined in the Service Level Agreement.

Sprint also issues performance statistics, which are posted on its public Web site, showing delay and packet loss for traffic on various global network segments. In addition to the actual metrics, Sprint lists its committed metric, which is written into its SLA, as shown in Table 12.4. If Sprint does not deliver on a performance metric, it offers to give customers a three-day credit, rather than the typical one-day industry standard credit.

In addition to availability, latency, and packet delivery guarantees, some carriers and service providers offer their Internet customers reporting guarantees. A network outage guarantee, for example, provides customer notification within 15 minutes after it is determined that service is unavailable. The standard procedure is to ping the customer's router every five minutes or so. If the router does not respond after two consecutive ping cycles, the service is considered unavailable, at which time the customer is notified by telephone, email, fax, or pager.

Another notification guarantee concerns scheduled maintenance. With this type of guarantee, the customer is notified 24 to 48 hours in advance of any maintenance activity that takes place at the hub to which the customer's circuit is connected. Usually the carrier will have a standard maintenance window, during which maintenance is performed during

TABLE 12.4   Delay and Packet Delivery Statistics, Along with the Committed Metric, Issued by Sprint and Posted on Its Public Web Site

Metrics	December 2002	January 2002	Committed metric
Transatlantic Delay	86 ms	84.2 ms	95 ms rt
Intra-Europe Delay	33.56 ms	32.12 ms	45 ms rt
USA to Japan Delay	100.26 ms	102.46 ms	120 ms rt
USA to Australia Delay	177.72 ms	181.70 ms	210 ms rt
USA to Hong Kong Delay	150.30 ms	166.68 ms	190 ms rt
Continental USA to Hawaii Delay	68.90 ms	69.56 ms	85 ms rt
Intra-USA Delay	45.10 ms	46.76 ms	55 ms rt
Transatlantic Packet Loss	0.01%	0.00%	≤ .3%
Intra-Europe Packet Loss	0.00%	0.04%	≤ .3%
USA to Japan Packet Loss	0.02%	0.01%	≤ .3%
USA to Australia Packet Loss	0.00%	0.00%	≤ .3%
USA to Hong Kong Packet Loss	0.02%	0.02%	≤ .3%
Intra-Asia Packet Loss	0.00%	0.00%	≤ .3%
Continental USA to Hawaii Packet Loss	0.01%	0.00%	≤ .3%
Intra-USA Packet Loss	0.04%	0.03%	≤ .3%

nonbusiness hours. Notice of scheduled maintenance is provided to the customer by telephone, email, fax, or pager.

If the carrier or service provider fails to meet its reporting guarantees, the customer's account is credited the prorated charges. Typically, this amounts to one day of the monthly fee for the service. The customer is responsible for providing the carrier or service provider with accurate and current contact information for his or her designated points of contact. Otherwise, as noted in the SLA, the carrier or service provider is relieved of its reporting obligations and any credits due the customer for nonperformance.

12.4.2 Collocation services

Many carriers and service providers offer their Internet customers collocation services. Under this arrangement, the customer leases floor space and/or rack space within the carrier's local network node for its Web servers, e-commerce servers, or application servers. The carrier or service provider may even offer equipment installation and ongoing management of these systems. By placing its equipment in the carrier's facility, a company can leverage the secure, fail-safe environment set up by the carrier while reducing operational costs, target its resources more effectively, and compete on a more equal footing with larger companies. The key advantage of collocation is cost savings on infrastructure, which frees the customer to target its resources in core business activities. To make the arrangement more attractive, carriers and service providers offer performance guarantees on such services as equipment installation and power availability.

A basic service associated with any collocation arrangement is the provision of on-site technical support, which may include some or all of the following operational functions:

- Power cycling of equipment to resume proper operation

- Securing cabling to prevent accidental detachment

- Setting switches to ensure normal operation

- Swapping backup tapes to protect data from loss

- Entering commands into server machines from a keyboard to perform administrative functions

- Limited diagnosis and repairs for select brands of equipment, per customer instructions

- Swapping hardware components with customer-supplied spares or upgrades

- Adding customer-supplied memory
- Conducting hardware diagnostics
- Returning defective equipment to the manufacturer or customer
- Restarting the system after the addition of hardware or software

There are some caveats that go along with these services. The carrier or service provider will not likely accept responsibility for lost revenue or service unavailability owing to delays or problems in repairing equipment or returning a system to full operational status. Nor will it be responsible for failure of equipment to boot, even if the equipment is not found to be defective. Likewise, the carrier or service provider will not be responsible for lost data or damage to a program as a result of entering commands or repairing the equipment. Another issue that merits consideration is whether having a third party perform diagnostics and maintenance of the equipment voids the manufacturer's warranty.

A collocation agreement may have provisions for on-site equipment installation by the carrier or service provider, saving a company the time and cost of dispatching a technician to a remote location. Installation activities usually consist of

- Taking inventory of equipment received
- Unpacking the equipment
- Assembling equipment based on information and instructions provided by the customer
- Labeling cables and hardware based on information provided by the customer
- Making Ethernet cable connections
- Connecting equipment to a power source
- Powering up equipment

Installations may require advance notice of up to 10 days, and the service is typically performed during normal business hours. In such cases, the carrier or service provider may require that the customer furnish appropriate systems documentation and a detailed cabinet diagram showing where each item of equipment is to be installed. The diagram should be supplemented with a detailed description of how each item is to be connected. All cables should be labeled for easy identification. Ideally, the equipment should arrive at the collocation site properly configured and with all software installed. If the customer fails to perform any of these steps, the carrier or service provider may decline to install the equipment.

Even when the customer takes care to perform all the steps necessary for a smooth installation, the Service Level Agreement may include caveats that exempt the carrier or service provider from providing the customer with remedies for

- Lost revenue or service unavailability owing to delay or problems in installing the equipment

- Failure of equipment to power up, even if the equipment is not found to be defective

- Loss of warranty owing to installation of the equipment

The SLA should completely describe the scope of the installation. For example, the SLA might specify that the carrier or service provider install network connections and all equipment within seven business days for any customer requesting 100 Mbps or less of connectivity or 20 business days for any customer requesting over 100 Mbps and up to 1000 Mbps of connectivity. The start date depends on the receipt of all contracts, forms, and documentation from the customer.

The SLA should also completely describe the remedies due to the customer upon failure of the carrier or service provider to perform. Upon installing the equipment and completing the network connections, the customer will receive from the carrier or service provider password and login information for the installed systems. The customer has a period of time, usually 10 days, within which to contact the carrier or service provider if it has failed to meet the installation guarantee. Interestingly, the carrier or service provider typically determines whether, in its reasonable commercial judgment, the service provider has failed to meet the guarantee. If the service provider agrees that it has violated the guarantee, the customer's account is credited a percentage of the standard installation fee.

Power availability is another component of equipment collocation that should be guaranteed by the carrier or service provider. The goal should be to have AC power to collocation cabinets available 100 percent of the time. Power unavailability would consist of the number of minutes that AC power was not available to the customer's collocation cabinet. Commercial power outages should be counted as power unavailability as well, since backup power in the form of batteries and diesel generators should be available to overcome such occurrences. Some carriers or service providers credit the customer's account automatically, while others force the customer to open a trouble ticket within a specific time frame after the outage in order to qualify for the credit.

In any case, power unavailability does not usually include occurrences resulting from customer circuits or equipment, customer applications or equipment, acts or omissions of the customer, or events beyond the reasonable control of the carrier or service provider, as defined in the SLA. If the guarantee is violated, the customer's account should be credited for every day the guarantee has not been met. Usually, the amount of the credit is equal to one day of the total monthly charge for every day that power was not 100 percent available to the customer's equipment.

12.4.3   Shared hosting SLA

Shared hosting refers to the arrangement whereby a company puts its Web site, e-commerce site, or enterprise applications on a server that it shares with other organizations. The server resides in a data center owned by a carrier or service provider that leases disk space and Internet connections to its customers. This arrangement relieves the customer of having to spend money on supporting infrastructure. To attract companies to shared hosting, the carrier or service provider offers a Service Level Agreement, which, among other things, contains a service availability commitment that seeks to limit server unavailability. Depending on the carrier or service provider, the goal for server unavailability can be one hour or less in any calendar month.

As with other types of service guarantees, some carriers and service providers will credit a customer's account automatically upon violation of the guarantee, while others insist that the customer open a trouble ticket within a specified time of the violation. Essentially, a server is deemed to be unavailable if it does not respond to HyperText Transfer Protocol (HTTP) requests issued by the carrier or service provider's monitoring software. But the carrier or service provider usually determines whether the server was unavailable for reasons that were within its span of control. Unless otherwise noted in the SLA, the records and data of the carrier or service provider may be the basis for all service availability calculations and determinations.

Of course, some caveats apply. Unavailability usually does not include taking the server off-line for scheduled maintenance. But customers will get ample advance notice, usually 48 hours, of when maintenance will be conducted. And usually maintenance will be scheduled to occur during nonbusiness hours. Unavailability also does not include occurrences owing to the customer's Web content or application programming, acts of the customer or its agents, network unavailability outside of the carrier or service provider's network, or unavailability owing to any event beyond its reasonable span of control, as described in the SLA.

Carriers and service providers have different ways of calculating the credit, if one is due the customer. If a customer's server was unavailable for one or more (but fewer than four) consecutive hours during a calendar month, some carriers or service providers issue a credit equal to the prorated charges for one day's service for that server. Or if the server was unavailable for four or more consecutive hours during any calendar month, the credit can be as much as the prorated charges for one week's service for that server. Credits may or may not apply to data transfer charges or to charges for services other than the monthly charge for the server for which the availability commitment was not met. In addition, there is usually a cap on the number of times per month the customer's account will be credited. Usually, customers will not be credited more than once per month, unless the SLA has been negotiated otherwise.

12.4.4   Managed hosting services

Managed hosting services entail leasing server space from an ISP or ASP for the purpose of establishing Web sites, engaging in e-commerce, or leasing enterprise applications. Subscribing to a managed hosting service is often preferable to a company setting up and managing its own infrastructure with dedicated staff. Enterprise-level applications, for example, entail having a staff of IT professionals, including systems administrators, database administrators, and help desk personnel. Applications need to be activated and then continuously monitored and supported. A flock of vendors needs to be managed. Integration with related applications can be a nightmare. Software upgrades imply another round of intensive effort. With leasing applications, businesses can achieve a 70 percent cost reduction the first year versus outright ownership, and a 30 to 50 percent cost reduction over a five-year period.

In turning over hosting to a service provider, however, there must be performance guarantees for such metrics as server installation time and server availability. And since the server or server array will be directly connected to the network at the service provider's data center, there must be network performance guarantees as well, including those for latency and packet delivery, as previously described.

Hosting providers typically offer a service availability commitment for a nonredundant server, that is, hosting services provided via a standard production server. The goal might be to limit server unavailability to less than 100 or 200 minutes or so in any week, for example, depending on the mission-criticality of the enterprise application and, consequently, the service plan selected by the customer. For an additional charge, the service provider might seek to limit unavailability to less than 25 or 50 minutes or so in any week. This might entail the use of an array, which consists of two or more servers configured for redundancy. The servers in the array load-balance HTTP traffic, which improves overall efficiency and application response time. When one server fails, the traffic load is redistributed among the remaining servers. When the failed server is restored to proper operation, the traffic load is redistributed among all the servers. The result is much less unavailability than for a nonredundant server.

Some service providers automatically credit customer accounts when the period of unavailability exceeds the number of minutes in the Service Level Agreement. Others require the customer to open a trouble ticket with the service provider's help desk or customer support organization within a stipulated time frame of the occurrence. A server or server array is usually considered unavailable if the service provider's hardware, software, operating system, or network is functioning in a manner that prevents HTTP, or SQL, access if applicable, to the server or array. If the service provider determines that the server or array was unavailable, that outage will be used to calculate a credit, which will be applied to the customer's monthly invoice. When the SLA provides for this level of service provider discretion, it is often helpful for the customer to provide supporting documentation from a recognized performance-monitoring tool. In any case, unavailability cannot arise as a result of the service provider's scheduled maintenance activities, or even emergency maintenance performed on an as-needed basis.

12.4.5   Web-enabled call centers

Call centers are specialized work environments that are equipped, staffed, and managed to handle a large volume of inbound or outbound calls. An inbound call center specializes in receiving calls via a toll-free number to take sales orders or provide customer assistance. An outbound call center is staffed with salespeople who make calls to sell a product or service. Of course, the distinction between inbound and outbound call centers is arbitrary; the same call center can do both. On the inbound side, calls can come in with customers placing orders for computers, for example. On the outbound side, calls can be placed to customers to confirm hardware and software specifications, notify customers of the shipping date, and determine satisfaction with sales assistance. And when inbound call volume drops, agents can do outbound calling.

One of the newest developments in electronic commerce is the integration of the Internet with traditional call centers to create Web-enabled call centers. This type of call center enables companies to personalize relationships with Web site visitors by providing access to a customer service agent during a critical moment—when the visitor has a question, the answer to which will influence the decision to buy. With the ability to intervene in the online purchase decision and influence the outcome, a company can realize several benefits:

- *Competitive differentiation.* By providing a convenient value-added service that improves customers' Web experiences via live agent assistance

- *Sales generation.* By removing obstacles in the buying process with immediate interaction between customers and knowledgeable call center agents

- *Increased customer satisfaction.* By delivering technical support and responding to customer needs quickly with personalized one-on-one service

For companies that do not want to create and manage the Web component but still want to receive calls from the Internet to their call center, there is the alternative of outsourcing the infrastructure to an ISP or carrier. A Service Level Agreement that guarantees performance and describes remedies for nonperformance encourages companies to entrust this part of their call center operation to an ISP or carrier. The SLA will apply only to inbound voice transactions, however, since this is where the ISP or carrier can exert its influence. Specifically, the ISP or carrier will make every effort to route calls, via its Web-based call center service, to the customer's call center. The maximum number of simultaneous in-bound voice calls will equal the number of agents for which the customer's call center is configured, plus 20 to 30 percent, which will wait in queue for the next available agent.

The SLA will define what constitutes service denial and describe the remedies that will be offered to the customer given such an occurrence. The number of agents used to calculate the maximum number of simultaneous transactions will be listed in the contract. To report service denial, the customer opens a trouble ticket with the ISP or carrier's help desk or customer support organization within a specified time frame of the occurrence. If the ISP or carrier determines that the number of simultaneous in-bound voice calls did not exceed the number of agents, that outage will be used to calculate whether a denial of service has occurred. Some ISPs and carriers will issue a credit, while others will not issue a credit, since they view the service as provided on a "best-effort" basis.

12.4.6   Managed firewall services

Companies, branch offices, and telecommuters increasingly are using "always-on" Internet access connections that provide the high-capacity bandwidth necessary to fully realize the benefits of the networked economy. On the downside, however, the benefits of always-on access come with security risks that originate from the public Internet. The risks go far beyond exposure to viruses and include serious hacker attacks that can bring down internal systems, the theft of sensitive information, or the corruption of mission-critical databases. The comprehensive solution most commonly applied to meet these threats is the firewall, which enforces security by regulating access to and from the Internet.

While firewall operation is fairly simple, complications are encountered during initial configuration of the device and in the course of managing the life-cycle issues of day-to-day operation. In a fairly benign environment, these are significant technical challenges for small and medium-size businesses. In the chaotic environment of the public Internet, with its resident hacker population eager to exploit any vulnerability, the challenge of maintaining a secure network may seem daunting.

After submitting the network to a battery of tests, the managed firewall service provider presents the customer with recommendations for fixing problems that have been identified. The recommendations will be codified in the form of rule sets—essentially security policies—that will be loaded into the firewall. Rule sets can be defined to regulate the passage of traffic according to its source and destination, specific applications and types of files, and users or groups of users; rule sets can even limit access to resources by time of day.

When a customer subscribes to a managed firewall service, the service provider should be able to take care of all configuration and design changes going forward. Any change to an existing firewall configuration, such as moving a user to a new subnet, adding a rule to the rule set, or changing an existing port configuration on the firewall, is a configuration change. Design changes are additions that must be engineered into the firewall, such as adding new hardware, software applications, or Internet services to an existing firewall.

Since the service provider prescribes a firewall rule set in consultation with the customer, there should be a grace period to allow for minor changes without charge to the customer. This period, as much as 14 days, is provided to allow for minor adjustments and testing of the new firewall rule set. Generally, customers may make unlimited changes during the grace period. After that, changes are typically billed on a time and materials basis.

The managed firewall service provider will have responsibility for maintaining backup copies of the rule sets for the firewall, at all customer locations, along with all the firewall passwords. A copy of the most recent router configuration should be kept as well. This information might be needed to reconfigure the firewall or router in case of any failures. It should take no more than four hours for the service provider to fully restore a firewall rule set and associated configuration files, assuming the dedicated connection is available or a dial-up modem link can be established.

Like the SLAs of other services, those for network security vary by carrier, but they make it easier for prospective customers to shop around for the best deal. Sprint, for example, offers multilevel IP security to customers of its Internet and VPN-managed services. Level 1 service is for customers who require the essential elements of an Internet security service at a more cost-effective rate, while Level 2 service is a more robust security solution for enterprisewide implementation. Sprint's IP security SLA provides the following performance guarantees:

- 100 percent firewall uptime and/or availability

- Four-hour response time after failure detection to fix hardware on the customer's site

- Next business day response to a "normal" firewall change management request

- Up to 25 two-hour "emergency" firewall change order responses per month for high-level service customers

- Critical-event notification within two hours for standard-offering customers; 30 minutes for high-level service customers

- Online performance reports within ten business days after the close of a month

All of Sprint's performance guarantees are supported with up to a 25 percent money-back guarantee for standard service customers and up to a 50 percent money-back guarantee for high-level service customers. In addition, Sprint's 24-hour security management center will notify customers of all problems rather than requiring customers to report a problem.

Table 12.5 provides some baseline performance benchmarks that small and midsize companies can use when evaluating carrier-provided managed firewall services. These performance benchmarks should be addressed in the Service Level Agreement, along with remedies for noncompliance.

In the event that a firewall-protected corporate network is attacked to the point that an intruder manages to damage or steal information, it must be determined if the intrusion originated from within the company or outside the company. Most damage and theft are perpetrated by employees of the company, and many managed firewall service providers will write SLAs that exempt them from liability in the event of such occurrences. If the intrusion originated from outside the company, it must be determined if there was a rule set in place to prevent such an occurrence. If the intrusion was made possible by an inadequate rule set that was approved by the customer, the service provider will not accept liability and will not provide a remedy. On the other hand, if the intrusion was made possible by a mistake in loading a rule set to the firewall, the service provider must accept liability and provide the customer with a remedy.

TABLE 12.5   Performance Benchmarks for Carrier-Managed Firewall Service

Description	Requirement, %
Network Security Operations Center (NSOC) availability	99.95
Report outage of monitored equipment within 30 minutes to customer point of contact	90
Help desk answers call from customer within eight rings	90
Call back from help desk (if required by customer) within 30 minutes of contact	90
Status reports of outage provided to customer every four hours until resolved (unless customer requests otherwise)	95
NSOC complies with established escalation requirements	95
Firewall configuration changes accomplished within 16 business hours	95
Mean Time to Response (MTTR) for troubleshooting is two hours or less	95
Firewall uptime, assuming firewall Customer Premises Equipment (CPE) is functional	95

The rapid pace of both technical and legal developments related to the Internet, combined with the current scarcity and consequent high cost of knowledgeable security personnel, makes it virtually impossible for all but large companies to attract seasoned talent to monitor their networks on a 24×7 basis, especially when multiple firewalls from different vendors are deployed. A carrier-managed firewall solution allows small and medium-size firms to implement best-of-breed security solutions at a reasonable monthly cost, but without the exorbitant expense associated with in-house implementation.

12.5   Issues to Consider

The key issue to consider with regard to SLAs is having the appropriate tools to monitor the performance of the service provider. A variety of performance metrics need to be measured on an ongoing basis to ascertain whether the carrier (or IT department) is delivering the grade of service promised in the SLA. These metrics differ according to the type of network service. A variety of vendors provide measurement tools, and each uses a different approach to address a narrow range of metrics. By being able to independently measure various performance metrics, companies have the means to effectively manage the SLA.

But taking measurements independently should not stop companies from preferring to do business with carriers and service providers that approach their SLAs proactively. After all, customers appreciate confirmation that the carrier or service provider is aware of the problem and that steps are being taken to achieve a prompt resolution. To maintain positive relations with their customers, most carriers and service providers understand the value of proactive network monitoring, which entails continuous surveillance to determine potential problems and take steps to prevent their occurrence. Unfortunately, many carriers and service providers, such as smaller Competitive Local Exchange Carriers (CLECs) and ISPs, cannot afford the tools for proactive network monitoring and do not have the mechanisms in place to warn their customers of possible service-affecting events in a timely fashion.

Although SLAs exact financial penalties from carriers and service providers when network performance does not meet expectations, it is important to remember that they do not indemnify the organization for lost business. Particularly, such guarantees do not help much if the organization's network were to go down for an extended period of time, as was demonstrated in April 1998 when AT&T's frame-relay network went down for 26 hours and in August 1999 when MCI WorldCom's frame-relay network went out of service for ten days. Both carriers tried to smooth things over with customers—AT&T by waiving frame-relay service charges until it could be sure the problem would never recur, and MCI WorldCom by doubling the credits to 20 days to compensate users for the network outage. In both cases, many customers were left to fend for themselves in dealing with lost revenues and business opportunities, unless they had the foresight to have backup services such as ISDN in place.

With regard to Internet performance, there are several reporting services available on the Web that periodically rank service providers according to various metrics. These services are not intended as SLA enforcement tools, but only to narrow down the choice of carrier or service provider. One reason that these services cannot be used for SLA enforcement is that the measurements represent an "outside-in" view of a third party and not the "inside" view of a customer. Thus the rating service may show a carrier or service provider's latency to be greater than 60 ms, while its SLA guarantees a latency *under* 60 ms.

Furthermore, ranking carriers and service providers by monthly latency shows which ones offer the fastest service, but may not take into account the consistency of this performance. For businesses that depend heavily on Internet access, it will not matter how fast the service is if the service is not available. This is why ranking service providers by "reachability" or "packet loss" is the best methodology for corporate decision making.

12.6 Conclusion

The intent of the Telecommunications Act of 1996 is to encourage competition. In order to meet the challenges of this new competitive era, carriers and service providers of all kinds are exploring new approaches to doing business. A major step has been taken with SLAs that offer service guarantees and credits to customers if various performance metrics are not sustained. In some cases, the customer need not even report the problem and provide documentation to support its claim. The carrier or ISP will report the problem to the customer and automatically apply appropriate credits to the invoice, as stipulated in the SLA. As noted throughout this chapter, however, not all carriers and ISPs follow this practice, preferring instead to put the burden on customers to report incidents that violate the SLA. As a result, the customer must invest in tools and personnel, driving up the cost of doing business.

At the same time, there is renewed use of SLAs for ensuring the peak performance of IT services within companies. Many IT departments are implementing SLAs to support the networking and system resource needs of various business units. The purpose of the intracompany SLA is to specify, in mutually agreeable metrics, what the various end-user groups can expect from IT in terms of resource availability and system response. Service Level Agreements also specify what IT can expect from end users in terms of system usage and cooperation in maintaining and refining the service levels over time. Service Level Agreements also provide a useful metric against which the performance of the IT department can be measured. How well the IT department fulfills its obligations, as spelled out in the SLA, can determine future staffing levels, budgets, raises, and bonuses. The biggest benefit of an SLA is that it raises awareness among all operating units of the business that there is a relationship between support and dollars.

Chapter 13

Technology Asset Management

13.1 Introduction

The migration of business applications from the data center to the desktop has been going on for almost 20 years and shows no signs of slowing down. After initial skepticism about the utility of Personal Computers (PCs) in the mid-1980s, virtually every company, regardless of type or size, has fully embraced desktop computers as the means to encourage enterprisewide information sharing via Local Area Networks (LANs) and Wide Area Networks (WANs). In turn, such information sharing empowers employees, improves customer response, and reduces operating costs.

Today PCs have become part of the business culture. Although hard numbers are difficult to come by, networked computers collectively are responsible for productivity gains and innumerable other competitive advantages. Yet trying to manage this diverse and growing assemblage of hardware and software has become a costly problem for many organizations. There are tens of thousands of different computer models, cards, peripherals, and software packages in use—most of them purchased from a multitude of vendors to suit the differing needs and preferences of individuals, work groups, and departments.

Unfortunately, most IT departments do not have the time, staff, and expertise to properly address the inventory and support needs of the growing population of PC users, especially when they are connected to LANs and WANs. Adding to this complexity is the growing number of notebook computers and Personal Digital Assistants (PDAs) that must be accounted for, as well as cell phones and hybrid wireless devices that combine the functionality of cell phones and PDAs.

This lack of attention can result in huge support costs. According to some industry estimates, the cost of operating and supporting a single workstation can reach $40,000 over five years. About 90 percent of this amount consists of hidden support costs, which remain unmanaged and unaccountable. A hidden cost can be anything a LAN or systems manager does not know about, such as having too many software licenses or overconfigured workstations that have more disk capacity or memory than is required for particular applications. With most PCs and workstations networked over LANs and/or accessible via remote access links, these hidden costs can be controlled through the use of asset management tools. U.S. corporations alone could save $20 billion a year by implementing asset management programs.

Another hidden cost is training. Nonstandardized configurations mean that training cannot be implemented on a corporatewide basis, which means that possible economies of scale are missed. Corporations spend about $200 billion per year on PC-related training.

Among other things, asset management tools provide the means to inventory the network. A central repository stores such information as equipment serial numbers, hard-disk configurations, and memory utilization. The tools track changes in hardware and software configurations and update the database in terms of moves, adds, and changes. They also monitor software usage, manage software licenses, handle software distribution, and implement security features.

In addition to reducing operating and support costs, this and related information can be used to more efficiently manage the network, standardize on particular hardware and software platforms, plan capacity, analyze costs and benefits, and assist in budget planning and technology migration.

The cost of asset management products differs widely, depending on the number of PCs/workstations, their features, and the level of integration. For a work group of ten users, for example, a stand-alone asset management product may cost less than $100 per machine. Enterprise asset management applications for the client-server environment can run as high as $100,000 to $400,000. The high initial cost may be recovered in a very short time, however, through savings in manual data entry, training, elimination of duplicate purchases, and gains in productivity and efficiency.

Not only does an asset management system automate inventory keeping, it also enables organizations to quickly identify underutilized equipment that can be transferred to departments requiring additional hardware. Thus existing assets can be leveraged, and new equipment expenditures reduced.

An asset management program can prove itself in a very short time. The most optimistic industry estimates claim that organizations implementing asset management can expect to reduce the total cost of system and network ownership by 10 percent in just six months. There are also appreciable increases in end-user productivity, since having the right information at hand allows help desks and on-site technicians to fix problems faster, minimizing system downtime.

13.2   Types of Assets

The major network management vendors have long neglected asset management as an integral feature, preferring instead to rely on third-party applications that work under their platforms. This neglect has resulted in not only a multiplicity of asset management products for companies to choose from, each with different features, but also interoperability problems between the different types of products. These products tend to specialize in hardware, software, or cable assets. Even when the management of two or more kinds of assets are combined into a single product, they may be better at one category than the other(s).

Compounding the problem is that the products in each category may not always work in mixed network operating system environments. A package that is adept at working with Novell's NetWare, for example, may not perform as well in other environments such as Microsoft's Windows NT, or the various flavors of UNIX. All of these variables enter into the purchase decision, with the inevitable result that network managers and administrators typically need several asset management programs to provide all the information they require.

13.2.1   Hardware assets

Recognition that desktop assets should be centrally managed and controlled is a relatively new trend. The reason is that the PC was conceived and designed as a stand-alone device. There was no need to treat the installed base in the aggregate. The means to track assets was not built into the system components. As the need developed to share data with or accept instructions from other machines, an overlay of an entirely new software and hardware system was required—a network. With PCs interconnected over the network, asset management became not only possible from a central location but also increasingly necessary.

Hardware inventory starts with identification of the major kinds of systems that are in use in the distributed computing environment—from the servers all the way down to the desktop computers as well as their various components, including the CPU, memory, boards, and disk drives. The utilities that come with servers generally provide this kind of information, along with various performance metrics. What is relatively new is the extension of these information-gathering capabilities to desktop systems. Nevertheless, asset management packages differ in their ability to correctly make such identifications. It is not uncommon for even the most sophisticated asset management products to misidentify key components or to omit certain information entirely.

Most asset management products provide the following basic hardware configuration information (see Fig. 13.1):

- *CPU*. Model and vendor
- *Memory*. Type (extended or expanded) and amount (in kilobytes or megabytes)
- *Hard disk*. Amount and percentage of disk space used and available, volume number, and directories
- *MAC address*. The address of the Network Interface Card (NIC)
- *Switch and port*. The name of the switch and the assigned port

Other products, such as Microsoft's Systems Management Server (SMS), provide very detailed asset information, which is accessible through the

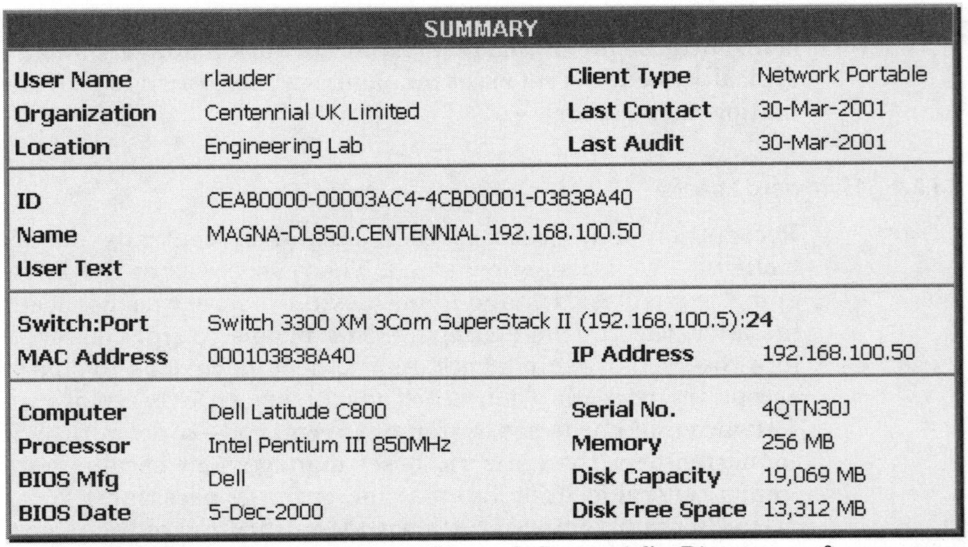

SUMMARY			
User Name	rlauder	**Client Type**	Network Portable
Organization	Centennial UK Limited	**Last Contact**	30-Mar-2001
Location	Engineering Lab	**Last Audit**	30-Mar-2001
ID	CEAB0000-00003AC4-4CBD0001-03838A40		
Name	MAGNA-DL850.CENTENNIAL.192.168.100.50		
User Text			
Switch:Port	Switch 3300 XM 3Com SuperStack II (192.168.100.5):24		
MAC Address	000103838A40	**IP Address**	192.168.100.50
Computer	Dell Latitude C800	**Serial No.**	4QTN30J
Processor	Intel Pentium III 850MHz	**Memory**	256 MB
BIOS Mfg	Dell	**Disk Capacity**	19,069 MB
BIOS Date	5-Dec-2000	**Disk Free Space**	13,312 MB

Figure 13.1   The system summary screen of Centennial's Discovery software usage monitoring tool.

familiar hierarchical tree menu (see Fig. 13.2). During hardware inventory, SMS collects a large amount of additional information about SMS clients and stores it in the SMS site database, where it can be used to create queries and collections. By default, SMS collects as many as two hundred hardware properties. If for some reason the client agent is unable to complete the inventory, the process ends, but it is retried in 24 hours. Systems Management Server also maintains a complete hardware history of all the information ever collected from the client.

Most hardware identification is based on the premise that if a driver is loaded, then the associated hardware must be present. Yet many of these drivers go unused and are not removed, resulting in inaccurate inventory information. This situation is remedied by industry standards, such as the Desktop Management Interface (DMI) and Plug and Play (PnP).

Hardware inventories can be updated automatically on a scheduled basis—daily, weekly, or monthly. Typically included as part of the hard-

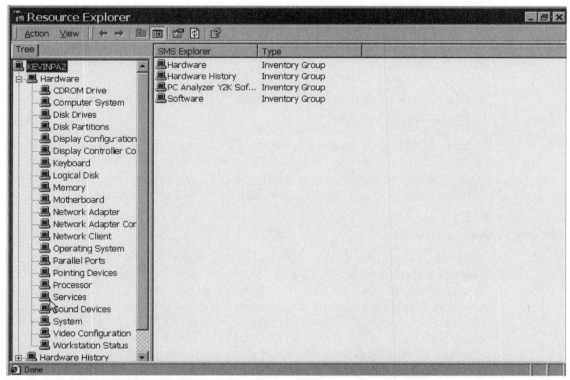

Figure 13.2 Microsoft's Systems Management Server (SMS) provides detailed configuration information about each user's computer.

ware inventory is the physical location of the unit, owner (work group or department), and name of the user. Other information may include vendor contact information and the unit's maintenance history. All of this information is manually entered and updated. When all resources are scanned, an inventory report can be printed.

In addition to providing inventory and maintenance management, some products provide procurement management as well. They maintain a catalog of authorized products from preferred suppliers, as well as list and discount prices. They track all purchase requests, purchase orders, and deliveries. With some products, even the receipt of new equipment can be automated, with the system collecting information from scans of asset tags and bar codes. Warranty information can also be added.

Still other asset management packages accommodate additional information for financial reporting, such as

- *Cost.* Purchase price of the unit and add-in components

- *Payment schedule.* Principal and interest

- *Depreciation.* One-time expense or multiyear schedule

- *Taxes.* Local, state, federal (as applicable)

- *Lease.* Terms and conditions

- *Charge back.* Cost charged against the budgets of departments, work groups, users, or projects

This kind of information is manually entered and updated in the asset management database. Depending on the product, this information can be exported to spreadsheets and other financial applications and used for budget monitoring, expense planning, and tax preparation.

13.2.2 Software assets

Software is another technology asset that must be tracked. Not only can software tracking (also called *applications metering*) reduce support costs it can also protect the company from litigation resulting from claims of copyright infringement, as when users copy and distribute software on the network in violation of the license agreement. Obtaining the initial license constitutes only 25 percent of the total life-cycle cost of software. About 75 percent of the total life-cycle cost of software is attributable to support, upgrades, and training.

Keeping track of software registration and licenses is not a simple task. Basically, tracking software assets involves assessing what products are owned and identifying their location. More sophisticated products also track software cost and depreciation; account for maintenance, training, and support costs; ensure that software is free from viruses; and manage and distribute software.

With an integrated asset management and software distribution solution, the required asset information can be used to quickly build a software distribution list. This means that the system administrator can quickly query all workstations to determine which ones have word processing or spreadsheet software installed, for example, and then create a distribution list that includes each of these workstations. The administrator can then schedule the delivery of any upgrades or security patches to each workstation.

Some management products not only automate network software distribution but also simplify the complex installation procedures, making it possible for the administrator to add conditional logic to customize mass installations across the network. Normally, it would take considerable effort for the administrator to develop a script to automate software installations because she or he would first have to figure out all of the changes that an application makes when it is installed. With the right installation package, however, the administrator can concentrate on adding the customization logic and let the installation software figure out the hundreds of files that are changed by an individual installation.

Asset management products that support software tracking automatically discover what software is being used on each system on the network by scanning local hard drives and file servers for all installed software, looking for the names of all executable files, and arranging them in alphabetical order. They determine how many copies of the executable files are installed and look into them to provide the product name and the publisher. Files that cannot be identified absolutely are listed as found but flagged as unidentified. Once the file is eventually identified, the administrator can fill in the missing information.

The administrator can monitor the usage status of all software on the network. Software is identified by total licenses, licenses used, licenses inactive, licenses free, and users queued. Such information is used to monitor software license compliance and optimize software usage for both network and locally installed software packages. Some products include filters, which let the administrator view installed software according to a specific manufacturer.

Some software metering products issue a warning if the limit on the number of legal copies in use has been exceeded. Depending on the product, the tracking program may even specify the directories in which these files were found. If not, you, as a network manager, may have to track down any illegal copies by visiting every machine to delete them.

Some inventory packages use a date and time stamp as well as file size to further identify .exe and .com files, but even this may not be enough.

Software-inventory packages should examine support files, such as .dll files, that may reside in the same directory.

From a troubleshooting perspective, some software tracking programs can alert you to changes in the user-defined configuration files, such as the config.sys, autoexec.bat, and win.ini files. Previous versions of these files can be called up to restore the unit to a functional level until in-depth troubleshooting can be performed. By backtracking to previous versions, it is often possible to determine what changes may have possibly caused the system to stop working. A history of changes can also be called up without having to visit the troublesome user and node.

In evaluating software inventory packages, it is important to determine how well the package identifies the versions of drivers and applications, the level of software use, and other such details. This is particularly important if the asset management product also automates software distribution and license management.

License management capabilities are important to have because it is a felony under U.S. federal law to copy and use (or sell) software. Companies found guilty of copyright infringement face civil penalties of up to $150,000 for each work infringed upon. In criminal cases, the maximum penalty for copyright infringement is up to $250,000 with a jail term of up to five years. The Software & Information Industry Association (formerly, the Software Publishers Association) runs a toll-free hotline and receives about 40 calls a day from whistle-blowers. The group sponsors an average of 250 lawsuits a year against companies suspected of software copyright violations. Having a license management capability can help the network administrator track down illegal copies of software and eliminate a company's exposure to litigation and financial risk.

Some software metering tools can designate applications as "portable" and then be checked out for use on nonnetworked computers. Program-specific time limits ensure that expired licenses are returned automatically. Users still have access to the programs they need at home or on a trip, while the IT department retains control of the software licenses without sacrificing security. Other metering systems either do not support mobile users, or they employ a façade that does not protect the organization from legal liability.

Application metering products not only help companies ensure license compliance but they also allow them to offset operations costs by charging groups for application usage. Charges can be assigned on the basis of general network use, such as time spent logged onto the network or disk space consumed. Reports and graphs of user groups or department charges can be printed out or exported to other programs, such as an accounting application. Although companies may not require departments or divisions to pay for applications or network usage, charge-back capabilities can still be a valuable tool for breaking down operations costs and planning for budget increases.

The charge-back feature is also useful for finding out who is using what. For example, if a company has accounting software running on six different servers, it might want to consolidate applications and reduce the number of servers. To do this properly, the company needs to know which servers are getting accessed the most. With this information, a decision can be made as to which servers can handle the load. Some other advanced capabilities of today's software metering products include

- *Custom suite metering with optimization capability.* Allows the administrator to monitor the distribution of software suites in compliance license agreements. This ensures that a suite will never be broken up illegally. The metering software automatically monitors the usage of individual components by end users and switches whole suite licenses to users working with more than one suite application at a time, leaving stand-alone licenses available whenever possible. With the optimization capability, single and suite licenses are used efficiently and legally—automatically.

- *Interactivity tracking and reminders.* Allows the administrator to track the amount of time open applications are inactive and reminds users to close inactive applications to make those resources available to others.

- *License allocation.* Allows the administrator to allocate licenses to an individual, group, machine, or any combination thereof. Access to applications is given on a priority basis to users who need it most. Overflow pool licenses can be created for common access.

- *Prioritized queuing.* Gives users the option of joining a queue when all eligible licenses for an application are in use. A queued user will be notified as soon as a license becomes available and will have a preset amount of time to access that license before it is released to other users. Different queuing arrangements and access limits can be implemented for each application.

- *License sharing across locally connected servers.* The metering software can be installed on any server for tracking applications across multiple servers. Access to licenses for a product installed on more than one server can be pooled together.

- *Enterprise management capabilities.* Allows licenses for applications to be transferred to remote locations across WAN connections, facilitating configuration changes and organizational moves.

- *Local application metering.* Tracks software usage and restricts access to unauthorized applications installed on local hard drives.

- *Enhanced application identification.* Allows the administrator to use a variety of categories to identify an application for metering, including filename, size, date, drive, and path, or any combination of these.

- *Dynamic reallocation of licenses.* Allows the administrator to transfer licenses between groups and users to accommodate emergency access needs.

Software metering products display information in a variety of ways, including

- *Graphs.* A graph is created daily for each application showing peak usage over a 24-hour day. Administrators can also view usage by group or user, as well as queuing patterns over time.

- *Error report.* An error report describes users who have been denied access to an application, attempted unauthorized access, or who have restarted their applications in mid-operation.

- *E-mail alerts.* Messages can be sent via any mail application to a designated address when an unauthorized user attempts to access an application, when a user has been denied access to an application, or when license limits have been exceeded.

- *Color-coded status screen.* A single screen displays the ongoing status of each metered application, and administrators may select different colors to indicate that a license limit has been reached or that users are in the queue.

- *Color-coded user screen.* This screen displays which users are active and which are inactive on any application.

13.2.3 Web-based asset management

For organizations that understand the value of asset management and just do not have the time to do it, there is a new option available—a hosted inventory service that provides instant, accurate inventory over the Web. Tally Systems, for example, offers WebCensus as a fast, centralized, low-maintenance way to audit enterprise PCs without the requirement to install or manage anything. The tool securely and efficiently audits computers when they access the network and returns hardware and software inventory results in minutes (see Fig. 13.3).

Information Technology managers collect inventory data by sending users an email with an embedded link to the service provider's Web site. When users click on the link, an agent is installed on the PC; the agent performs an inventory and sends it to a database on the service provider's site before uninstalling itself. All data is encrypted during

Details for Workstation BBBouchard

User	Login	Group	
Bruce Benjamin Bouchard	BBBouchard	Finance	
MAC Address	IP Address	Serial Number	
00C04F8B3677	192.169.4.654	SN16173	
Total Disk Space (MB)	Free Disk Space (MB)	Total Memory (MB)	Last Inventory
7683	3438	64	12/14/1999 12:43:29 PM

System

Dell OptiPlex G1 266MTbr+

Operating System

Microsoft Windows 95 4.00

Hardware Components

Category	SubCategory	Manufacturer	Product
			System Board
		Novell	NetWare Shell Driver 3.26-0
			Memory Module
			Memory Module
			Memory Module
BIOS	BIOS	Phoenix	ROM BIOS
CD/DVD	CD-ROM		CD-ROM Drive
Diskette	Diskette		Diskette Drive
Hard Drive	Hard Drive	Quantum	Fireball ST2.1A
Keyboard	101/102 key		101/102 keyboard
LAN Adapter	LAN Adapter	Intel	EtherExpress PRO/100B
Logical Drive	Logical Drive		FAT-16 Partition - Big DOS
Logical Drive	Logical Drive		FAT-16 Partition - Big DOS
Logical Drive	Logical Drive		FAT-16 Partition - Big DOS
Monitor	VGA	Dell	D825TM
Monitor	VGA	Gateway	CrystalScan 1572DG
Mouse	Serial Mouse	IBM compatible	PS/2 Mouse
Parallel Port	Parallel Port		Parallel Ports
Processor	Pentium II	Intel	Pentium II
Serial Port	Serial Port		Serial Ports
Video Adapter	VGA	ATI	3D RAGE IIC Controller

Software Components

Category	SubCategory	Manufacturer	Product
Asset Mgmt	PC Inventory	Tally Systems	NetCensus Win32 Collector Unknown
Comm Software	Integrated	Microsoft	Exchange Client for Win32 5.0
Comm Software	Internet Tools	Microsoft	NetMeeting for Win32 2.1
Comm Software	Internet Tools	Netscape Communications	Netscape Communicator for Win32 4.5
Comm Software	Fax	Cracchiolo and Feder	RightFAX Client for Windows 5.0
Database	DB Managers	Sybase	Sybase SQL Anywhere for Win32 5.0
Database	DB Managers	Microsoft	Access 97 8.0 SR-2
Games	Misc	Microsoft	Minesweeper 95
Games	Misc	Microsoft	Hearts Network for Win32 95
Games	Misc	Microsoft	FreeCell 95
Games	Misc	Microsoft	Solitaire 95
Graphics	Presentation	Microsoft	PowerPoint 97 8.0 SR-2
Integrated	Integrated	Microsoft	Office 97 Taskbar 8.0
Multimedia	Integrated	Macromedia	ShockWave 7 7.0
Spreadsheet	Spreadsheets	Microsoft	Excel 97 8.0 SR-2
Utility	PIM/Cont. Mgr.	Microsoft	Schedule+ for Win95 7.5
Utility	PIM/Cont. Mgr.	Microsoft	Outlook 98 8.5
Utility	Data Compress.	Nico Mak Computing	WinZip for Win32 7.0 SR-1
Word Processor	Word Processors	Microsoft	Word 97 8.0 SR-2

Open in Excel

Figure 13.3 Workstation Details is one of about 20 reports offered by Tally Systems'
WebCensus, an online inventory service.

transport. Information Technology administrators then log on to the site and run reports against the database. The reports can be used to determine such things as how many copies of Microsoft Office are deployed and which versions. The reports can be saved as an Excel spreadsheet. Aside from tracking inventory, the data can be used to plan for operating system and application upgrades. The necessity for end users to initiate the inventory collection is a weak link in the process. If users ignore the link in the email, the inventory process is not initiated.

If everyone cooperates, the advantage of a subscription-based inventory service is that it allows you, as a network manager, to offload management costs and pay only for the services you use, rather than invest in a software product and incur its associated installation and maintenance costs. A hosted service completely eliminates the need for installing, configuring, and updating complex software, thus reducing stress on IT staff, protecting IT infrastructure, and eliminating the need for in-house upgrades. Hosted inventory services are available in one-, three-, and twelve-month subscriptions and are priced from $3 to $15 per PC based on length of subscription and type of service.

Of course, there are vendors that offer editions of their inventory products that are designed to operate over the Internet. In this case, the IT department takes responsibility for ensuring security by setting up the security features during installation of the inventory software.

For instance, you can force all users to go through the HyperText Transfer Protocol Secure sockets (HTTPS) protocol to access the Web-based control center where inventory information resides. This option can be enabled only after a Secure Sockets Layer (SSL) certificate has been installed on the company's Web server. Once the option is enabled, all access to the Web-based control center is made using the HTTPS connection.

If anonymous access is allowed, then read-only access is supported without logging on to the Web site. To change any data on the site, the user must still log on to the network. If a user "guest" exists, then its configuration is used for all anonymous access, allowing the site administrator to restrict anonymous users to a specified subset of the collected inventory data.

When the inventory software is set up with the log-on cookie enabled, the user is given the option to automate the log-on process. Once the cookie is enabled and the user enters the correct username and password, he or she is then automatically recognized on subsequent visits. It is not advisable, however, to enable this feature when the Web-based control center is accessed via the public Internet. Rather, this feature is intended for use over a company's private intranet.

13.2.4 Network assets

Network management systems have always been better at accounting for network assets, rather than the hardware and software assets at the desktop. The kinds of network elements that must be monitored, controlled, and accounted for in inventory include repeaters, routers, gateways, hubs, and switches. These types of equipment are usually designed to be centrally managed and controlled from a Network Operations Center (NOC) and may be part of or added to a larger enterprise management platform such as Hewlett-Packard's OpenView.

Although network devices are often purchased as stand-alone products, they can also be packaged in the form of modules that plug into intelligent wiring hubs or switches. Since hubs and switches consolidate the LAN backbone into a tight, easily managed package, it is a natural point where additional network elements can be added to facilitate local and wide area networking, as well as asset management.

There are even asset monitoring tools for smaller networks. Cisco, for example, offers a version of CiscoWorks that simplifies the administration of small to medium-size business networks and work groups with 20 or fewer switches, routers, hubs, and access servers. The network management solution operates on Windows-based servers and includes a set of Web-based Simple Network Management Protocol (SNMP) management tools for auto-discovery and mapping of network devices as well as graphical real-time monitoring for proactively troubleshooting network issues and analyzing historical network trends. In addition, the network management solution provides inventory and device change management, network configuration and software image management, network availability, and system log analysis. A Web-based tool graphically provides real-time status of the network devices, a drill-down capability to display monitoring information on interfaces, and access configuration functions.

13.2.5 Cable assets

Also included among the assets that must be managed is the cabling that connects all of the devices on the network. There are a number of specialized applications available that keep track of the wiring associated with connectors, patch panels, cross-connects, and wiring hubs. These applications draw upon a graphical library of system components to display a network on both a geographic and a hierarchical structure. Clicking on any system component brings up the entire data path, with all its connection points. These cable management products offer color maps and floor plans that are used to illustrate the cabling infrastructure of one or more facilities. Managers can create both logical and physical views of their facilities; they can even view a complete data path by simply clicking on a connection.

Some products can calculate network load statistics to facilitate proactive management and troubleshooting and generate work orders for moving equipment or rewiring, complete with a picture of the connections. With this information, the network administrator knows where the equipment should go, what needs to be disconnected, and what should be reconnected. This function is particularly useful, since studies show that companies relocate on average 30 percent of their staff each year, often with the need to reconfigure the networks to reflect the moves. With printed work orders that include cable information, technicians can zero in on the exact connections that need to be modified.

The use of cable management tools allows administrators to plan, organize, and seamlessly execute these moves in a nondisruptive way. By being able to coordinate the changes in the cable infrastructure in real time, companies can also experience much less network system downtime, halve the average service call duration, decrease the number of interventions, and find extra capacity on the network.

Some cable asset management products automatically validate the cabling architecture by checking the continuity of the data paths and the type of network and application for every wire. The statistics that are generated about the load of the cable system or any of its branches are useful for problem tracking.

Like other types of asset management applications, cable management applications can be run as stand-alone systems (see Fig. 13.4) or may be integrated with help desk products, hub management systems, and major enterprise management platforms. The minimum requirements for a cable management system should include capabilities to

- Record the equipment, cables, and pathways for the cable plant

- Define the connectivity and circuit routes

- Search and sort cable and phone switch data

- Design, print, and export reports and move, add, and change forms

- Identify space capacity while planning major moves and changes

- Design and print floor plans

- Interface data with floor plans

- Import bitmap files of CAD drawings

- Interface with the phone switch for programming it with script files

- Synchronize management data with changes made at the phone switch

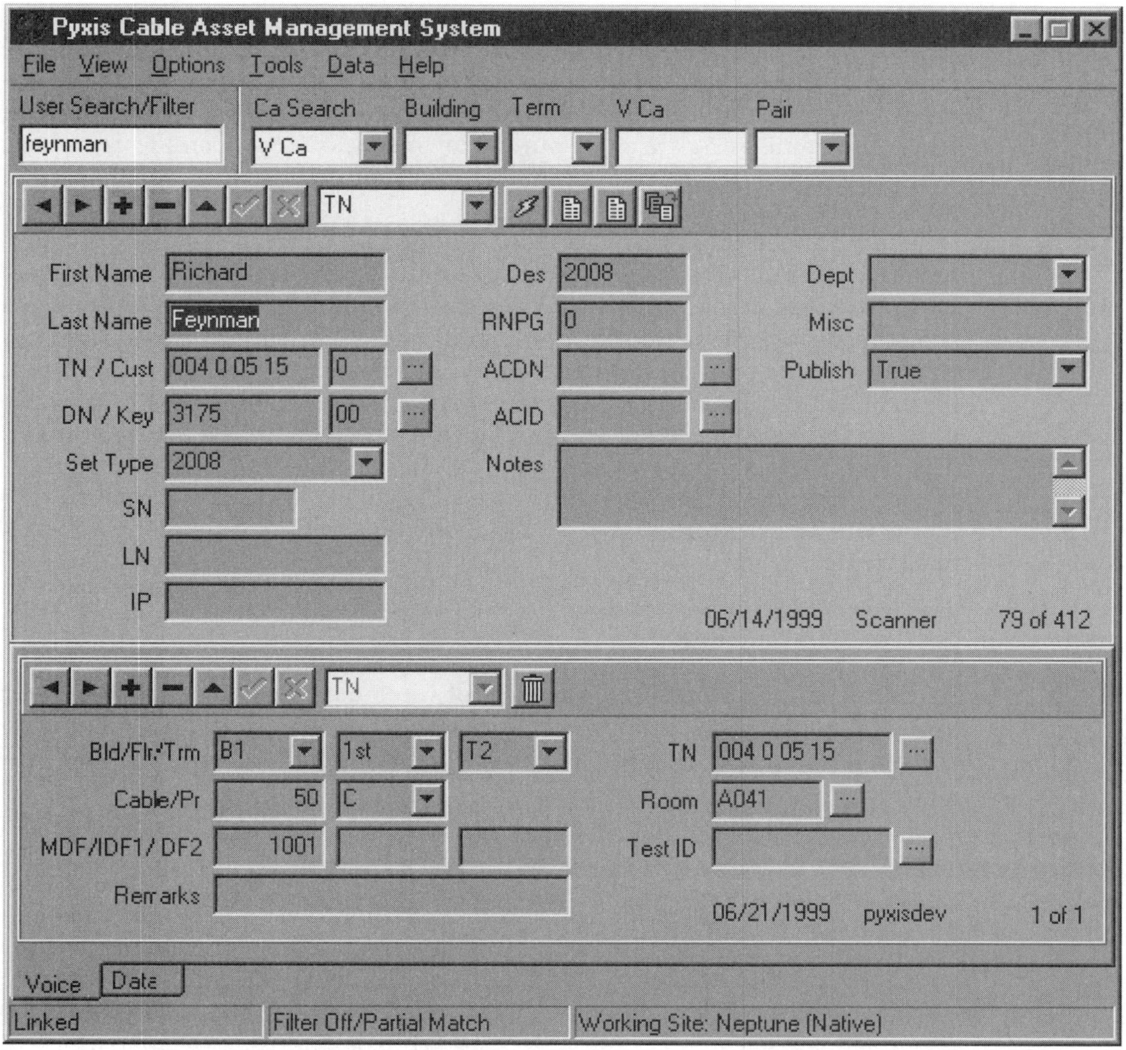

Figure 13.4 The Pyxis Cable Asset Management System from Neptune Technologies provides a view of Private Branch Exchange (PBX) cable information.

- Export and import data between text files
- Maintain permanent records of physical location and connections for every circuit
- Provide service request and trouble ticket management
- Store information for multiple buildings or campuses and show views by floor, closet, and zone

When integrated with a hub management system, for example, a cable management system's value comes in its capabilities for documenting and managing the equipment and cable plant inventory. It can identify equipment locations and actual cable runs through a CAD interface and can track the cable runs through punch-down blocks, multiconductor cables, and cable trays. It can also produce bill-of-materials reports for new and existing installations.

When a cable management system is tightly coupled with a hub management system and trouble ticket system, information can be exchanged and navigated across systems. When the hub management system detects a media failure, for example, the actual cable run can be extracted from the cable management system and submitted along with the trouble ticket for problem resolution tracking. When the trouble ticket is sent to the help desk, as network manager, you can receive a notification by email or pager. The hub management system, in turn, can be integrated with an enterprise management platform, such as HP's OpenView, allowing all of this activity to be monitored from a single management console at the network operations center.

13.3   Methods of Implementation

As noted, there are several methods of implementing asset management systems: running the application as a stand-alone application, integrating the application with help desks or network management systems, and subscribing to a third-party service.

13.3.1   Stand-alone applications

The advantage of stand-alone asset management systems is that they can be unwrapped and used immediately. This makes them very attractive for work groups, departments, and small businesses that want to get started quickly. Full-function demonstration packages are even available for download at some vendors' Web sites, thus enabling administrators to assess the look and feel of the product, as well as to evaluate its capabilities, before committing to a purchase decision.

There are over a hundred asset management products available today; the majority are for tracking computers and metering software. Some use Terminal Stay Resident (TSR) programs, while others use a disk-based utility that allows data to be collected from the stand-alone machine and then imported into a central database. Still others perform periodic scans of the network from a server that collects data from each computer's hard drive. The data is then compared with that collected from previous scans to track changes to the installed asset base.

Such products usually include a variety of preformatted reports to minimize the learning curve. These preformatted reports can even be customized to fit particular needs. To facilitate report writing, the product should also support third-party SQL knowledge bases such as Microsoft SQL Server and Sybase SQL Server for NetWare. The report generator should aid the tracking of asset movement by allowing the network administrator to specify queries, sort fields, add graphics, and save reports for later review and follow-up. Other useful features include printing bar-code labels, maintaining inventories on a room-by-room basis, and providing reports that detail missing or unidentified items.

The obvious disadvantage of stand-alone asset management products is that the data they accumulate about hardware, software, and cabling cannot always be shared with help facilities and other management or accounting applications. Therefore, to avoid having to start from scratch in the future, it is advisable to choose a product with an open architecture so as to eliminate the need for tedious manual data reentry in case integration with the existing management platform and/or third-party management applications is desired.

Such integration must be accomplished via an Application Programming Interface (API), which allows data to be transferred and used across applications instead of merely sharing a menu. An integrated solution is important because it permits all the necessary management information to be consolidated for use at a single management console.

13.3.2 Help desk

Help desks have evolved to the point where they are not just for answering hardware and software questions from end users. With the introduction of new integrated software, they have become endowed with features that make them veritable information clearinghouses about network performance and the status of desktop and cable assets.

Many callers are not familiar with the configuration details of their hardware and software, and this information is often essential for problem resolution. By combining traditional help desk functions with an asset management database, operators and technicians can have easy access to configuration information. With this information readily available, the time the help desk personnel will have to spend with any single caller will be greatly reduced.

In the course of resolving problems, operators and technicians can also keep the asset management database current as new software versions and hardware components are installed. Since the help desk may also play a role in administering moves, adds, and changes, this function also becomes the means of keeping the asset management database updated with regard to the current user and location of equipment.

13.3.3 Network management systems

The major network management platform vendors have formal programs that encourage the development of third-party management applications. When choosing a third-party application, it is important to determine what level of integration the application provides. Such products offer differing levels of integration: data sharing, menu bar, or Graphical User Interface (GUI).

The highest level of integration involves *data sharing* with the platform in key application areas. In other words, data from one application can be collected and manipulated by another application, thereby simplifying the database structure and conserving resource utilization. The application should have passed interoperability testing conducted by the platform vendor.

Menu bar integration means that the network manager does not have to close one application to open another, and possibly have to change various operational settings before continuing. Menu bar integration typically involves the use of a macro or script program that comes with the third-party application, enabling it to show up on one of the platform's pull-down menus. Whatever data sharing takes place is very limited and must be specifically written into the macro.

The lowest level of integration involves only the use of the platform's *GUI*. Third-party applications can be launched from the platform by clicking on the appropriate icon. This level of integration provides separate, but complementary, management functionality. No data sharing takes place between applications.

The management systems of intelligent wiring hubs usually include tools for reporting the identity and attributes of various nodes on the network, as well as their configuration and operational status. These management systems also provide a graphical map depicting the arrangement of modules in the hub chassis. A zoom feature allows you, as the network manager, to focus in on a particular module. The module is identified by model—including model name, IP address, and security string—and by device name, type, location, and firmware version. The specific configuration of that module, including ports and connections, can also be called up for viewing or printing. The operational status of the module is indicated by the Light Emitting Diodes (LEDs) that are also displayed graphically. In addition, performance statistics are available for each module.

13.3.4 Third-party subscription services

Another method of implementing asset management is by subscription through a third-party service. Tally Systems, for example, offers an online asset management system called WebCensus, which is available in 1-, 3-, and 12-month subscriptions. It requires no installation, training, or maintenance. Reports are accessible over the Web, and inventory accuracy is achieved with a unique algorithm that, combined with hardware and software "fingerprints," ensures comprehensive, accurate, and automatic inventory results.

Tally Systems works with PC manufacturers to develop hardware fingerprints that enable its recognition algorithm to identify manufacturer, model, serial number, and other details of their hardware during a scan. Instead of simply reporting "Pentium III," the scan can report "Compaq Proliant, Pentium III, Serial number 123456." If these initial tests are inconclusive, System Management Basic Input/Output System (SMBIOS), DMI, and other data can be compared and contrasted to report on discovered hardware.

To create software fingerprints, each application is installed individually and then analyzed for recognition variables such as filenames, sizes, and combinations that will consistently set it apart from other applications and other versions. By analyzing file combinations in fully installed products, software fingerprints identify vendor and product names, as well as specific version numbers, serial numbers, and foreign language versions.

Some computer leasing companies offer asset management services that go beyond the traditional lease versus purchase analysis—encompassing cost of funds, useful lives of equipment, and residual expectancy—to include the analysis of the complete life-cycle costs of technology assets, from initial acquisition through disposal, weighing financial implications with such issues as usage, software, support, maintenance, and risk of interruption.

Not only do these asset management services help track systems, software, and personnel resources; some may include consulting and management services as well. Under the service, companies can contract with the service provider to gather information about hardware, software, and other resources in a customer-specific relational database, which may contain over six hundred potential data elements for each asset.

In tracking software, for example, such systems identify each license by asset ID, product, version, manufacturer, contract number, license fee, renewal fee, validation date (from and through), license code, license key, number of installations allowed, and number of users allowed. Typically, the asset management system also accommodates free-form comments.

The information collected can help IT departments make decisions on such issues as software licensing agreements, hardware and software standardization, technology upgrades and migration, maintenance, and budgeting. Reports can be obtained that are categorized by user and/or location and that graphically illustrate asset information.

After the initial inventory, the service will update the asset information to include ongoing changes. To maintain database accuracy, a mixture of internally developed and third-party tools may be used, which reside on network servers to automate reporting of configuration changes. Updates can be loaded into the customer's asset database over public and private networks, the Internet, or dial-up serial connections. The service provider usually offers an 800 number, online forms, customized fax service, and email services to facilitate customer reporting of moves, adds, and changes. The service provider may also offer periodic physical inventories to ensure that all changes have been properly recognized and that the database is current and accurate.

Access to information from the asset management service is usually available to network managers through an enterprise management platform. OpenView users, for example, can read information from the service provider's database when and if that information is pertinent to systems management routines. At the same time, the asset management service database can retrieve data from the customer's on-site OpenView database to provide more complete asset reporting. In addition, the service database can share asset information with third-party network management systems that interface with OpenView.

Other asset management services approach asset management from the perspective of the help desk, which, in addition to providing traditional support functions, acts as a gateway to a coordinated suite of complementary services that also include maintenance, configuration, procurement, and multivendor network integration.

With services oriented for use with the help desk, the service provider offers 24-hour operational guidance and troubleshooting assistance to users. Among other things, this service entails providing immediate telephone support for industry-standard PC and LAN software and hardware. When problems arise, the end user dials a dedicated 800 number to reach the service provider's central support center. The technical specialist who takes the call answers the end user's questions. If the problem is due to faulty equipment or software, a technician may be dispatched to the customer site. If nonsupported equipment or software is involved, the help desk will route the call to the right vendor.

With the help desk as the central point of all transactions, the service provider can offer value-added services. If an end user needs more memory or hard disk capacity, for example, it can be ordered through the help desk. The service provider will even distribute and upgrade software electronically in accordance with each vendor's license agreement. Other services may include trouble ticket tracking, management reports, and multivendor service contract management.

13.4   The Cost of Asset Management

Most asset management offerings consist of software products, which range in price from approximately $10 to $55 per desktop. Turnkey enterprise systems may cost $200,000 or more and may entail additional charges for systems integration and customization. And there is also the cost of staff time spent collecting data, maintaining the database, and defining and generating the various reports. A do-it-yourself asset management program requires not only a commitment of time and staff but also a hefty budget.

Pricing for third-party asset management services is typically based on the service level desired by the customer and the corresponding level of effort required on the part of the vendor. The cost factors include

- The number of assets tracked

- The amount of information required for each asset

- The geographical distribution of assets, which can affect physical inventory collection

- Reporting requirements in terms of the number of reports and their frequency

- The system and/or LAN environment, which affects the use of automated data collection tools

- The kind of online access and/or system linkage required by the customer

Typically, the service firm charges a nonrecurring setup cost, which can be billed separately or amortized with ongoing operating costs over the life of the contract on a dollar-per-asset-per-month basis. This boils down to two payment options for users:

- *Pay as you go.* With this payment option, customers pay a nonrecurring setup charge for data collection and database loading, and a monthly charge for database maintenance and standard management reports. Under this payment plan, customers can expect to pay $5 to $10 a month per asset over a three-year period for a no-frills basic asset management service.

- *Monthly installment.* With this payment option, customers pay a fixed monthly charge for a term of at least three years, with the setup cost spread over the monthly installments. Under this payment plan, customers can expect to pay $15 to $20 a month per asset over a three-year period.

An asset management service should provide an effective way to track hardware and software assets and control costs in a way that will not unduly burden internal staff. After the three-year commitment with a service firm, the fully functional asset management program can be brought in-house with much less disruption to daily business operations than would be created by starting a program from scratch.

13.5   Standards

Most asset management tools for the distributed computing environment are proprietary in nature. The emergence of standards such as the Desktop Management Interface (DMI), developed by the Desktop Management Task Force (DMTF), and Plug and Play (PnP), developed by a consortium of computer equipment and software vendors that includes Microsoft, Compaq, and Intel, may help broaden usage of asset management systems.

13.5.1   Desktop Management Task Force

The Desktop Management Task Force (DMTF) was formed in 1992 to develop and deliver the enabling technology for building a new generation of PC systems and products that will make them easier to manage and configure. The goals of the DMTF were to design a programming interface for easier desktop management, to specify a method for making desktop components manageable, to define a way for management applications to gather information, and to simplify implementation of network management. These goals, now largely achieved, extend to platforms, add-ons, peripherals, management applications, and management consoles.

Desktop Management Interface gives network administrators a window into their PCs, allowing them to monitor, manage, and configure any DMI-compliant product either locally or remotely. Desktop Management Interface defines two application programming interfaces: the Management Interface (MI), which provides management applications with a common method of querying and controlling network resources; and the Component Interface (CI), which lets software and hardware components tell the network about themselves.

Desktop Management Interface's Management Information Format (MIF) defines the standard manageable attributes of PCs, servers, printers, LAN adapters, applications, and various other peripheral devices. These attributes describe a product's identifying characteristics such as name, manufacturer, version, serial number, and capabilities such as speed. The MIF will eventually include attributes to support installation, de-installation, verification, and maintenance.

When an application or hardware product is installed, its MIF is passed to the Service Layer, a local resident program that passes the information to various management applications via the Management Interface. The Service Layer resides in the operating system and acts as a traffic controller, handling all requests for data in the MIF. The Service Layer dynamically notifies management applications of the new device and makes information about that device available to other products, even if they are from different manufacturers.

13.5.2 Plug and Play

Plug and Play (PnP) identifies components in Extended Industry Standard Architecture (EISA) and Micro Channel Architecture (MCA) machines at start-up via the Basic Input-Output System (BIOS). Plug and Play makes use of these architectures' built-in ID codes to provide the inventory. The older Industry Standard Architecture (ISA) does not include such codes. Users will have to rely on asset management packages to take inventory of the components, or they will have to buy new ISA components that comply with the PnP standard.

In addition to facilitating asset management, PnP is designed to self-configure PCs, without the user worrying about such things as device conflicts, jumper settings, and modifying the config.sys file. Today, without PnP, switches and jumpers have to be set for each installed device. Often the user does not know there is an address conflict until after the device is installed and the system is rebooted. At that point, a network connection might be lost or, worse yet, the system will not reboot.

Operating systems that support PnP pass on the configuration information to any application. This means that asset management packages are able to make use of this information without having to gather it themselves. They can focus instead on creating analyses and reports for managing the hardware.

Plug and Play is backed by many of the same vendors who are behind the DMTF. Plug and Play is similar to but less extensive than DMI in its information-gathering capabilities. Plug and Play is compatible with DMI but functions only at boot time. Desktop Management Interface is also compatible with the SNMP, which will allow network managers to retrieve the kind of information DMI collects.

13.6 Conclusion

Asset management is a methodology, combined with one or more software products, which helps IT departments gain control over what they own and operate. Asset management administers every piece of the technology puzzle—from users and IT staff, to the procurement process, to specific pieces of hardware and versions of software. An effective asset management program optimizes the use, deployment, and disposal of all assets. It can ease the record-keeping burden of routine moves, adds, and changes. In addition to containing the cost of technology acquisitions and reining in hidden costs, such programs can improve help desk operations, enhance network management, assist with technology migrations, and provide essential information for planning a reengineering strategy.

For some organizations, asset management is a form of protection that helps guard against fraud, waste, and abuse. The failure to implement an asset management program can actually encourage theft or "asset shrinkage." If employees know that assets are not accounted for and that they will not be held accountable for lost or stolen equipment, they are more inclined to carelessness when entrusted with company assets. Although all companies are potentially vulnerable to employee theft, those with no way to track assets suffer greater losses than do companies that can track assets.

Keeping the Network Healthy

Maintenance and Support Planning

14.1 Introduction

Voice and data communications networks have increased in functionality and complexity, catapulting maintenance and support into the forefront of issues that you, as a network manager, must grapple with. Unfortunately, the importance of maintenance and support is not always recognized until problems arise. While managers have become more attuned to the importance of evaluating vendors in terms of their maintenance and support capabilities before making major purchase decisions, it is still easy to overlook the costs and commitment associated with these activities over the life expectancy of telecom equipment and information systems. It is also easy to overlook the effect of these considerations on the price-performance equation.

Communications equipment rarely stands alone in the business environment; this equipment is most likely part of an expansive, complex network that requires continuous maintenance and support. Whether problems are revealed through alarms, diagnostics, predictive methods, or through user notification, the need for timely and qualified maintenance and support services is of critical importance. For some organizations, continued survival and competitive advantage hinge on the proper functioning of communication systems and networks. For all organizations, a problem-free environment ensures full return on investment for all technology acquisitions. Recognizing these concerns, many traditional third-party maintenance firms have expanded their services from simple repair-and-return operations to on-site maintenance in multivendor environments.

Manufacturers of data communications equipment also offer maintenance services, providing customers with a broad range of plans that encompass systems that are in or out of warranty. Some of these firms even support mixed-vendor installations. With carriers bundling hardware with their services—such as videoconferencing systems with Integrated Services Digital Network (ISDN) or Integrated Access Devices (IADs) with voice, data, and Internet over the same access line—they too have begun to offer maintenance plans, buttressed by Service Level Agreements (SLAs). The competition between traditional maintenance firms, hardware vendors, and carriers over the years has had at least three important ramifications:

- The cost of maintenance and support services has dropped dramatically; in the case of high-end Private Branch Exchanges (PBXs), by as much as 50 percent in the last decade.

- Organizations now have more choice in the selection of service firms; they need not be pressured into making decisions at the time of equipment purchase, nor must they be boxed in by single-source procurements.

- Organizations now have more leverage; that is, continuation of the service agreement may be contingent upon good vendor performance. The user may even elect to use several service firms at once, making each compete for a bigger slice of the pie.

As an aid to planning, this chapter will outline the general service approaches that have become available in this highly competitive marketplace and explore some of the issues you, as a network manager, should consider when choosing maintenance and support services.

14.2   Service and Support Offerings

The service and support offerings of carriers, vendors, and third parties may encompass dozens of individual activities. Generally, service and support activities include, but are not limited to

- Site engineering, utilities installation, cable laying, and rewiring

- Performance monitoring of the system or network, alarm interpretation, and initiation of diagnostic activities

- Identification and isolation of system faults and degraded facilities on the network

- Notification of the appropriate hardware vendor or carrier for service restoration

- The repair or replacement of the faulty system or component by a service technician dispatched to the site

- Monitoring of the repair and/or replacement process and the escalation of problems that cannot be readily solved

- Testing of the restoration action to verify proper operation of the system or network

- Trouble ticket and work order administration, inventory tracking, maintenance histories, and cost control

- Administration of moves and changes

- Network design, tuning, and optimization

- Systems documentation and training

- Preventive maintenance

A variety of other types of support are also available, such as 24-hour telephone ("hot line") assistance, short-term equipment rental, fast equipment exchange, and guaranteed response time. Customized cooperative maintenance plans qualify the organization for premium reductions if an internal help desk is established to weed out routine problems, most of which are applications related. An increasingly popular support offering is remote diagnostics and network management from the vendor or carrier's Network Operations Center (NOC), which perform continuous surveillance of equipment and access lines on a 24 × 7 basis.

There are several distinct methods of delivery for maintenance and support services, each with its advantages and disadvantages:

- In-house staff

- Equipment vendor or communications carrier

- Cooperative arrangement between user and equipment vendor or communications carrier

- Third-party maintenance firm

Considering that annual service charges may amount to 4 to 10 percent of the equipment's list price (sometimes more), making the wrong decision can prove to be quite costly. For example, a maintenance program for a $750,000 router network may cost $30,000 to $75,000 annually, which may or may not include certain types of spare componentry, repairs, and after-hours service.

14.3   In-House Approach

When it comes to the provision of maintenance and support, many large organizations have become more self-reliant out of dissatisfaction with the quality of service and response time provided by vendors and other maintenance providers. Although the vendor community is becoming more attuned to the service needs of users and is attempting to meet these needs with new and flexible programs, backed by SLAs, many are simply too small to have elaborate service programs that can respond quickly when customers experience trouble with their networks. Even more vendors lack sufficient field service personnel to staff their customer support operations. In addition, most equipment vendors lack a nationwide—let alone international—service presence. Consequently, they must enter into strategic relationships with third-party maintenance firms to ensure that an acceptable level of service and response is available to their customer base.

Many large organizations provide their own maintenance services, if not for high-end systems such as backbone routers and PBXs, then for equipment attached to a Local Area Network (LAN), such as microcomputers, printers, and other peripherals. All such equipment is very sensitive to heat, static electricity, dampness, dust, grease, food and drink, and smoke. An in-house preventive maintenance program can mitigate the damage done by these contaminants, thereby prolonging the useful life of equipment and saving money on vendor-provided maintenance services.

Scheduled preventive maintenance every hundred hours or so can spot such common problems as misaligned heads on floppy disk drives before extensive damage is done to the disks, rendering stored data irretrievable. Preventive maintenance also involves checking air filters and ventilation systems in "clean rooms" and wire closets, testing Uninterruptible Power Supplies (UPSs) for proper operation, and periodically inspecting for proper electrical grounding.

By performing their own preventive maintenance, many organizations are discovering that they can reap substantial savings in both time and money. Furthermore, the ready availability of technical training, certification programs, and comprehensive troubleshooting guides, as well as relatively inexpensive test equipment, makes faulty subsystems, boards, and chips fairly easy to isolate and fix. Of note is that the Federal Communications Commission's (FCC's) Part 68 regulations prohibit repair or modification of registered equipment, except by the manufacturer or its authorized service agent. Failure to comply with this rule may void the warranty and possibly FCC registration.* Spare parts can be ordered by phone and, in most cases, shipped on the same day for delivery the next day. In extreme cases, in-house technical staff can ship the faulty unit to a third-party maintenance vendor, who can provide a loaner until the original equipment is repaired.

With in-house technical staff and a help desk, most problems can be diagnosed and fixed in a short amount of time, especially if they are applications problems. Users can be up and running again in a matter of minutes instead of waiting hours (or until the next business day) for outside help to arrive. But in deciding whether to handle maintenance in-house, the organization must determine if it has the resources, time, and the commitment to handle the job. Corporate management must become "diagnostic aware" and demonstrate support for the program with a realistic budget and charge network managers with responsibility for carrying out the program.

Many companies perceive no risk at all in providing in-house maintenance services, mainly because advances in technology and production processes have combined to greatly increase the reliability of today's communications products. In addition, many products are now modular in design, permitting fast isolation and easy replacement of faulty components from inventory. Some products even come standard-equipped with redundant power supplies, backplanes, and control logic so that when subsystem A fails, subsystem B takes over with minimal disruption in performance. While these factors contribute to the timeliness and quality of in-house maintenance, this is not to say that an in-house maintenance program will come cheaply. In fact, an in-house maintenance program requires a substantial investment in technical staff, among other things.

14.3.1 Staffing requirements

As noted in Chapter 1, a self-sufficient communications department is usually divided into several specialized areas of service. Among them, the

---

*As of July 2001, the FCC no longer accepts applications for certification of terminal equipment under 47 CFR Part 68. Part 68 was originally designed to protect wire lines and telecommunications employees from harm, while fostering competition. It called for the phone companies to allow other manufacturers' terminal equipment to be attached to their wire lines if the terminal equipment complied with Part 68, ensuring that the equipment would not damage the wire lines or harm personnel. The FCC decided that with the speed of technological changes and advances in the telecommunications industry, it would no longer micro-monitor each piece of equipment. The Administrative Council for Terminal Attachments (ACTA) is now the body responsible for establishing and maintaining a database of equipment found to be compliant with industry-established technical criteria, establishing numbering and labeling requirements, and establishing filing requirements for certification.

help desk, technicians, and operations management play key roles in providing maintenance services, and each function has its own staffing requirements. While these three functions are mandatory for in-house programs without vendor support, the help desk and operations management can also be well utilized in conjunction with traditional vendor maintenance plans. The reasons for such specialization in the delivery of maintenance services are as follows:

- First, the communications department must strive to provide the best possible response to its user community.

- Second, the department must work toward developing quick and cost-effective troubleshooting procedures to minimize downtime and conserve expensive labor resources.

- Third, the department should provide effective systems and network management—a necessity in today's multivendor voice and data environment.

These areas of expertise virtually demand specialized staff, who must possess an up-to-date set of skills.

14.3.2 User support

The first level of operations support is the user support position, which is essentially the help desk. The responsibilities of this individual or group fall into six categories: troubles; knowledge base; moves, adds, changes; training; record keeping; and interpersonal communications skills. The user support position serves as the first point of contact for users who are experiencing problems. This position requires individuals with demonstrated interpersonal communications skills, as well as technical competence in the use of applications, databases, systems, and networks. Usually there will be several people with different skills staffing the help desk to provide appropriate assistance.

Troubles. The user support position handles troubles by

- Screening the problem to ascertain whether the trouble is with the user, application, database, system, or network to determine whether a technician should be dispatched

- Instructing users on the proper operation of hardware and software to prevent the recurring problems

- Preparing and administering trouble tickets to identify, organize, and track problem resolution activities

- Initiating and maintaining log entries to provide statistics on troubles by category on a daily, weekly, or monthly basis

- Monitoring trouble report escalation, according to defined response times and escalation procedures

- Verifying cleared troubles with users to determine level of satisfaction and service quality

Help desk operators are usually able to answer from 50 to 70 percent of all calls without having to pass them on to another authority.

Knowledge base.   With expert systems, organizations can basically discern what application and communication experts know, encode it, store it in a database, and make it accessible when needed. This way, whenever the resident expert is unavailable, that accumulated knowledge and expertise can still be applied, translating users' problem descriptions into workable solutions.

Moves, adds, changes.   The user support position handles moves, adds, and changes by

- Processing move, add, and change requests from users, work groups, and departments

- Assigning effective implementation dates to moves, adds, and changes

- Providing move, add, and change information to technicians so they can go to the right location and be properly equipped to resolve problems

- Monitoring move, add, and change requests for daily work scheduling and to ensure the completion of work by due dates

- Updating the help desk database to facilitate problem resolution

- Handling reconfigurations such as feature, ports, and password assignments

- Creating service orders for the repair or replacement of components and subsystems

- Maintaining order and receiving logs to track the movement of new equipment and software

- Preparing summary reports of move, add, and change activities on a daily, weekly, or monthly basis

As noted in Chapter 1, if these responsibilities overwhelm the help desk, they can be escalated to a higher level of support, perhaps to you, as the network manager, who can interface with a vendor, carrier, or integrator to find the cause of a problem.

Training.   The training function of the user support group includes

- Providing on-the-spot user assistance and training via the help line or through a remote-control "show me" connection whereby the help desk operator takes over the operation of a remote desktop computer to show the user how to navigate through a database or application

- Conducting scheduled training sessions for multiple users, in either a classroom environment or a distance learning arrangement

- Interfacing regularly with various department managers to ascertain the training requirements of their staff

- Preparing training summary reports on a daily, weekly, or monthly basis

Record keeping.   This function involves maintaining a central library of all operations logs and reports. This work usually is accomplished with the aid of software associated with other help desk functions. Typically, data is entered into various online forms. The forms provide the raw information used by the report generator to issue a set of standard reports. Most software of this type accommodates Structured Query Language (SQL), so reports can be customized to meet organizational needs. In addition, the information can be exported to other applications such as spreadsheets for further analysis and be displayed in various presentation formats, including pie charts, bar charts, and histograms.

Interpersonal communications skills.   The user support position requires individuals with demonstrated communications and organizational skills. A sense of diplomacy and urgency is also very important. As the primary contact point for the user community, the support staff must have well-developed "people" skills so that users feel their problems are being given the attention that is required. The staff must also be self-motivated and self-directed to a large extent since the workload is not set by a schedule but rather by the ringing of the phone.

Beyond the capacity for self-direction and the possession of strong interpersonal communications skills, the support staff should have hands-on experience with a variety of hardware and software. After all, users require that the support person be able to effectively handle their problems, and most hardware and software products have their little quirks and nuances that are only appreciated through experience. Although this kind of information can be shared among support people, and may even be available in a problems-solutions database (i.e., a knowledge base), the need for formal cross training may be warranted as organizations increasingly move toward multivendor and multiplatform environments.

14.3.3 Technical support

The next level of operations support is the technician. The technician's key function is to provide routine and remedial service for systems and networks through quick, efficient, and cost-effective diagnosis, troubleshooting, and service restoration procedures. Problems that must be fixed on-site are assigned to a technician by the help desk or NOC. The current shortage of technical personnel makes technical support a competitive field. Individuals with internetwork and systems integration experience are especially valuable. Special incentive programs should be devised to keep them and preserve staff continuity. Following is a summary of the typical job responsibilities of a technician:

- Support of voice and data systems and networks in multivendor, multiplatform environments

- Routine monitoring, which involves daily testing and adjustments to keep systems and networks optimized to handle applications efficiently

- Remedial maintenance, which involves problem identification, testing, and system or network restoration in response to trouble reports

- Preventive maintenance, which involves performing various maintenance procedures on a scheduled basis per manufacturer recommendations

- Interface with multiple vendors and carriers to isolate problems to specific products or services

- Maintainance of the spares inventory and keeping test equipment properly tuned

- Implementation of moves, adds, and changes, especially when they involve system disassembly and/or reassembly and special handling

- New installations, especially when they involve cable and equipment in wiring closets, hubs, and network nodes

- Documentation, which involves maintaining the alarm log, preventive maintenance log, vendor notification and escalation log, repair-replace log, and shipping-receiving log

Depending on the size and physical layout of an information system or network, the number of technicians and specialists required will vary, as well as their respective skill levels. With regard to information systems, for example, among the factors affecting staffing requirements is the sophistication of such systems:

- Are the systems largely defined in software?
- Are users continually demanding advanced capabilities?
- Are many types of communications interfaces, lines, and protocols required?
- Are the applications distributed via client-server networks?
- Are the applications classified as "mission-critical," requiring a high level of security?

Such requirements affect the hiring of technicians, making the candidate's level of formal training, work experience, and demonstrated expertise all the more important. A PBX technician, for example, should have prior experience with data communications because most troubles encountered in today's integrated voice-data PBXs are based in software rather than in hardware. The technician must feel equally at ease in chasing 0s and 1s throughout the latest generation of switches as in measuring the standard analog parameters of any legacy switches that may be on the network. In addition, as data applications are added to the PBX, the possible points of trouble increase dramatically. The technician must know how to troubleshoot the various data components and be able to check the integrity of various communications protocols. This requires knowledge and expertise in the use of digital test equipment and protocol analyzers.

Training is an ongoing activity; in many cases, equipment warranties will not be honored unless in-house technicians have been trained by the vendor and achieve certification on specific products. The amount of time required to complete programs at such schools varies according to the type of product. Training in the use of a central office digital switch adapted for use on a private network, for example, may require up to 16 weeks of school at the vendor's location. For a PBX, the training period may last up to 10 weeks; for a key system, four weeks; for most desktop computers, a full week. Not only must the technician's travel and living expenses be factored into the cost of in-house maintenance but the organization must also be prepared to go without the benefit of that person's services until training is completed.

Ongoing training is also made necessary by the sheer number of equipment types that must be interconnected over the communications network and by the fast pace of technological innovation among vendors. If the organization's communications needs are continually changing, additional training may be necessary to keep the technician(s) current on engineering changes, new types of devices added to the network, proper maintenance procedures, and the use of more sophisticated diagnostic tools. Depending on the nature of these changes, people with specific expertise may have to be hired to augment existing personnel.

The value of the in-house maintenance program can be increased dramatically by devising a cross-training program whereby a senior technician teaches other technicians to become proficient in servicing multiple equipment types that are out of warranty. This kind of program will provide the flexibility the organization needs to make future product selections. An in-house certification program can even be implemented to motivate higher levels of performance among technical staff, completion of which may weigh heavily in future raises and promotions. Instituting such a program assumes that a technician who is both technically competent and reasonably skilled at teaching others is available. Even so, the time spent away from mission-critical activities must be weighed against the benefits of the training effort itself.

14.3.4 Operations management

The third major position in the in-house maintenance organization is that of the operations manager. This person is responsible for ensuring system integrity and an optimum grade of service. The key function of this position is to oversee all operations procedures and resources to ensure maximum system availability and accountability. This person is the highest point of escalation for problems that cannot be solved at the user support or technician levels. Typical job responsibilities include the following:

- Oversight of all maintenance and operations functions including fault detection; service restoration; vendor relations; software and database modifications; installation activities; moves, adds, and changes; and inventory and spares kit levels

- Evaluating system performance to identify any weak links or quality control problems

- Evaluating vendor performance to determine if established standards, service levels, and other contractual obligations are being met

- Determining maintenance and equipment budgets in consultation with department heads and executive staff

- Determining required equipment to be ordered and any necessary system reconfigurations to accommodate corporate growth and expansion into new markets

- Providing technical, administrative, and policy guidance for technical staff
- Planning, testing, and implementing mission-critical operating procedures, such as disaster recovery scenarios
- Establishing required service levels and response times in consultation with other departments

In large organizations, more than one operations supervisor or manager may be required—one for telecom and another for data networks and perhaps others for information systems and database management. For instance, in a large campus setting, the physical plant activities alone can consume the efforts of one person. This position requires a balanced blend of technical expertise and management skills.

14.4 Reporting Requirements

To justify system upgrades and network expansion, and to determine the requirements for maintenance and support staff, an understanding of the basic maintenance reporting requirements is necessary. The variables that need to be tracked on an ongoing basis include

- Network equipment replacement and repair
- Station equipment replacement and repair
- Software, firmware problems and versions
- Trunk-related transmission problems on the Wide Area Network (WAN)
- Wiring problems on the LAN
- Database updates: nature, scope, and frequency
- Moves, adds, and changes
- Vendor and carrier performance
- User-related problems

The information that is derived from these variables includes the time spent performing the following tasks:

- Repairing and/or replacing faulty components and subsystems
- Testing and debugging software
- Monitoring quality of service provided by telephone companies and interexchange carriers
- Testing and/or installing wire and cable

- Moves, adds, and changes
- Database modifications
- Vendor response and reliability
- User training

The collected information can be used in a variety of ways, such as

- Determining the spares inventory—types of components and quantities
- Predicting the frequency and level of parts replacement and/or repair
- Predicting the turnaround time for parts replacement and/or repair
- Determining the Mean Time between Failures (MTBF) and Mean Time to Repair (or Response) (MTTR) of critical components and subsystems
- Determining the required number and type of labor hours for an appropriate staffing level
- Determining required support levels from vendors and carriers
- Determining vendor and carrier performance and reliability in terms of response times, speed of problem resolution, and number of calls requiring escalation

A series of daily and monthly logs should be maintained so that periodic analyses can be performed. These logs should be structured to incorporate all the factors previously discussed. On a daily basis the following logs should be run:

- Trouble activity
- Software bugs and alarms
- System maintenance
- System utilization
- Moves, adds, and changes
- Vendor notification and escalation

On a monthly basis the following reports should be run:

- Trouble activity summary
- Software status
- Parts replacement and repair
- Preventive maintenance

- Vendor performance
- Move, add, and change activity summary
- Training activity summary

The proper maintenance of systems and networks requires accurate record keeping. Documentation procedures should be strictly enforced regardless of maintenance program—vendor or in-house.

14.5   Equipment Requirements

Along with a well-qualified staff and accurate record keeping, the task of in-house maintenance will require an inventory of parts, tools, and test equipment.

14.5.1   Spares inventory

Consideration must be given to stocking the appropriate types and quantities of components for both spares kits and inventory. Spares kits are kept on hand to make possible on-site replacement of failed components, and their contents are derived from both the user's experience with hardware problems and any mandatory spares requirements of a vendor service contract. There are several ways to contain inventory costs:

- Only order spares that cannot be readily purchased within the time frame dictated by system and/or network uptime requirements. For example, if routers can be bought on-line from a reliable local source for next-day delivery, there may be no need to stock them as spares.

- Spares can be stored centrally, where possible, so that they can be made available to neighboring locations on short notice. That way a single spare unit can cover a wider area. Alternatively, an account with an overnight courier enables a single spare unit to cover locations nationwide.

- Avoid purchasing equipment that uses hard-to-find or outdated components, or components that are hard to repair, whenever possible.

- Have a disaster recovery plan in place, which allows faulty systems or lines to be bypassed, so that mission-critical operations can continue until the fault can be repaired or replaced.

Instead of trying to keep pace with new products and technologies, many organizations implement in-house maintenance only for the installed base of older products. The savings can be enormous. Not only are the older technologies stable and the need for continuous training eliminated, but spare parts may be purchased from the used equipment market at very economical prices.

14.5.2   Test equipment

Test equipment also should be available at each location. A basic starter approach might include the following items, some of which are bundled into the same device:

- Breakout box
- Tone generator
- Portable dB meter
- Bit error-rate tester
- Data line monitor or protocol analyzer
- Power meter with graphical output
- Volt ohmmeter
- Time Domain Reflectometer
- Portable printer

Large maintenance organizations might keep all test equipment in a central location. Technicians could sign out equipment from the pool, as they need it. To account for all test equipment, a standard nomenclature is applied for describing the performance of each item and how it is tuned. All items are bar-coded for easy identification and for tracking the movement of items in and out of the pool. Some companies have even set up a reservation system to schedule the use of equipment in advance. If a technician knows a specific oscilloscope will be needed a month from now to do scheduled maintenance, he or she can reserve it in advance to be sure it is available.

Information about the movement of equipment in and out of the pool can be tracked in an asset management database. The operations manager has online access to the database so information on daily equipment usage can be obtained. This information can be used to determine whether new equipment should be added to the pool to keep up with demand. It can also be used to charge back equipment costs to appropriate departments, work groups, or projects.

14.5.3   Technical references

In-house technical staff should be equipped with a full reference library of product-specific information regarding all aspects of system operation and service. Generally, the vendor or manufacturer provides a series of manuals documenting system architecture, operation, procedures for installation and service, and other technical and procedural information. Some vendors supply this information on CD-ROM, which can substantially shorten look-up time and, consequently, speed up the fault isolation and restoration process. Supplementary information may be available on the vendor's secure Web site.

It is very important that arrangements be made to receive all updates and revisions as they are issued to the vendor's field engineering staff. Some vendors provide this as a value-added service at an extra charge. On a quarterly basis, the vendor provides the most up-to-date technical product information on maintaining system or network efficiency and reliability. Written by engineers and field service personnel, with an emphasis on how to more effectively operate and manage the vendor's products, this information might take the form of technical bulletins, product application notes, software release notes, user guides, and field bulletins.

Other sources of technical reference information, software bug fixes and upgrades, and troubleshooting advice include user forums on such services as America Online and newsgroups on Usenet and other news feeds.

14.6   Vendor and Carrier Services

From an administrative perspective, the easiest way to obtain service and support is to let the equipment vendor or carrier handle it. There are some key advantages in doing so, which may override cost concerns.

The vendor can bring more resources to bear on a problem with its specialized staff of hardware and software engineers, who are experienced in solving a broad range of problems for an entire installed base of customers—domestic and international. The largest equipment vendors, and some third-party maintenance vendors, are able to expedite problem-solving with dial-up links from a central service center to remote customer sites. Using advanced predictive maintenance tools, these vendors can monitor and analyze user systems to identify potential problems before they affect system performance. When something goes wrong with a system that is being monitored, the vendor's technicians are dispatched to the scene immediately with the right spare parts and information to correct the problem. Thus problems can be identified and corrected before customers are even aware that a problem exists. Such an arrangement is roughly equivalent to having the vendor on call 24 hours a day, seven days a week. This capability is especially important to large companies with international locations.

Expert systems are being applied to network management, specifically for diagnostic and restoration applications. Basically, expert systems rely on massive databases that house the accumulated solutions to a multitude of problems. The technician can input a symptom, and the expert system will propose a course of action based on the stored knowledge gleaned from the past experiences of the vendor's entire customer base. Some expert systems are capable of accepting plain language input from nontechnical people, who will use the plain language output to restore malfunctioning systems.

Most vendors offer a range of basic service agreements. Forty-hour-a-week service is sufficient for most companies, with a time-and-materials provision for emergency service after normal business hours. There are also standard agreements for 24-hour, seven-day-a-week service. The price differential between the two plans may be quite high, as much as 30 percent. Most firms offering maintenance and support services will customize a service agreement to include performance guarantees, specifying such things as response time to trouble calls and penalties for given levels of downtime. Typically, penalties take the form of credits on future maintenance service billing.

Depending on the scope and complexity of service needs, it might be advantageous to retain a consultant to perform a needs assessment of maintenance and support requirements and, based on the findings, issue a Request For Proposal (RFP) to ensure that the most qualified firm is selected. Another technique entails paying service firms a fee to perform the needs analysis with the understanding that, in the process, they are competing for the contract.

There is another way to obtain the level of maintenance and support services economically, but this usually involves a user taking more responsibility for diagnosing problems before calling in the vendor or third-party maintenance firm. Such services usually require the customer to set up a help desk, which is staffed by personnel who can determine the cause of routine problems and offer solutions that will get the user up and running quickly. This arrangement minimizes unnecessary equipment downtime, saves the vendor an unnecessary trip to the customer site, and eliminates unnecessary maintenance charges. As a result the help desk can reduce overall maintenance charges by as much as 25 percent. In standardizing reporting procedures and structuring maintenance charges uniformly over its entire customer base, the vendor or maintenance service provider can reduce its own costs and pass the resulting savings on to its customers.

Another way that maintenance and support services can be obtained economically is through a master contract. Instead of negotiating each service option and getting the paperwork approved at several points within the vendor organization, the user signs one master contract. Each option is selected off the master list and initialed. This simplifies service administration, which benefits both the vendor and the user.

14.7 Third-Party Maintenance Firms

Despite these innovative services, many organizations are not willing or able to increase their involvement in maintenance and support activities, even if doing so qualifies them for substantial discounts. Nor can they afford to completely turn over management of their networks to a single vendor. For such firms, bargains abound among third-party maintenance providers, and many do not base discounts on user participation via help desks. Instead they provide whatever services customers need when the trouble call is placed.

Price and breadth of equipment coverage are the principal reasons users opt for third-party maintenance vendors. Third-party vendors are more open to negotiation on such matters as the amount of coverage, price, and the types of equipment to be serviced. But this option entails the greatest amount of risk to uninformed users.

While cost savings and equipment coverage are important, care must be taken not to sacrifice service quality. Anything less is not only shortsighted but will end up costing more in the long run in terms of lost production owing to downtime from inoperable equipment and/or unusable software. Before choosing this type of vendor, it is prudent to inquire about response times, particularly for the organization's remote locations. It is also necessary to check into how many customers the service firm supports at these locations, ascertain the staffing levels, and determine the locations of the nearest spare parts inventories. All of this information can be used to confirm vendor statements about response times.

The more types of equipment the vendor supports, the better. The user can then have more flexibility in future equipment selections, since it will not have to contract with other service firms. In a mixed-vendor environment, a third-party maintenance provider who can handle everything eliminates finger-pointing among competing vendors. Before making a decision, however, it is a good idea to check the third-party vendor's references and make inquiries about the ability of the vendor to work with prime vendors, who may be servicing mission-critical systems that are still under warranty.

A vendor's capability to perform remote diagnostics will limit unnecessary downtime and save the user money over the life of the contract. The vendor should also have a computerized spare parts management system. After all, qualified technicians are of little value unless extensive spare parts inventories and a parts distribution network are in place to ensure fast delivery of the right parts on a moment's notice.

If the third-party maintenance vendor can provide computer-generated reports on equipment service histories and reliability factors, by product line as well as by customer location, this information can be used to guide future equipment selections. And if the vendor's computerized inventory and dispatch system is also used for billing, so much the better because there will be less chance of errors that tie up the user's staff time with invoice reconciliation.

Third-party maintenance vendors should be screened for organizational depth to be sure that they have sufficient resources to provide comprehensive support. Screening entails checking into the number and qualifications of technical people who service particular types of equipment to determine whether they have the required expertise to assist with upgrades, reconfigurations, and relocations, as well as cabling and rewiring.

The third-party maintenance vendor should also have regional repair and refurbishment facilities, where equipment and printed circuit boards can be fixed or rebuilt. Oftentimes using such a facility is a more economical alternative than purchasing new units. The company should also offer a warranty on items sent in for repair. A limited 90-day warranty is typical.

Some third-party maintenance vendors even stage new equipment and perform thorough systems-level testing before installing it on customer premises. If any problems are revealed, the maintenance vendor can interface directly with the equipment vendor to resolve them. Meanwhile, the customer is saved from protracted dealings with a recalcitrant equipment vendor, while being spared the burden of having to deal with malfunctioning equipment.

If the organization depends on its communications system or network for its continued survival, it should verify the capability of the third-party maintenance vendor to implement a disaster recovery plan, which protects key sites against catastrophic loss. Many large primary vendors offer such services, under a separate annual fee, as a supplement to the force majeure clause within the standard maintenance contract. The type of disasters covered under such programs include loss by fire, water, vandalism, theft, power surge, or air-conditioning malfunction.

A key feature of a disaster recovery plan is the replacement of nonrepairable equipment with like equipment within a specified time frame, which varies among vendors from 48 to 72 hours. This provision is particularly significant when an organization relies heavily on products that are discontinued or in long-term production cycles. The vendor typically ships replacement parts for repairable equipment within 24 hours, whereas a technician is dispatched to the customer location within four hours.

Although an insurance policy might cover the replacement costs of equipment, the user may be responsible for such things as on-site installation fees and rush delivery charges for parts and/or system equipment. But many third-party maintenance vendors are ill-prepared to address critical problems on short notice. Others are slow in responding to trouble calls after normal business hours. Such behavior should disqualify them as providers of disaster recovery services.

Some users may have doubts about the service claims of a third-party vendor but cannot resist the promise of big cost savings. One way to deal with this dilemma is to use the service firm on a trial basis at select sites to determine its response times and quality of service. When the service firm demonstrates satisfactory performance, the user may gradually turn over more sites, rather than give out all of the available service business at once. The rationale of this strategy is to keep rewarding the vendor for good performance, and, in the process, provide more incentive to perform. Many times such strategies will not be workable. The alternative is to stick with short-term contracts with third-party maintenance vendors to minimize risk until a consistent level of high performance is demonstrated.

14.8   Cooperative Arrangements

The benefits of a cooperative approach to maintenance, whereby in-house staff work right alongside primary vendors and third-party maintenance firms, are quite compelling. Oftentimes such arrangements produce a synergistic effect among the various players that manifests itself in performance of the highest quality, resulting in maximum equipment uptime which, in turn, ensures a high level of satisfaction among the numerous users and work groups of the organization.

Such arrangements usually require that the customer establish a help desk to minimize trouble calls for routine problems. In turn, the customer can keep maintenance costs in line by using experienced staff to perform first-level maintenance. Staff members are typically certified to assume responsibility for primary equipment service, to respond to service calls, and to perform first-level troubleshooting and repair. The vendor acts as a support backup and as the primary escalation point for service.

There is another kind of cooperative arrangement whereby companies share computer and network resources in times of disaster. When one company experiences a prolonged power outage, for example, it uses the spare capacity of its partner company. Of course, such arrangements must be worked out in advance by negotiation. Once the arrangement is in place, it must be tested to see what technical problems surface and to find out how well designated staff work together before a real disaster strikes.

The following issues should be discussed when negotiating this type of cooperative arrangement:

- Identify the staff members in each company who need to be involved and make sure they all participate in the planning process.

- Thoroughly define the purpose of the agreement and the specific circumstances that will trigger its implementation.

- Create links among staff members via email and monthly meetings.

- Limit the layers of bureaucracy in each organization to expedite decision making regarding the planning and implementation of the agreement.

Although any number of companies can participate in the cooperative agreement, it is best to achieve a track record of success before widening the circle. As more partners are included and all existing members make the necessary adjustments, additional members can be brought into the arrangement.

There are firms that specialize in offering such disaster recovery services, even across national borders. They supply computer disaster recovery services and network facilities in the event of fire, flood, power outage, or any other type of disaster. In the event one of its customers experiences a disaster, the firm reroutes voice and data traffic to prearranged alternative facilities.

14.9 Cabling and Rewiring Considerations

Another aspect of maintenance and support is cabling and rewiring. Although not the most glamorous aspect of maintenance and support, cabling and rewiring are probably two of the most planning intensive, especially when multiple floors are being tied together in city office buildings, or multiple buildings are being tied together in a campus environment.

When a new system or network is installed, invariably some wiring already exists, and there is strong economic incentive for its continued use. When new wiring is installed, the desire to make it as useful as possible in future expansions must be balanced against the initial installation cost.

14.9.1 Premises wiring

There are typically several hierarchical layers of wiring that merit attention. With PBX installations, for example, these layers include

- The telephone carrier's distribution frame or demarcation point, which is the termination of the carrier circuits on the premises, and also the termination of carrier responsibility for the wiring. Often, this is simply a series of terminal blocks through which user Customer Premises Equipment (CPE) is attached to carrier circuits.

- The user's incoming circuit distribution frame is the place where carrier lines and user CPE cross-connections can be made. Cut-throughs or PBX bypass lines used to answer or originate calls when a PBX fails may be attached here, in front of the PBX. This is normally placed with the carrier's demarcation point, and may be omitted if there is only one CPE destination for all lines.

- The user's telephone equipment, which is linked to the incoming circuit distribution frame on one side, and to the private network or intermediate distribution frame on the other. The station wire pairs from the telephone system exit this equipment.

- The intermediate private-network wiring distribution frame is where the telephone system station wire pairs are connected to terminal blocks or panels for matching with the building wiring.

- Riser cables terminate in the intermediate distribution frame and link it to, for example, horizontal distribution panels on floors.

- Horizontal distribution, or wiring-closet, panels take riser connections and distribute them to the actual station wiring.

- Outside wiring serves the combined functions of riser cables or horizontal distribution cables where the run must exit a premises and transit an outdoor space.

- Station wiring links the horizontal distribution panel with the instruments.

The structured nature of the wiring process is designed to achieve an important goal—permit restructuring of the system without actually stringing new wiring or performing extensive rewiring to accommodate moves, adds, and changes. Virtually the same considerations apply to hub-based LANs.

14.9.2 Treating the cable plant as an asset

In a properly designed wiring system, each cable pair should be viewed as a manageable asset, which can be manipulated to satisfy any user requirement. Individual cable pairs should be color coded for easy identification. Cables should be identified at both ends by a permanent tag with a serial number. This permits the individual pairs to be selected at either end with a high degree of reliability and connected to a patch panel or punch-down block as appropriate to implement moves, adds, or changes.

Patch panels and punch-down blocks should be designed to segregate different functions or circuit types to expedite the easy location of pairs. Data and voice connections, for example, can be terminated in different areas of the panel or block to avoid confusion. Many panels and blocks come in colors or have color-tagging capabilities, which can assist in locating pairs later.

The cabling used in wiring a telephone system or LAN depends on the requirements of the devices being used and the formal distribution plan that the telephone system and/or computer vendor provides. Formal plans, such as IBM's Cabling System or AT&T's Premises Distribution System, compete with similar plans that are available from nearly every major vendor. All of these plans have the common goal of establishing a wiring strategy, which will support present needs and future growth. The in-house technician (or maintenance firm) should be familiar with these cabling schemes.

There are a number of asset management applications available that keep track of the wiring associated with connectors, patch panels, and wiring hubs. These cable management products offer color maps and floor plans that are used to illustrate the cabling infrastructure of one or more offices, floors, and buildings. Managers can create both logical and physical views of their facilities, and even view a complete data path simply by clicking on a connection.

Some products provide complete cable topologies, showing the locations of the cabling and connections, providing views of cross-connect cabling, network diagrams by floor, and patch panels and racks (see Fig. 14.1). Work orders can be generated for moving equipment or rewiring, complete with a picture of the connections. With this information, the network administrator knows where the equipment should go, what needs to be disconnected, and what should be reconnected. The technician can take this job description to the location and perform the changes.

Other products provide a Computer Aided Design (CAD) interface, enabling equipment locations and cable runs to be tracked through punchdown blocks, multiconductor cables, and cable trays. In addition, bills-of-material reports can be produced for new and existing cable installations.

Cable management applications can be run as stand-alone systems or may be integrated with help desk products, hub management systems, and network management platforms. When coupled with a hub management system and help desk, a high degree of automation can be brought to bear on the problem resolution process. When the hub management system detects a media failure, the actual cable run can be extracted from the cable management application and submitted along with a trouble ticket generated by the help desk. And when the hub management system is integrated with a network management platform such as Hewlett-Packard's OpenView, all of this activity can be monitored from a single management console, thereby expediting problem resolution.

14.9.3 Cable planning

Determining if existing wiring can be used to support new system installations requires the preparation of a complete wiring plan, preferably before the start of system installation. The plan acts not only as a guide to

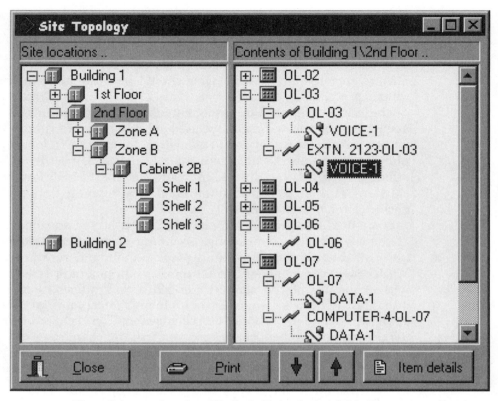

Figure 14.1   View of site topology from Unylogix Technologies' Cable Management System.

the installation process, but also as a check on capacity and planning, which can be carried out before installers appear and begin working. In the case of a PBX, for example, the plan should associate each instrument with a complete path back to the station wiring on the PBX, through all panels, horizontal feeds, and risers. During this process, the capacity of each trunk or riser should be checked one last time, and additional cabling run as needed. Spare pairs on each cable can be identified for possible future use.

As with station cabling, there are many factors to consider in deciding whether to purchase new or continue to use existing house cabling. The high cost of large-paired distribution (300-, 400-, 600-, or 900-pair) cabling and its installation make the lease or purchase of existing cabling cost-effective. As with station cabling, there is a labor cost for identifying, reter-minating, and documenting the existing cable. This cost increases as the number of pairs increases, and the probability of an error does likewise. Because existing cabling often introduces unknown factors, such as pair counts, condition, and destination, many interconnect companies prefer to install new cabling. This cabling is much easier to document, install, and cut over because it is not being used.

A building under construction is the ideal environment for cable planning because the cable can be installed according to the company's needs without concern for the requirements of the existing cable plant. Moreover, factors that hamper installation in an existing building are not present in a building under construction (i.e., cosmetic concerns, disruption of office personnel, and inaccessible areas), allowing attention to be focused on meeting the needs of each possible telephone instrument or terminal location.

Many companies specialize in cable installation and may be contracted on a per project basis through a bidding process. Some cabling contracting companies only install cabling; others both install and maintain cable networks. Because of the complexity of a cable network, it is convenient to deal with only a single company—one that endeavors to become familiar with the organization's current and future cabling requirements.

If information systems and communications networks constitute a strategic resource to a company, so must its wiring. The proper planning for each wiring and rewiring of a facility can preserve these resources intact for the support of future applications. Improper wiring and poor record keeping often leave an organization no recourse but to fully rewire a facility—an undertaking that is often far more expensive than a new installation.

14.10   Other Planning Activities

In addition to the hands-on issues of managing maintenance and support services, there is the job of providing reports to top management, which address costs and benefits and account for allocated resources. Such information is typically used to validate the current approach to maintenance and support, to make changes that will bring about additional efficiencies and economies, and/or to expand the nature and scope of maintenance and support activities.

Toward these ends, it is necessary to provide top management with annual and long-range plans, which generally involve the following activities:

- Maintaining records that provide concise information about the current status of information systems and networks

- Auditing the progress and performance of vendors on major projects

- Keeping personnel records up-to-date, including all pertinent information about technical and management skill levels, continuing education, and incentive plans

- Gathering information about the current and future equipment and applications requirements of corporate divisions, departments, and work groups

- Assisting department heads with advice on the alternatives for achieving near- and long-term objectives

- Keeping track of developments in information systems and networking technologies and their potential impact on the organization's competitive position

14.11   Conclusion

Sometimes problems in multivendor environments prove so intractable that no single vendor can offer a solution. Because such problems were once so common, the Technical Support Alliance Network (TSANet) was created in 1993. The goals of the 100-member group are to expedite resolution of shared customer problems and to avoid having vendors pass the problem around until the customer finally gives up. TSANet also provides a method by which vendors can exchange selected information and share their common experiences in providing customer support.

As members of TSANet, competing vendors are willing to put aside issues among themselves in order to learn better ways to serve their customers in a post-sale situation. Vendors work together to solve the customer's multivendor problems and agree to accept ownership of customer calls, which eliminates finger-pointing at each other. As a result customer satisfaction levels increase because problems can be resolved more efficiently and with less impact on the customer.

When buying computer products and information systems, companies should give preference to vendors that belong to TSANet. In large procurements, this preference should be specified in the RFP, to provide the assurance that vendors will work cooperatively with each other and eliminate finger-pointing, thus improving the speed of problem resolution between vendors.

15

Network Monitoring and Testing

15.1 Introduction

As networks become larger and more complex and expansive, an increasing amount of corporate resources goes into monitoring and testing. Precise information about network performance must be gathered and interpreted properly to ensure high levels of availability and reliability. Among the many challenges faced by network managers is the growing number and types of equipment and lines found on today's networks, which slows problem isolation. For example, if a user is experiencing performance problems—in that the applications run too slow—the cause could be the user's computer, the application software, a server, the Local Area Network (LAN) cable, the interconnection devices (i.e., hubs, switches, or routers), the carrier lines, or the remote server. To complicate matters, the network may include international locations, which may use equipment and lines that adhere to different standards.

The value of network managers to the Information Technology (IT) organization is in their ability to minimize the disruption caused to the enterprise when technical problems arise. To aid in identifying and correcting problems, various test equipment is available that provides the following basic functions:

- Specify the type of data to be collected and when.

- Set time intervals for data collection over a specified time frame.

- Gather specified performance data for analysis.

- Summarize gathered data into a variety of graphical forms.
- Generate traffic to simulate loads and inject errors to test their impact on the network.
- Sort gathered data by multiple criteria.
- Edit gathered data.
- Program various monitoring tasks and the sequence of their execution.
- Play back recorded information for later analysis.

The test instruments that provide these and other functions range from sophisticated protocol analyzers to simple, hand-held line testers. There are software-based collection tools as well, which are referred to as *agents*. These are software programs that reside on remote devices to collect performance data. This data is sent to a network management system or protocol analyzer when requested. Sometimes the agents have enough intelligence to act on the data without requiring human intervention.

15.2   Protocol Analyzers

Protocols are the rules about how communications equipment should format information for transmission on a particular network. When types of equipment from different manufacturers adhere to the same protocols, they have the ability to communicate with each other. Typically, data is prepared for transmission by formatting it into packets and adding a header and trailer, which forms an envelope for the message while it traverses the network. Devices on the network using the same protocol know how to read the envelope so it can be routed over the appropriate link for delivery to the proper addressee.

In addition to source and destination addresses, a packet might include other features such as error correction. A protocol violation occurs when these procedures are not properly followed. Then the cause of the problem must be found so that corrective action can be taken to restore communication. Important clues can be found by reading the headers and trailers attached to the messages, or by opening the envelope to read the messages themselves. The protocol analyzer is the tool used to take these components apart and to look inside them to see what is wrong.

The protocols themselves can be categorized as either byte-oriented or bit-oriented. *Byte-oriented protocols* have been around for many years and include bisync, HASP, poll-select, and many others. Byte-oriented protocols are relatively simple to decode to determine protocol violations. This does not require a very sophisticated protocol analyzer. *Bit-oriented protocols,* on the other hand, are more complicated and require a more sophisticated analyzer. Common bit-oriented protocols include Ethernet, token ring, Transmission Control Protocol/Internet Protocol (TCP/IP), Integrated Services Digital Network (ISDN) Q.931, Systems Network Architecture/Synchronous Data Link Control (SNA/SDLC), X.25, and DECnet.

Diagnosing problems in an environment where such protocols are used can be very tedious, if not impossible, without the proper analytic tools. Among these tools is the protocol analyzer, which connects directly into the LAN as if it were just another node, or to the port of a communication device [Data Terminal Equipment (DTE) or Data Communications Equipment (DCE)] under test. As part of its analysis, this type of equipment can be used to display a line of information describing the packet, including the type, protocol, and its function. By drilling down, the technician can read progressively more detailed information about the packet.

Since some protocols like SNA are not routable, they are often encapsulated by other protocols that are routable. For example, TCP/IP can be used to route SNA over the Wide Area Network (WAN). Other times, LAN protocols must be emulated to run over Asynchronous Transfer Mode (ATM) networks. Whenever a native protocol must be manipulated in such ways to run over another type of network, protocol analysis becomes more complex. In the case of routing SNA over TCP/IP, examining the TCP/IP protocol suite will not reveal a problem with SNA unless the SNA "envelope" itself is opened. In the case of LAN traffic over ATM, the protocol analyzer must include the ability to penetrate the ATM layer to read the LAN protocol. Analysis is further complicated by the fact that the larger LAN packets have been fragmented into smaller ATM cells and a large amount of data may have to be captured to isolate the problem. Many companies with SNA networks are using IBM's Data Link Switching (DLSw) technology to interconnect their network segments. But SNA networks are fragile because timing is critical. A single delayed or out-of-sequence frame can cause a connection to shut down.

When a network problem occurs on-line, a single analyzer should be able to narrow the problem down to a certain device through a process of elimination. In a real-world situation where other traffic is present, however, only by watching multiple segments at the same time will it be possible to identify when non-SNA traffic is causing a problem. By observing synchronized time stamps from multiple segments, the technician can identify the frames causing the problem. Real-time analysis of multiple segments can be performed simultaneously by linking protocol analyzers (see Fig. 15.1).

By comparing monitored traffic from each segment, the technician can see what happens as non-SNA traffic is mixed in. For example, heavy File Transfer Protocol (FTP) traffic may overload a router, slowing throughput, and causing a time-sensitive SNA session to be dropped. Because the same traffic from each link using the synchronized time stamps is being

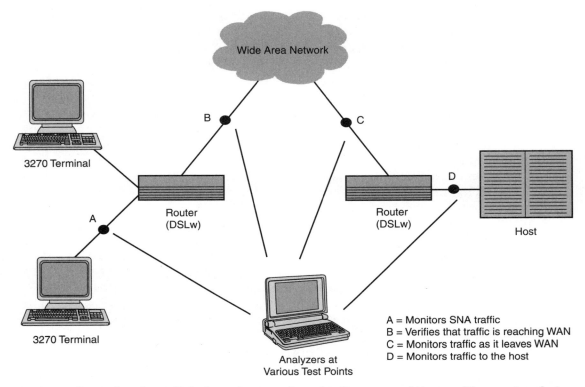

3270 Terminal

Router (DSLw)

Router (DSLw)

Host

3270 Terminal

Analyzers at
Various Test Points

A = Monitors SNA traffic
B = Verifies that traffic is reaching WAN
C = Monitors traffic as it leaves WAN
D = Monitors traffic to the host

Figure 15.1 Protocol analyzers linked together to perform simultaneous, real-time, multisegment analysis.

compared, the technician can identify precisely when the overload begins so that adjustments can be made, such as changing the router's protocol prioritization.

There are certain features of protocol analyzers that can greatly shorten the time it takes to find the causes of problems. These include monitoring and simulation features, as well as a variety of settings for configuring counters, timers, traps, and masks.

Monitoring and simulation modes. Protocol analyzers can be used in either of two modes: passive monitoring or active simulation. Passive monitoring is used to collect information on network performance and active simulation is used to mimic a network node under a variety of conditions to see what impact it may have on the network (see Fig. 15.2).

In the monitoring application, the analyzer sits passively on a ring or segment and monitors both the integrity of the cabling and the level of data traffic, logging such things as excessive errors that can tie up a token ring LAN, for example. In this application, the protocol analyzer merely displays the protocol activity and user data that traverses the ring, providing a win-

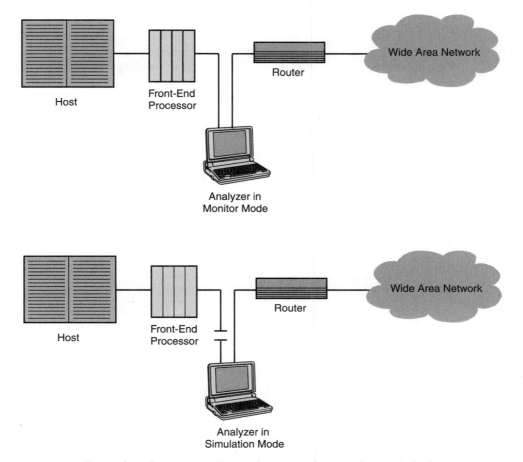

Figure 15.2   Protocol analyzer in monitor mode (top) and in simulation mode (bottom).

dow into the message exchange between network nodes. The collected information is then retrieved by the network management system.

In addition to monitoring network performance for purposes of problem isolation, the analyzer's monitoring mode can be used to gather data required for planning. For example, to manage WAN bandwidth for optimum throughput requires that you, as the network manager, know exactly how much LAN traffic is traversing the WAN. To improve bandwidth efficiency or plan for increased bandwidth demand, you must know what is meaningful LAN payload and what is extraneous overhead. Because you can monitor WAN links and collect data to provide detailed analysis of upper-layer LAN traffic on each link—displaying station-level statistics and WAN payload versus overhead efficiencies—you can obtain the information necessary for optimizing WAN bandwidth utilization.

Protocol analyzers are not limited to monitoring one link at a time. This is actually determined by the number of ports a protocol analyzer has. On a legacy WAN, for example, one port could monitor a gateway's X.25 interface while the other port monitors the gateway's SNA interface. A multiport protocol analyzer can collect information from two or more sources simultaneously, permitting the operator to view each decoded protocol in its own window.

In the simulation mode, the protocol analyzer is programmed to exhibit the behavior of a network node, such as a gateway or communications controller. This makes it possible to test the impact of more traffic on the entire network, before actually buying a new gateway or controller. By simulating a specified device, the network manager can also find out how the network will respond to its failure, which can aid disaster recovery planning. Protocol simulation is most often used to verify the integrity of a new installation before it is put into operation. Some manufacturers offer prewritten emulations, while others provide a programming language that lets users create their own test routines.

Trapping.   With the trapping function, the protocol analyzer records specified events into its buffer or disk. For example, the protocol analyzer could be set to trap the first errored frame it receives. In this way, only essential information is captured. Some protocol analyzers allow the user to set performance thresholds according to the type of traffic on the network. When these performance thresholds are exceeded, an alarm message is triggered, indicating that there is a problem.

Filtering.   The protocol analyzer's filtering capability provides the operator with the means to include or exclude certain types of protocol data that require analysis, such as:

- Destination and source addresses

- Protocol type

- Errored packets

Filtering narrows down the number of packets that need to be captured, so that the technician does not have to waste time with extraneous information when trying to isolate the cause of problems. With X.25, for example, the operator can specify a Logically Connected Node (LCN) and only packets on that LCN will be captured. In isolating an X.25 virtual circuit, the operator can specify a network address of interest. The analyzer monitors traffic to or from that address and automatically filters on LCNs selected for the connection. Similar capabilities are supported for Physical

Unit (PU) and Logical Unit (LU) addresses in IBM's SNA environment, letting the user focus on a single device in the network.

If the test technician suspects that errors are being generated at the data link layer, network layer packets can be excluded from collection. At the data link layer, the analyzer will track information such as where the data was generated and whether it contains errors. If no problems are found, the technician can set the filter to include only network layer packets. At this layer, the protocol analyzer tracks information such as where the data is destined and the type of application under which it was generated. If the technician has no idea where to start looking for problems, then all of the packets may be captured and written to disk. A variety of filters may be applied later for selective viewing.

Other filtering possibilities include collecting and displaying only packets that are going to or from specific nodes, packets that are formatted according to specific protocols, or those packets containing only certain kinds of errors. Where voice and data are integrated over the same digital line, voice messages can be eliminated so that only data packets of interest show up for viewing. Some protocol analyzers allow users to set multiple filters so that various types of relevant data may be collected and displayed at the same time. Others even allow users to view decoded data in real time while the protocol analyzer continues monitoring and capturing more data.

Information may be displayed in terms of bar charts that depict packets per second, bytes per second, errors, collisions, and other relevant information. Some analyzers support multiple windows, allowing the operator to view many types of information at once. With the right windowing software, users can customize the presentation of information, as well as size and position graphs and charts on the screen for maximum clarity.

Packet generation. In being able to generate packets, the protocol analyzer can be used to test the impact of additional traffic on the network. Among the packet parameters that can be set for this purpose are the following:

- The source and destination addresses

- The minimum and maximum frame size

- The spacing between the packets, expressed in microseconds

- The number of packets sent out with each burst

The technician can also customize the contents of the data field section of the packets to simulate real or potential applications. When the packets are generated, the real-time impact of the additional traffic on the network can be observed. Packets can also be generated to force a suspected problem to recur so that possible solutions can be devised and tested.

Load generation. A related capability is load generation, whereby varying traffic rates can be put onto the network. By loading the network with extra traffic, various network devices such as bridges and repeaters can be stressed for the purpose of identifying potential points of weakness. The information can be used to identify the need for more bandwidth or equipment upgrades.

Timing. Protocol analyzers can measure the time interval between events. By setting up two traps—one for the transmit path and one for the receive path—the technician can verify if a handshake procedure has exceeded its maximum time interval, for example. After the relevant information is captured and sent to the analyzer's buffer or disk, the operator can place markers between any two events that are viewed on the display screen to determine the elapsed time between these events.

Although this type of measurement is most often better handled from the central network control point or host location, a protocol analyzer that supports simple timing measurements between data events can often isolate the problem without having to invoke these resources.

Some analyzers have a review mode that lets the operator cursor between events in the capture buffer to measure the time interval between events. With this feature, the technician can measure the time between when data was sent to the host and when the corresponding response was received. At the same time, other activity on the link can be observed to determine if the slow response is network or host related.

Captured data can also be related to alarms generated by the management system to give the technician various reference points. Over extended monitoring periods, various events can even be correlated with end-user trouble reports and system logs.

Text editing. Some analyzers come with text editors that can be used on captured data. This allows the technician to enter comments, delete unimportant data, print reports, and save the data to a particular database format.

Terminal emulation. Some analyzers support asynchronous terminal emulation. If there is a requirement to communicate with network devices for configuration management, for example, or to access remote databases, this feature can eliminate the need for the technician to carry a separate terminal. Implementations of terminal emulation can vary considerably among analyzers, with some supporting complete 24 lines $\times$ 80 characters on a single display, and others requiring windowing to access an entire page. Support may be limited to dumb terminal emulation, or VT100/200 emulation may be included.

Cable testing. Many protocol analyzers include the capability to test for cable faults, breaks, and improperly terminated connections using Time Domain Reflectometry (TDR). This technique is especially useful for pin-pointing problems caused by shorts, crimps, and water. The test procedure involves sending a signal down the cable and then receiving and inter-preting its echo. Depending on the specific TDR device in use, the status of the cable and connections may be reported simply as no fault detected, no carrier sense, open on coax, or short on coax. The distance to the problem is also reported and the test results can be printed.

There are optical TDRs designed specifically for testing optical fiber. Among the advanced features of Optical Time Domain Reflectometry (OTDR) is one-button operation, which allows a technician to characterize a link merely by connecting the fiber under test, switching on the unit, and pressing RUN. This feature is particularly helpful for infrequent users of OTDRs. Of course, OTDRs have features that allow a more skilled user to set up measurement parameters in advance and program complete proce-dures, including printing and saving. An installation team can select the required setting from the OTDR's internal memory or from a floppy disk to run through the entire process. This is important to users who must gen-erate many measurements and reports for new installations or acceptance tests that entail a range of OTDR measurements, all of which must follow the same guidelines and protocols. To reduce setup time for different appli-cations, some devices even allow the user to set the OTDR to achieve the best possible performance for a given parameter.

Remote fiber-testing systems improve network quality because they reg-ularly conduct automatic tests on fiber links, and store and evaluate the test information centrally. Such test systems can be used to identify link degradations over time and issue an alarm detailing the faulty fiber and its physical location. This feature allows maintenance personnel to take action in time to prevent disruption in network service. Because the sys-tem also identifies the exact location of cable breaks, repair crews can be dispatched quickly to the right location. Repair time and cost are reduced, thereby minimizing loss of revenue and customer dissatisfaction due to service outages.

When integrated with an open management platform, such as Hewlett-Packard's OpenView, centralized control can be exercised over the entire remote-testing configuration and can interface with the operations-sup-port system of the network. The management system can perform such functions as auto-configuring measurement hardware, storing network maps, controlling access levels, providing a software-development envi-ronment, and linking to Structured Query Language (SQL) relational databases.

Plain-language translation. Some protocol analyzers have the capability to decode packets and display their contents in plain language notation, in addition to hexadecimal and binary code. Further details about a specific protocol may be revealed through the analyzer's zoom capability, which allows the technician to display each bit field, along with a brief explanation of its status. This feature may be applied to any protocol. Some vendors have quite extensive libraries of software that can identify and symbolically decode just about any currently used protocol. Installing the software is often as simple as copying it to a subdirectory.

Storage. Protocol analyzers typically come with Random Access Memory (RAM), which is used as a capture buffer. Some capture buffers can be quite large. In the case of protocol analyzers that are used on high-speed ATM virtual circuits, for example, the buffer can be quite large, starting at 32 MB.

RAM permits the temporary storage of recorded data. When RAM is filled, the operator can scroll forward or backward to find specific information within the captured data. Using an integral text editor, the captured data can be revised, comments can be added, and selected material can be printed. Alternatively, all of the data can be written to disk for later analysis. Compared to a standard Personal Computer (PC), the analyzer's RAM and disk storage are usually quite meager. However, the use of trapping and filtering capabilities minimizes the collection of irrelevant data, thus conserving these resources.

Remote operation. When there are a number of remote locations on the network, it is often useful to control the protocol analyzer from a central site. Via a dial-up modem link or dedicated connection through the WAN, the network manager can assist field personnel in such matters as verifying proper procedures and, if appropriate, by transferring buffer contents or other types of files.

The remote control capability is a virtual necessity when the remote network locations are unmanned. A remote protocol analyzer connected to a test port of a remote data switch or wiring hub, for example, can monitor any of the remote links as if they were local.

Mapping. The mapping capability of some protocol analyzers automatically documents the physical locations of LAN nodes in graphical form, assigning appropriate icons for servers and workstations, and allowing you, as the network manager, to name each node. The icon for each station also provides information about the type of adapter used, as well as the node's location along the cable. When problems arise on the network, you can quickly locate the problem by referring to the visual map. Some protocol analyzers can depict network configurations according to the usage of network nodes, arranging them in order of highest to lowest traffic.

Programmability. The various tasks of a protocol analyzer may be programmed, allowing performance information to be collected automatically. While some analyzers require the use of programming languages, others employ a setup screen, allowing the operator to define a sequence of tests to be performed. Once preset thresholds are met, a sequence of appropriate tests is initiated automatically. This capability is especially useful for tracking down intermittent problems. An alternative to programming or defining analyzer operation is to use off-the-shelf software that can be plugged into the data analyzer in support of various test scenarios.

15.3 Breakout Boxes

Some types of testing are intrusive in nature, requiring that you first notify users of impending downtime before cables can be disconnected to accommodate the testing device. Other types of testing are nonintrusive and can be run without interrupting user traffic—if a connection can be made to the network. It makes sense to implement various test access points in the communications system into which testing devices can be inserted without interrupting users.

Patch panels constitute a convenient and nondisruptive method of test access. Since the panel is already wired into the network, the test device is merely plugged into it, thereby eliminating the need to momentarily break network connections to accommodate a test set. Patch panels are built with every conceivable type of interface. The more common interfaces include RS-232, which is used for relatively low-speed applications, as well as V.35 and RS-449, which are used for higher-speed applications.

In many cases, a more convenient method of checking device interconnection and conducting rudimentary performance tests entails the use of an inexpensive, hand-held breakout box. By connecting the breakout box between two devices, a technician can determine if the interface leads are properly connected. With the breakout box, leads may be opened, closed, or crossed in any pattern required. Some breakout boxes also test cable continuity. There are breakout boxes for each type of interface; the breakout box that physically conforms to the interface on the device under test must be used.

The ability to cross-connect the leads in the breakout boxes means they can also be used to connect devices whose interfaces are not identically configured. This is particularly the case with DTE, and is not uncommon with DCE.

15.4 Bit Error Rate Testers

Although many protocol analyzers feature an integral Bit Error Rate Tester (BERT), this type of device is also available as a separate unit. BERTs can also be integrated into other types of equipment such as T1 multiplexers and intelligent hubs.

BERTs are used to determine whether data is being passed reliably over communications links. They send and receive various bit patterns and data characters so a comparison can be made between what has been transmitted and what has been received. The bit error rate is calculated as a ratio of the total number of bit errors divided by the total number of bits received. Any difference between the two is displayed as the error rate.

Low-end BERTs may only indicate that an error has occurred, but not how many errors, or what kind. High-end BERTs display a real-time cumulative total of bit errors, as well as a real-time calculation of the bit error rate itself. Additional information that may be presented on high-end BERTs include sync losses, sync loss seconds, errored seconds, error free seconds, time unavailable, elapsed time, frame errors, and parity errors. Should an error condition be identified, the technician can pursue the fault by testing various portions of the circuit and devices in the network.

15.5 Analog Line Impairment Testers

Analog lines are not usually used for LAN interconnection because of their inability to support high data rates. The highest data rate currently supported on analog lines without compression is 56 kbps, assuming ideal line conditions. While this is fine for occasional dial-up PC access to a remote LAN, it would cause serious delay if used for frequent transfers of large files.

However, dial-up lines can be used to support certain LAN interconnection devices. For example, you, as network manager, can access a router isolated on the network by digital line failure via its integral modem. The cause of the problem can be determined, and a workaround performed. Also, the router's accumulated statistics can be downloaded to a central management facility via the integral dial-up modem to aid in problem resolution.

Because there is still a large installed base of analog lines in the local loop, and they can be used for such applications as disaster recovery, remote reconfiguration, and statistics gathering, the network manager's arsenal of test equipment might include devices that are often referred to as Transmission Impairment Measurement Sets (TIMS). These test sets are used to measure various types of impairments—noise and distortions—that affect analog lines. Whether analog circuits are switched or dedicated, conditioned or not, they are all supposed to meet some very basic performance guidelines, which are published by Bellcore, now known as Telcordia

Technologies. By obtaining this information and comparing it with impairment measurements, users can determine whether carriers are complying with their stated levels of performance and, if not, get help in resolving line problems.

The ability to test for impairments is especially important when conditioned leased lines are being used. A conditioned line is one that has been selected for its desirable characteristics or treated with equalizers to improve the user's ability to transmit data at higher speeds than would normally be possible over analog private lines. Since conditioning is provided by the carrier as an extra cost option, periodically testing these facilities with a TIMS allows users to verify that they are indeed getting the level of performance for which they are paying.

Such test equipment can make a few very basic measurements by passively bridging into a circuit. However, since all of its sophisticated testing is intrusive, testing must be coordinated so that it will not affect users. These measurements require sending reference tones or combinations of frequencies and receiving them back. By analyzing the difference between what was sent and what is received, a TIMS is used to determine the extent of impairment. TIMS can measure a variety of Voice Frequency (VF) impairments, including:

- Overall signal quality
- VF transmit level
- VF receive level
- Data carrier detect loss
- Dropouts
- Signal-to-noise ratio
- Gain hits
- Phase hits
- Impulse hits
- Frequency offset
- Phase jitter
- Nonlinear distortion

The measurements for selected parameters (listed above) can then be compared against the performance thresholds set by you, the network manager. If these thresholds are exceeded, data traffic may have to be rerouted to another facility until the primary line can be brought back into specification, or down-speeded to avoid the corrupting effects of the line impairment on data.

Low-end impairment sets may only measure the Decibel (dB) level and frequency. More sophisticated units measure noise, noise with tone, noise to ground, signal-to-noise ratio, as well as other noise measurements with various notching filters, phase jitter, envelope delay distortion, impulse noise, dropouts, and Peak to Average Ratio (PAR).

15.6   DSL Testing

Digital Subscriber Line (DSL) is a digital service that is provisioned over analog lines to provide large amounts of bandwidth for Internet access. The growing popularity of DSL has brought about the need to test phone lines to determine if they can support this high-speed service. Until mid-2000, prospective customers, business as well as residential, had to look up a service provider's Web page and enter their phone number and address to determine what flavor of DSL service was available in their area and what speed their POTS line could support.

This qualification process, however, is based on line-of-sight distance estimates and is calculated by maps and telephone company cable plant records. Many of these records are often incomplete and inaccurate. Consequently, the process fails to provide an accurate picture of who can qualify for the service, as demonstrated by the large number of failed installations and dissatisfied users.

Now there are automated systems designed to streamline the process of accurately testing and qualifying phone lines for DSL service. Such Web-based tools empower IT managers desiring DSL service for their organizations to test and qualify their own lines in a matter of minutes using a downloaded browser plug-in. With the plug-in, IT managers can calculate the actual distance to their company's local central office by launching a battery of tests designed to determine the electrical characteristics of the telephone lines. DSL inhibitors like load coils, fiber in the loop, Digital Loop Carriers (DLC), and bridge taps can even be detected.

Automated line-testing services are more accurate than using conventional static maps accessed through Web forms. However, the major problem for many business users is that the actual copper loop used for the DSL service is currently provisioned as an Unbundled Network Element (UNE) purchased from the Incumbent Local Exchange Carrier (ILEC) by a third-party DSL service provider. Because it is provisioned after the customer has actually ordered the service from a DSL service provider, automated test tools are of little value. Once the industry moves fully into the "line-sharing" mode of provisioning DSL, in which DSL service is overlaid on the customer's active telephone line, these test tools can be used effectively.

15.7   T-carrier Testing

Certain network devices as T1 multiplexers and Data Service Units/Channel Service Units (DSU/CSU) offer various levels of T-carrier testing. For example, they usually include basic diagnostic capabilities such as detection of Bipolar Violations (BPVs) or frame errors. However, such capabilities do not always help technicians and network managers determine what is wrong with the circuit or where to find the fault. In such cases, more sophisticated fault isolation capabilities are provided by portable test sets.

As with most other types of products, the capabilities of T-carrier test equipment differ according to vendor. There are devices suitable for testing a single DS1 signal, as well as multifunctional devices capable of selecting and testing a single DS1 signal from within a DS3 bit stream, or a single DS0 from within a DS1 bit stream.

Complicating the testing process are the different frame formats the carriers support on the T1 facilities offered in their service areas. These capabilities include D4 and ESF* DS1 signal frame format, and in some cases, clear channel capability. This means the test set must be able to run and recognize a variety of test patterns, specifically quasi-random signal source, "all ones," and others. Carriers also differ in the type of line coding technique they use: specifically, Binary Eight Zero Substitution (B8ZS) or Alternate Mark Inversion (AMI). Many test sets support both, allowing the user to select the appropriate set of diagnostic routines.

T-carrier testing can be performed in two ways: in-service (nondisruptive) testing or out-of-service (disruptive) testing. Each needs different types of equipment, and users can perform different types of tests with the equipment.

For out-of-service testing, a BERT sends out a quasi-random pattern of bits—combinations of ones and zeros—on a T1 line in place of the 24 channels of information. With this, technicians can precisely measure the performance of the line in terms of how many bit errors are received at the remote BERT.

It is standard procedure among carriers to use bit-error-rate tests upon installation of T1 lines. But once a span is up and running, most users are

---

*Extended Super Frame (ESF) format is a carrier-provided capability for nonintrusive circuit testing and diagnostics. ESF diagnostic information, often displayed on a terminal or personal computer, supports circuit troubleshooting by the carrier, but is also useful to the user. The supervisory terminal provides information about T1 link performance over a long period of time, furnishing a historical record of circuit performance. Performance statistics are compiled every 15 minutes. This information is typically saved for a full 24 hours so that a complete one-day history can be assessed by the service provider The carrier polls the Channel Service Units (CSUs) attached to the network to retrieve data on a demand basis before clearing the storage registers. CSUs with dual registers allow both the carrier and user to access the performance history.

very reluctant to take the whole line down just for testing, unless there is spare capacity available elsewhere. If not, in-service testing is a viable alternative.

In-service testing uses the traffic on the T1 line. Two basic types of tests can be performed. One is performance monitoring in which the equipment searches for errors such as BPVs. Users can also look for Cyclic Redundancy Check (CRC) errors on ESF lines and can check for framing bit errors on any line (superframe or ESF). Another type of in-service test is channel access testing. The test set extracts a single channel from the T1 line, such as voice, data, a modem tone, or a Digital Data Service (DDS) circuit, and examines it. Since a T1 line may be running flawlessly except for a problem on one channel, each channel may be tested in turn until the problem is found.

Historically, a problem with in-service testing was that every error could not necessarily be detected. For example, if a data bit gets corrupted, it might not show up as a BPV. A reason for this is that BPVs are cleaned up at every network element, such as a Digital Cross-connect System (DCS) or a high-speed multiplexer. Framing bit errors incur similar problems; often, when the data goes through the DCS, framing patterns on T1 lines are replaced. This precludes test equipment from performing a true end-to-end check of the T1 line.

Today's T-carrier test equipment supports out-of-service testing, which allows the user to select the bandwidth increment for analysis. The test set then provides results in such terms as bit error rate, error-free seconds, and percent error-free seconds for that increment of bandwidth. This capability is especially important for users with Fractional T1 and generic digital services. Most of the test sets currently available allow both performance and channel access testing.

Fractional T1 is slightly more complicated to test. Essentially, FT1 utilizes a full T1 from the customer's premises to the local central office. From there, the T1 goes to an access tandem, where the channels are delivered to the interexchange carrier's Point of Presence (PoP). There, a DCS routes the various channels to their appropriate destinations. The user pays only for the number of Inter Office Channels (IOCs) ordered. The carrier manages to fill up partially used T1s by inserting channels from other customers, and dropping others off at intermediate locations on the DCS network.

While this can save on monthly line charges, it also complicates the testing of Fractional T1, since more sophisticated devices are needed to follow the signals along their various paths in the network. The testing process is further complicated by the various forms of carrier-offered FT1, which include both 56 and 64 kbps provided over contiguous or noncontiguous channels.

15.8   ISDN Testing

ISDN support for Basic and Primary Rate Interfaces (BRI and PRI) has become a key feature of many protocol analyzers. The use of such protocol analyzers enables ISDN users to verify that they are getting the kind of service for which they are paying.

Monitoring, analysis, and simulation over B and H are done via access to the D channel, which carries the signaling and control information concerning what to do with the information on the B and H channels. Once access to the D channel at Open Systems Interconnection (OSI) Layers 2 and 3 is established, further testing on the B or H channels can be performed. The B or H channels may carry anything, including LAN, X.25, or SNA traffic.

15.8.1   Physical link problems

ISDN standards and specifications at Layer 1 of the OSI reference model define mechanical, electrical, functional, and procedural considerations of network operation. Layer 1 standards describe the protocol that activates and deactivates the physical connection between terminals, network terminations, and ISDN switches. Any of the following can cause Layer 1 problems:

- A break in the physical connection

- A faulty digital subscriber line

- Improper or incorrectly implemented cabling

- Failure to plug into the correct jack

In addition, the physical-level interface of the Customer Premises Equipment (CPE) or network switch may not be operating correctly, or the physical-level "handshaking" may not be operating properly to establish the communications link. Other Layer 1 problems include power sources that may not be working or may have the wrong polarity.

To track down problems, it might be necessary to isolate the suspect device and use a protocol analyzer to simulate the Layer 1 functionality of the terminal or network termination. The analyzer may also be used to monitor the status of the physical-level handshaking process, activity on B, H, and D channels, and status of the power states.* It is sometimes desirable to measure bit-error-rate performance of the entire digital subscriber line or of a single

---

*H channels are high-capacity ISDN channels of 384 kbps and 1.536 Mbps. They are referred to as H0 and H11 channels, respectively. H0 consists of a group of six B channels, with the resulting 384 kbps of bandwidth used as a single channel. H11 consists of a group of 11 B channels, with the resulting 1.536 Mbps of bandwidth used as a single channel.

channel. To gain access to the channel for testing purposes, a call is established via the D channel signaling protocol with the protocol analyzer.

15.8.2 Data link problems

Layer 2, the link- or frame-level interface, is responsible for the reliable transfer of information across the physical links. Its functions include synchronization, error control, and flow control. The most basic Layer 2 tests look for physical-level problems that did not show up in Layer 1 testing. Layer 2 information, such as bad frame-check sequences, may indicate bit errors during transmissions. Frame-reject reports of an error condition indicate poor digital subscriber line quality.

Layer 2 tests also locate problems caused by configuration errors, such as when terminal service access point identifiers assigned by the network switch do not correspond to those of the CPE.

Another set of Layer 2 tests also applies to Layer 3. These are timing measurements designed to verify response time, check for premature time-outs, and determine whether a particular vendor implements handshaking sequences the same way as another vendor.

Finally, Layer 2 tests include protocol tests designed to verify that the proper Q.921 (LAP-D) procedures are followed for such functions as link setup, frame transfer, and link disconnection.

The tasks associated with performing these tests with the protocol analyzer include monitoring the line and decoding Layer 2 information, focusing on the suspected problem, and decoding and verifying frame types and responses, as well as checking their timing relationships.

15.8.3 Network layer problems

Although the core of Layer 3 is defined by the international Q.931 standard, some ISDN equipment manufacturers have gone beyond the basic definition and implemented different extensions to Layer 3. With a protocol analyzer, the user can verify that the proper procedures for Q.931 are occurring. This level of testing reveals incompatible implementations of Layer 3 message interactions, including those for call establishment, message transfer, and call disconnect.

Another type of Layer 3 testing is timing. If a particular response, such as alerting, is not received after call setup within the required amount of time, the call may be disconnected. The protocol analyzer displays the time stamps, along with the decoded messages. The location of the problem can be deduced from this information, possibly to Layer 3 software.

With a high performance protocol analyzer, several verification tests of the various ISDN channels may be performed, including:

- Verification that each voice connection on the B channel is working in both directions

- Verification that B or H channel circuit-switched data transfer is functioning properly

- Verification that all Layer 2 and 3 packets are being processed properly when the link is being used to send D-channel packet data

- Verification that LAP-B (Layer 2) and Layer 3 messages are correct when the link is being used to send B or H channel packet data

Among the ISDN-specific features a protocol analyzer should include are the following:

- Full drop-and-insert access to both ISDN interfaces, BRI, and PRI

- Full Q.921 and Q.931 support

- Simultaneous B, H, and D channel support, including rate adaptation

- Flexible programming language and libraries

- Analysis of B, H, and D channels

- Simulation of ISDN network elements, including Terminal Equipment (TE) and Network Terminations (NTs).

15.9 Digital Data Service Testing

Most testing of DDS-type services by network managers consists of protocol testing done within the CSU/DSU, or with stand-alone protocol test sets. Some test equipment vendors also provide end-to-end diagnostic capabilities. This test capability may also be provided by the carrier, offered as an extra-cost option to the digital service. DDS-specific test equipment is available that makes use of the secondary channel available on some DDS circuits.

The secondary channel provides users with either a secondary data channel for in-service testing or the capability to perform end-to-end diagnostics without taking the DDS line out of service. The channel is provided through the use of a time-sharing scheme that allows the control bit in one out of every three information bytes to carry user information. This information can be user data or diagnostic information.

15.10 Testing Frame Relay Networks

Frame relay is a popular method of carrying LAN traffic between corporate locations due to its ability to support intermittent or "bursty" transmissions at up to the T3 rate of 45 Mbps. Many test equipment vendors offer frame relay software upgrade packages for their protocol analyzers. The frame relay software generates simulated frames for testing with a separately available analysis product. The simulation software tests endpoint devices such as switches and routers in frame relay networks. Vendors also provide tools to enhance the user interface for these tests. Network evaluation systems, for instance, provide real-time statistics; protocol monitoring development systems let users modify decoding programs to accommodate variations of the emerging frame relay specifications.

Multiprotocol analyzers decode packets and provide statistics on frame relay as well as X.25, SNA, and ISDN data streams. Easy-to-use menus allow the user to specify triggers and filters for different protocols. Screens include performance analysis, statistics, and multilayer decodes and data presentation.

15.11   X.25 Testing

Although X.25 is not the best choice for interconnecting LANs, because of its store-and-forward nature, it is done for routine applications that are not time-sensitive. When applied to upper-level protocols, the trace and statistics capabilities of protocol analyzers permit the decoding of encapsulated LAN protocol data transported over X.25 packet-switched networks at data rates up to 2.048 Mbps.

An analyzer's trace features can decode and display data in three modes: single-line trace, multiline trace, or raw data (undecoded) form. In single-line mode, decoded X.25 packet summaries appear in sequence. Information presented includes address, logical channel number, frame type, send-receive frame sequence numbers, and packet type. The multiline trace decodes each field and subfield of an X.25 packet, including encapsulated upper-level LAN protocols. Fields are displayed line by line. Raw data mode shows all frame data in hexadecimal form with ASCII equivalents.

The analyzer can be configured to filter real-time X.25 data or data captured in the buffer. Data can be filtered by such parameters as discrepancy condition, port, side of line, or specific string or ASCII/HEX character. An operator can also create multiple independent, conditional triggers. Trigger parameters can be based on frame type, Logical Channel Number (LCN), packet type, and field value.

Some analyzers can also support statistical performance analysis. Statistics can be generated for frame counts and packets (by type), clear and diagnostic causes, bad frames, aborted frames, and total packets. An LCN statistics report can provide rates for each side of the line for several ports and the average call duration for each LCN.

15.12   Testing ATM Networks

As ATM moves into the corporate environment, you, as a network manager, must pay attention to the specific testing and monitoring issues associated with ATM. For example, in the switched environment of ATM, despite the availability of a centralized database of addresses, it is more difficult to figure out who is on what link at any given time. Testing an ATM network is more complex because it can support many more types of interfaces and services. The interaction of different services running over ATM—such as voice and video in packet format—still has to be analyzed. As a result, the performance and quality of service issues get more complicated with ATM, and the speeds are higher.

There is now test equipment that combines WAN, LAN, and ATM analysis functions over full DS1 (1.5 Mbps), DS3 (45 Mbps), OC-3 (155 Mbps), and higher line rates. The key to seeing all of the traffic in real time on the link—not just the lower-layer protocols—is the ability of some analyzers to perform real-time AAL5 reassembly, which allows technicians to monitor and decode cell-encapsulated frames like Ethernet or IP running across high-speed ATM links.

With regard to finding ATM channels, many analyzers require the technician to enter the Virtual Path/Virtual Circuit (VP/VC) addresses of channels under test before any testing can start. Some analyzers have a VP/VC bandwidth discovery feature that automatically finds the active channels on a monitored ATM link so that various tests can be performed, such as measuring traffic volume per channel and the bandwidth usage for each.

Such systems can also analyze ATM cell errors, throughput, and quality of service via simulation tests for cell loss, delay, and delay variation. In addition, customized traffic patterns can be created, including cells, Payload Data Units (PDUs), and Operations, Administration, and Management (OAM) traffic. The simulated traffic can be captured as "live" traffic for analysis. Technicians can define and customize filters, as well as set counters to review trends. The throughput for each virtual circuit setup can be tested—some devices can monitor as many as 1023 channels in real time. In addition to ATM-specific tests, such analyzers provide for bit-error-rate testing to establish line quality.

Some ATM analyzers have an on-line help facility to automatically provide status information and potential solutions to errors. In addition to predefined tests for cell-loss and cell-delay measurements, these "smart" troubleshooting techniques allow technicians to measure the quality of a new service accurately by automatically setting up simulation and monitoring tests for which the user sets the pass-fail criteria. Depending on the vendor, ATM support can be added to existing test equipment as an upgrade.

Among the statistics that ATM analyzers are capable of collecting and displaying are the number of cells transmitted and received on a designated path, cell loss, and cells received with errors. The types of errors include ATM Header Error Check (HEC) errors, CRC errors, and header errors. A graphical, cell delay report dynamically displays the variance between the number of expected empty cells for every full cell received and the total number of cells received.

The physical interface options for ATM analyzers include Synchronous Optical Network/Synchronous Digital Hierarchy SONET/SDH, multimode and single-mode dark fiber, DS3, E3, and STS-1. In addition to ATM, broadband analyzers can be used to test other protocols, including ISDN and frame relay.

15.13   Conclusion

The trend in the telecommunications industry is to integrate as much functionality and intelligence as possible into a single test set, with the IT department selecting appropriate application software to suit the present and emerging needs of the entire organization. This is important because field service, product development, engineering, and technical support all have different test needs. Selection of the right equipment for the field technician or analyst, for example, can mean the difference between solving the problem quickly and having the situation degenerate into a finger-pointing contest among vendors, carriers, and users.

As a network manager who is contemplating the purchase of test equipment, you should give due consideration to ease of use, reliability, upgradability, portability, as well as the availability of test applications and vendor support. The right product will provide a growth path so that, as testing needs expand with the adoption of new facilities and services, the capabilities available through the test device can expand accordingly.

Network Management

16.1 Introduction

The task of keeping multivendor networks operating smoothly with a minimum of downtime is an ongoing challenge for the network managers of most organizations. While many companies prefer to retain total control of their network resources, others rely on computer vendors and carriers to find and correct problems on their networks, or depend on third-party service firms. Wherever these responsibilities ultimately reside, the tool set used for monitoring the status of the network and initiating corrective action is the Network Management System (NMS).

With an NMS, technicians can remotely diagnose and correct problems associated with each type of device on the network. Although as a network manager, you are primarily concerned with diagnosing failures, the likelihood of problems can also be predicted so that traffic may be diverted from failing lines or equipment with little or no disruption to applications and without unduly inconveniencing users.

Network management begins with such basic hardware components as modems, data sets Channel Service Units/Data Service Units (CSUs/DSUs), routers, switches, multiplexers, and servers (see Fig. 16.1). These and other devices typically have the ability to monitor, self-test, and diagnose problems regarding their own operation and are capable of reporting their status to a central management station, either periodically or on request. The management station operator can initiate test procedures on systems at any point in the network. On more complex multipoint and multidrop configurations, the capability to test and diagnose problems from a central location greatly facilitates problem resolution. This capability also

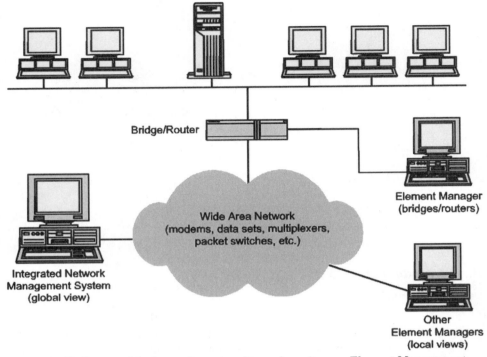

Figure 16.1 Each type of device on the network may have its own Element Management System (EMS), which reports to an Integrated Network Management System (INMS).

minimizes the need to dispatch technicians to remote locations and reduces maintenance costs.

A minimal network management system consists of a central processing unit, system controller, operating system software, storage device, and an operator's console. The central processor may consist of a minicomputer or microcomputer. The system controller, the heart of the network management system, continuously monitors the network and generates status reports from data received from various network components. The system controller also isolates network faults and restores segments of the network that have failed, or which are in the process of degrading. The controller usually runs on a powerful platform such as UNIX or Windows NT/2000.

16.2   Network Management System Functions

Although differing by vendor, the basic functions that most network management systems have in common include topology mapping, administration, performance measurement, control and diagnostics, configuration management, applications management, and security. Some network management systems include other functions such as network modeling, for example, which would enable the operator to simulate aggregate, node, or circuit failures to test various disaster recovery scenarios.

16.2.1   Topology mapping

Many network management systems have an automatic discovery capability that finds and identifies all devices or nodes connected to the network. Based on the discovered information, the NMS automatically draws the required topology maps. Nodes that cannot be discovered automatically can be represented by manually adding custom or standard icons to the appropriate map views. Or they can be represented by using the network management systems' Simple Network Management Protocol (SNMP)-based Application Programming Interfaces (APIs) for building map applications without having to manually modify the configuration to accommodate non-SNMP devices.

A network map is useful for ascertaining the relationships of various equipment and connections, for keeping accurate inventory of network components, and for isolating problems on the network. The network map is updated automatically when any device is added or removed from the network. Device status is displayed via color changes to the map. Any changes to the network map are carried through to the relevant submaps.

16.2.2   Administration

The administration capability allows you, as a network manager, to take stock of the network in terms of what hardware is deployed and where it is located. It also tells the user what facilities are serving various locations, and what lines and equipment are available with which to implement alternative routing. The vehicle for storing and using this information is the Relational Database Management System (RDBMS).

For administrative tasks, multiple specialized databases are used, which relate to each other. One of these databases accumulates trouble ticket information. A trouble ticket contains such information as the date and time the problem occurred, the specific devices and facilities involved, the vendor from which it has been purchased or leased, and the service contact. It also contains the name of the operator who initially responded to the alarm, any short-term actions taken to resolve the problem, and space for recording follow-up information. This information may include a record of visits from the vendor's service personnel, dates on which parts were returned for repair, serial numbers of spares installed, and the date of the problem's final resolution.

A trouble ticket database can be used for long-term planning. As network manager, you can call up reports on all outstanding trouble tickets, trouble tickets involving particular segments of the network, trouble tickets recorded or resolved within a given period, trouble tickets involving a specific type of device or vendor, and even trouble tickets over a given period not resolved within a specific time frame. The user may customize report formats to meet unique needs.

Such reports provide network managers with insight on the reliability of a given network management station operator, the performance record of various network components, the timeliness of on-site vendor maintenance and repair services, and the propensity of certain segments of the network to fail. And with information on both active and spare parts, network managers can readily support their decisions on purchasing and expansion. In some cases, cost and depreciation information on the network's components is also provided.

16.2.3   Performance measurement

Performance measurement refers to network response time and network availability. Many network management systems measure response time at the local end, from the time the monitoring unit receives a start-of-transmission (STX) or end-of-transmission (EOT) signal from a given unit. Other systems measure end-to-end response time at the remote unit. In either case, the network management system displays and records response time information, and generates operator-specified statistics for a particular terminal, line, network segment, or the network as a whole. This information may be reported in real time, or stored for a specified time frame for future reference.

Network personnel can use this information to track down the cause of the delay. When an application exceeds its allotted response time, for example, network personnel can decide whether to reallocate terminals, place more restrictions on access, or install faster communications equipment to improve response time.

Availability is a measure of actual network uptime, either as a whole, or by segments. This information may be reported as total hours available over time, average hours available within a specified time, and Mean Time between Failure (MTBF).

With response time and availability statistics calculated and formatted by the NMS, managers can establish current trends in network usage, predict future trends, and plan the assignment of resources for specific present and future locations and applications.

16.2.4 Control and diagnostics

With control and diagnostic capabilities, the NMS operator can determine from various alarms (i.e., an audio or visual indication at the operator's terminal) what problems have occurred on the network, and pinpoint the sources of those problems so that corrective action can be taken. Alarms can be correlated to certain events and triggered when a particular event occurs. For example, an alarm can be set to go off when a line's Bit Error Rate (BER) approaches a predefined threshold. When that event occurs and the alarm is issued, automated procedures can be launched without operator involvement. In this case, traffic can be diverted from the failing line and routed to an alternative line or service. If the problem is equipment-oriented, another device on "hot" standby can be placed into service until the faulty system can be repaired or replaced.

16.2.5 Configuration management

Configuration management gives the NMS operator the ability to add, remove, or rearrange nodes, lines, paths, and access devices as business circumstances change. If a T1 link degrades to the point that it can no longer handle data reliably, for example, the network management system may automatically reroute traffic to another private line or through the public network. When the quality of the failed line improves, the system reinstates the original configuration. Some integrated network management systems—those that unify the host Local Area Network (LAN) and Wide Area Network (WAN) carrier environments under a single management umbrella—are even capable of rerouting data, but leaving voice traffic where it is.

Voice and data traffic can even be prioritized. This NMS capability is very important because failure characteristics for voice and data are very different—voice is more delay sensitive and data is more line error sensitive. On networks that serve multiple business entities and on statewide networks that serve multiple government agencies, the ability to differentiate and prioritize traffic is very important.

On a statewide network, for example, state police have critical requirements twenty-four hours a day, seven days a week, whereas motor vehicle branch offices use the network to conduct relatively routine administrative business only eight hours a day, five days a week. Consequently, the response time objectives of each agency are different, as would be their requirements for network restoral in case of an outage. On the high-capacity network, there can be two levels of service for data and another for voice. Critical data will have the highest priority in terms of response time and error thresholds, and will take precedence over other classes of traffic during restoral. Since routine data will be able to tolerate a longer response time, the point at which restoral is implemented can be prolonged. Voice is more tolerant than data with regard to error, so restoral may not be necessary at all. The capability to prioritize traffic and reroute only when necessary ensures maximum channel fills, which affects the efficiency of the entire network and, consequently, the cost of operation.

Configuration management not only applies to the links of a network, but to equipment as well. In the WAN environment, the features and transmission speeds of software-controlled modems may be changed. If a nodal multiplexer fails, the management system can call its redundant components into action, or invoke an alternative configuration. And when nodes are added to the network, the management system can devise the best routing plan for the traffic it will handle.

16.2.6 Applications management

Applications management is the capability to alter circuit routing and bandwidth availability to accommodate applications that change by time of day. Voice traffic, for example, tends to diminish after normal business hours, while data traffic may change from transaction-based to wide band applications that include inventory updates and remote printing tasks.

Applications management includes having the ability to change the interface definition of a circuit so that the same circuit can alternatively support both asynchronous and synchronous data applications. It also includes having the ability to determine appropriate data rates in accordance with response time objectives, or to conserve bandwidth during periods of high demand.

16.2.7 Security

Network management systems have evolved to address the security concerns of users. Although voice and data can be encrypted to protect information against unauthorized access, the management system represents a single point of vulnerability to security violations. Terminals employed for network management may be password protected to minimize disruption to the network through database tampering. Various levels of access may be used to prevent accidental damage. A senior technician, for example, may have a password that allows her to make changes to the various databases, whereas a less experienced technician's password allows him to review only the databases without making any changes. Other possible points of entry such as gateways, bridges, and routers may be protected with hardware or software-defined partitions that restrict internal access.

Individual users, too, may be given passwords, which permit them to make use of certain network resources, and deny them access to others. A variety of methods are even available to protect networks from intruders who may try to access network resources with dial-up modems. For instance, the management system can request a password and hang up if it does not obtain one within 15 seconds. Or it can hang up and call back over an approved number before establishing the connection. To frustrate persistent hackers, the system can limit unsuccessful call attempts before denying further attempts. All successful and unsuccessful attempts at entry are automatically logged to monitor access and to aid in the investigation of possible security violations.

16.3   Network Management System

Simple Network Management Protocol is one of three components comprising a total network management system (see Fig. 16.2). The other two are the Management Information Base (MIB) and the Network Manager (NM). The MIB defines the controls embedded in network components while the NM contains the tools that enable network administrators to comprehend the state of the network from the gathered information.

16.3.1   Simple Network Management Protocol

Since 1988, the *Simple Network Management Protocol* (SNMP) has been the de facto standard for the management of multivendor Transmission Control Protocol/Internet Protocol (TCP/IP)–based networks. SNMP specifies a structure for formatting messages and for transmitting information between reporting devices and data-collection programs on the network. The SNMP-compliant devices on the network are polled for performance-related information, which is passed to a network management console. Alarms are also passed to the console. There, the gathered information can be viewed to pinpoint problems on the network or stored for later analysis.

SNMP runs on top of TCP/IP's datagram protocol—the User Datagram Protocol (UDP)—a transport protocol that offers a connectionless-mode service. This means that a session need not be established before network management information can be passed to the central control point. Although SNMP messages can be exchanged across any protocol, UDP is well suited to the brief request/response message exchanges characteristic of network management communications.

SNMP is a very flexible network management protocol that can be used to manage virtually any object. An *object* refers to hardware, software, or a logical association such as a connection or virtual circuit. An object's definition is written by its vendor. The definitions are held in a Management Information Base (MIB), which is often thought of as a database. In reality, a MIB is a list of switch settings, hardware counters, in-memory variables,

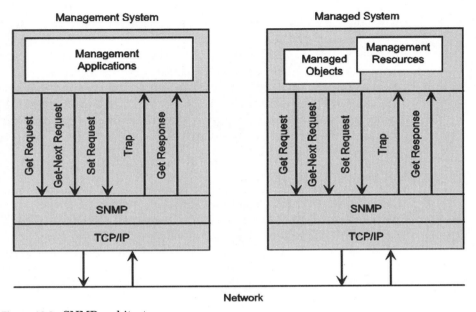

Figure 16.2   SNMP architecture.

or files which are used by the network management system to determine the alarm and reporting characteristics of each device on the network. The network can be any LAN or WAN or service over which the TCP/IP suite of protocols is run.

All of the major network management platforms support SNMP, including those of Computer Associates, Hewlett-Packard, IBM/Tivoli, and Sun. In addition, many of the third-party systems and network management applications that plug into these platforms support SNMP. The advantage of using such products is that they take advantage of SNMP's capabilities, while providing a Graphical User Interface (GUI) to make SNMP easier to use (see Fig. 16.3). Even MIBs can be selected for display and navigation through the GUI.

Another advantage of commercial products is that they can use SNMP to provide additional functionality. For example, Hewlett-Packard's OpenView is used to manage network devices that are IP addressable and run SNMP. OpenView's automatic discovery capability finds and identifies all IP nodes on the network, including those of other vendors that support SNMP. Based on discovered information, the management system automatically draws the required topology maps. Nodes that cannot be discovered automatically can be represented in either of two ways: first, by manually adding custom

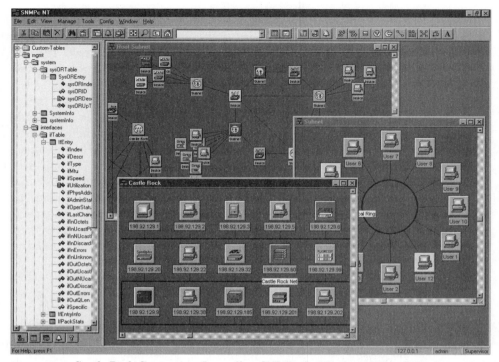

Figure 16.3 Castle Rock Computing, Inc., offers SNMPc for Windows NT and Windows 2000, which, among other things, provides a graphical display that supports multilevel hierarchical mapping.

or standard icons to the appropriate map views, and second, by using SNMP-based APIs for building map applications without having to manually modify the configuration to accommodate non-SNMP devices.

The SNMP system retrieves information from the agents through "get" and "get-next" commands, making SNMP basically a request-response protocol. The "get" request retrieves the values of specific objects from the MIB. The MIB lists the network objects for which an agent can return values. These values may include the number of input packets, the number of input errors, and routing information. The get-next request permits navigation of the MIB, enabling the next MIB object to be retrieved, relative to its current position. A "set" request is used to request a logically remote agent to alter the values of variables. In addition to these message types, there are "trap" messages, which are unsolicited messages conveyed from management agent to management stations. Other commands are available that allow the network manager to take specific actions to control the network. Some of these commands look like SNMP commands, but are really vendor-specific implementations.

16.3.2   Management Information Base

The *Management Information Base* (MIB) is a listing of information necessary to manage the various devices on the network. The MIB contains a description of SNMP-compliant objects on the network and the kind of management information they provide. An object refers to hardware, software, or a logical association such as a connection or virtual circuit. The attributes of an object might include such things as the number of packets sent, routing table entries, and protocol-specific variables for IP routing.

The first MIB was primarily concerned with IP routing variables used for interconnecting different networks. There are 114 objects that form the core of the standard SNMP MIB. The latest generation MIB, known as MIB II, defines 185 objects. It extends SNMP capabilities to a variety of media and network devices, marking a shift from Ethernets and TCP/IP WANs to all media types used on LANs and WANs. Many vendors want to add value to their products by making them more manageable, so they create private extensions to the standard MIB.

Many vendors of SNMP-compliant products include MIB tool kits that generally include two types of utilities. One, a MIB compiler, acts as a translator that converts ASCII text files of MIBs for use by an SNMP management station. The second type of MIB tool converts the translator's output into a format that can be used by the management station's applications or graphics. These output handlers, also known as MIB editors or MIB walkers, let users view the MIB and select the variables to be included in the management system. Some vendors of SNMP management stations do not offer MIB tool kits, but rather an optional service whereby they will integrate any MIB into the management system that a user requires for a given network. This service includes debugging and technical support.

There are also MIB browsers (see Fig. 16.4) that allow network managers, technicians, and engineers to query a remote device for software and hardware configurations via SNMP and make changes to the remote device. The remote device could be a router, switch hub, server, firewall, or any other device that supports SNMP. Another common use for a MIB browser is to find out what MIBs and Object Identifiers (OIDs) are supported on a particular device.

16.3.3   Network Manager

The *Network Manager* (NM) is a program that may run on one host or more than one host, each of which manages a particular subnet. SNMP communicates network management data to a single site, called a Network Management Station (NMS). Under SNMP, each network segment must have a device, called an *agent,* that can monitor devices (called *objects*) on that segment and report the information to the NMS. The agent may be a passive monitoring device whose sole purpose is to read the net-

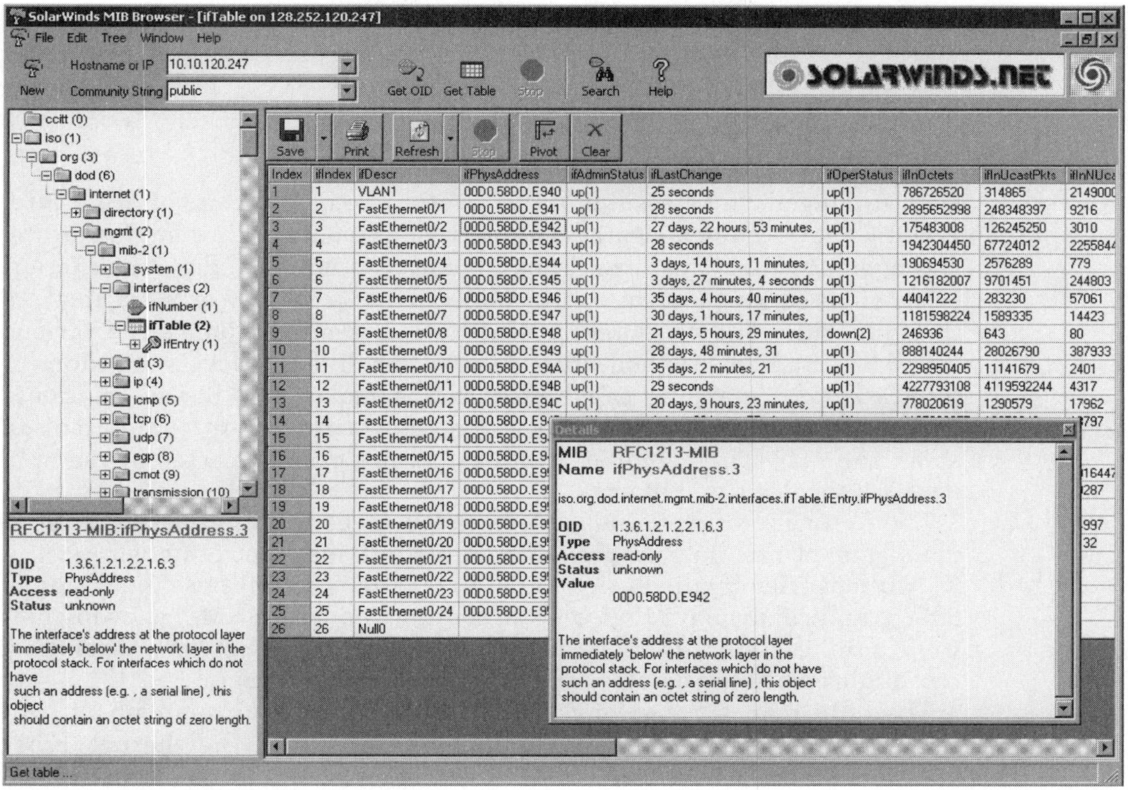

Figure 16.4   The MIB browser from SolarWinds.Net, Inc., is capable of reading over a thousand standard and proprietary MIBs and 110,000 unique OIDs.

work, or it may be an active device that performs other functions as well, such as bridging, routing, and switching. Devices that are non-SNMP compliant must be linked to the NMS via a proxy agent.

The NMS provides the information display, communication with agents, information filtering, and control capabilities. The agents and their appropriate information are displayed in a graphical format, often against a network map. Network technicians and administrators can query the agents and read the responses on the NMS display. The NMS also periodically polls the agents, searching for anomalies. Detection of an anomaly results in an alarm at the NMS.

16.4   SNMP Integration

Simple Network Management Protocol's continued popularity stems from the fact that it works, it is reliable, and it is widely supported. The protocol itself is in the public domain. SNMP capabilities have been integrated into just about every conceivable device that is used on today's LANs and WANs, including intelligent hubs and carrier services.

16.4.1 Intelligent hubs

With intelligent hubs assuming the pivotal role in LAN and/or WAN integration, they occupy a strategic position on the network for implementing management functions. The hubs usually contain modules for different types of LANs and interfaces for connection to a variety of WAN services through bridge and/or router cards. SNMP is used to manage the various interconnected devices and links. Many offer different levels of graphical interfaces and mapping capabilities. Some make good use of relational databases. The relational database provides a repository of information about the network, including what devices are interconnected at the hub, network topology, and alarms.

The first generation of hubs offered a low-level, terminal-based network management system that provided an instant snapshot of network activity without friendly displays and data retention capabilities. Today's hubs offer graphical displays that work in conjunction with SNMP, allowing network administrators to more easily view and change network topologies, reconfigure network devices, and relocate individual users.

The hubs offer a range of SNMP-based network management applications that integrate Asynchronous Transfer Mode (ATM), Ethernet, Fiber Distributed Data Interface (FDDI), and token ring management from a single management station platform. The hub's internal supervisor module communicates with other modules and relays network management information to the management console. Each module communicates with the supervisor module. The supervisor can then use SNMP to communicate with the management console. The agent software in the supervisor could be downloaded from the management console, an arrangement that permits easy upgrades.

Enterprise hub management systems run on UNIX or Windows platforms using GUI and SNMP as the element command protocol. Large networks are segmented into domains to more effectively gather, analyze, and present only salient data through the management station. Each domain may use a different protocol and support a different set of applications. Segmenting networks in this way also enhances security.

The management data is processed locally in the hub to build the information presented through the GUI at the management station. SNMP support allows basic management tasks for third-party devices that are tied into the hub, including graphical maps and gives users access to database information through the use of icons. Various network views are available through the manager, providing both static mapping capabilities, which are drawn by the user, and dynamic network representations, which are provided automatically by the network management system. These network views include:

- *Global Network Views.* User-generated, high-level representations created to provide the overall context for network operations; both static and dynamic subviews can be nested below the Global View to allow rapid and logical navigation through both local and remote hierarchical networks.

- *Flat Network Views.* Dynamic (automatically drawn and updated) view of a flat Ethernet or token ring network (bounded by routers) that provides a picture of the physical relationship among network segments.

- *Segment Views.* Dynamic topology representations of Ethernet or token ring concentrator segments bounded by bridges.

- *Expanded Views.* Dynamic, real-time graphical representations of the intelligent hubs in the network, with all of their modules and their associated ports and diagnostic Light Emitting Diodes (LEDs); information about the status of any point on the network can be obtained through the Expanded View capability.

The ability to filter information and only present what is needed by the LAN administrator is becoming more important as LANs grow from tens of attached devices to hundreds and even thousands. The hub network management station gathers, analyzes, and reduces management data from throughout its domain, which is an arbitrarily defined portion of a network consisting of intelligent hubs and other network devices. In addition, management information can be stored or logged locally and sent to a higher-level management system for subsequent analysis, so it does not traverse the backbone of the network until it is needed.

Some hubs include an integral protocol analyzer capability that is used to capture specified types of packets, automatically disassemble packets, detect nodes generating excessive packets, and debug protocols. Load profiling enables the administrator to observe the behavior of every station on a network. Network load selection parameters include time intervals, source nodes, destination nodes, protocols, applications, and packet sizes. Network load can be monitored over extended periods of time to determine how the load varies during the day; which stations interact with each other; how much of the load is generated by a specific node; or how much is generated by XNS, SNMP, NFS, and other types of traffic. Issues of performance or diagnostics not relevant to the overall management of the system are dealt with at the source, and a set of information—analyzed and reduced to only that which is needed—is sent up to the central station.

16.4.2 Carrier services

Carrier services can also be made manageable through SNMP, since they consist of definable objects such as network devices, circuits, and communication protocols. Frame relay, for example, is a carrier-provided service whose protocol is specifically optimized to support LAN interconnection. Some carriers offer users the option of managing frame relay services with the ubiquitous SNMP, rather than burden them with a separate, proprietary management system.

In such cases, SNMP is used to monitor the service in real time, receive alarms, and keep a history of service performance. All this is done by the customer at an on-premises management console—the same console used to manage their existing LANs and WANs.

With frame relay service, for example, agents on each switch in the network pass performance information back to customers' SNMP management stations. Configuration data lists individual ports and port speeds on each device, virtual connections to each device, and the utilization percentage of each Committed Information Rate (CIR). Alarm information includes notification when any link in the network becomes congested or goes down. It also provides notification when any port is lost. Statistics provided include the number of frames per second transmitted over each link in the frame relay network. The software archives configuration data, alarm information, and usage statistics.

16.5 Remote Monitoring Management Information Base

As networks expand, the ability to perform remote monitoring becomes more important. Problems can be identified and resolved from a management console, rather than by sending a technician to remote locations, which is expensive and time-consuming. The ability to monitor the performance of remote LAN segments (Ethernet) and rings (token ring) is made easier with SNMP's Remote Monitoring Management Information Base (RMON MIB) standard.

RMON provides a common platform from which to monitor multivendor networks. Hardware- and/or software-based RMON-compliant devices placed on each network segment monitor all data packets sent and received. Although a variety of SNMP MIBs collect performance statistics to provide a snapshot of events through the use of agents, RMON enhances this monitoring capability by keeping a past record of events that can be used for fault diagnosis, performance tuning, and network planning.

16.5.1 Advantages of RMON

The RMON MIB is a set of object definitions that extend the capabilities of SNMP. RMON is not used to directly manage the devices on the network; instead, monitors or probes equipped with RMON agents passively monitor data transmitted over LAN segments or rings. The accumulated information is retrieved from the probe or monitor, or another SNMP agent that reports to another central network management system, using SNMP commands. The RMON standard accomplishes several highly worthwhile goals.

Offline operation. There are sometimes conditions when a management station will not be in constant contact with its remote monitoring devices. This occurs by design in an attempt to lower communications costs, especially when communicating over a WAN or dial-up link, or by accident as network failures affect communications between the management station and the probe. For this reason, the RMON MIB allows a probe to be configured to perform diagnostics and to collect statistics continuously, even when communication with the management station may not be possible or efficient. The probe may then attempt to notify the management station when an exceptional condition occurs. Thus, even in circumstances where communication between management station and probe is not continuous, information on fault, performance, and configuration may be continuously accumulated and communicated to the management station conveniently and efficiently.

Proactive monitoring. Given the resources available on the monitor, it is potentially helpful for it to continuously run diagnostics and to log network performance. The monitor is always available at the onset of any failure. It can notify the management station of the failure and can store historical statistical information about the failure. The management station can play back this historical information in an attempt to perform further diagnosis to determine the cause of the problem.

Problem detection and reporting. The monitor can be configured to recognize conditions, most notably error conditions, and to continuously check for them. When one of these conditions occurs, the event may be logged, and management stations may be notified in a number of ways.

Value added data. Because a remote monitoring device represents a network resource dedicated exclusively to network management functions, and because it is located directly on the monitored portion of the network, the remote network monitoring device has the opportunity to add significant value to the data it collects. For instance, by highlighting those hosts on the network that generate the most traffic or errors, the probe can give the management station precisely the information it needs to solve a class of problems.

Multiple managers. An organization may have multiple management stations for different units of the organization, for different functions such as engineering and operations, or to provide continuous coverage across time zones in different countries. Because environments with multiple management stations are common, the remote network monitoring device has to deal with more than one management station, potentially using its resources concurrently.

RMON enhances the management and control capabilities of SNMP-compliant network management systems and LAN analyzers. The probes view every packet and produce summary information on various types of packets, such as undersized packets, and events, such as packet collisions. Intelligent probes can also capture packets according to predefined criteria set by the network manager or test technician. At any time, the RMON probe can be queried for this information by a network management application or an SNMP-based management console so that detailed analysis can be performed in an effort to pinpoint where and why an error occurred.

The RMON MIB defines objects broken down into nine functional groups. Some of those functional groups—the statistics and the history groups—have a view of the data-link layer that is specific to the media type and require specific objects to be defined for each media type, such as Ethernet and token ring. A map of the RMON MIB, showing objects defined for Ethernet and token ring, is provided in Fig. 16.5.

16.5.2   Remote Monitoring applications

Control, visibility, and easy-to-read information are essential characteristics of tools for internetwork monitoring and analysis. Vendors have added these characteristics to various network management applications that use data collected by the RMON MIB.

A management application that views the internetwork, for example, gathers data from RMON agents running on each segment in the network. The data is integrated and correlated to provide various internetwork views that provide end-to-end visibility of network traffic, both LAN and WAN. The operator can switch between a variety of views.

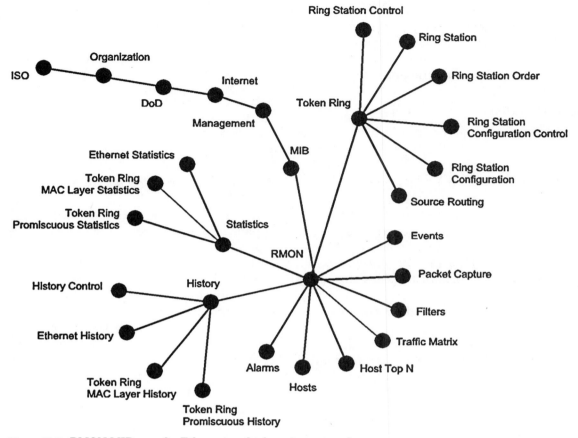

Figure 16.5 RMON MIB map for Ethernet and token ring networks.

For example, the operator can switch between a Media Access Control (MAC) view (which shows traffic going through routers and gateways), a network view (which shows end-to-end traffic), or apply filters to see only traffic of a given protocol or suite of protocols. These traffic matrices provide the information necessary to configure or partition the internetwork to optimize LAN and WAN utilization.

In selecting the MAC-level view, for example, the network map shows each node of each segment separately, indicating intrasegment node-to-node data traffic. It also shows total intersegment data traffic from routers and gateways. This combination allows the operator to see consolidated internetwork traffic and how each end node contributes to it.

Another application of RMON is that it allows the network manager to consolidate and present multiple segment information, configure RMON alarms, provide complete token ring RMON information, as well as perform baseline measurements and long-term reporting. Alarms can be set on any RMON variable. Notification via traps can be sent to multiple management stations. Baseline statistics allow long-term trend analysis of network traffic patterns that can be used to plan for network growth.

16.5.3   Ethernet object groups

As noted, the RMON specifications describe object groups for Layer 2 networks, including those for Ethernet and token ring.

Ethernet Statistics Group.   The Statistics Group provides segment-level statistics (see Fig. 16.6). These statistics show packets, octets (or bytes), broadcasts, multicasts, and collisions on the local segment, as well as the number of occurrences of dropped packets by the agent. Each statistic is maintained in its own 32-bit cumulative counter. Real-time packet size distribution is also provided. Table 16.1 lists the statistics available from the Ethernet Statistics Group and their definitions.

Ethernet History Group.   With the exception of packet size distribution, which is provided only on a real-time basis, the History Group provides historical views of the statistics provided in the Statistics Group. The History Group can respond to user-defined sampling intervals and bucket counters, allowing for some customization in trend analysis.

The RMON MIB comes with two defaults for trend analysis. The first provides for 50 buckets (or samples) of 30-second sampling intervals over a period of 25 minutes. The second provides for 50 buckets of 30-minute sampling intervals over a period of 25 hours. Users can modify either of these or add additional intervals to meet specific requirements for historical analysis. The sampling interval can range from one second to one hour.

Host Table Group.   A host table is a standard feature of most current monitoring devices. The RMON MIB specifies a host table that includes node traffic statistics such as packets sent and received, octets sent and received, as well as broadcasts, multicasts, and errored packets sent. In the host table, the classification "errors sent" is the combination of undersizes, fragments, Cyclic Redundancy Check (CRC)/Alignment errors, collisions, and oversizes sent by each node.

The RMON MIB also includes a host timetable, which shows the relative order in which each host was discovered by the agent. This feature is not only useful for network management purposes but it also assists in uploading those nodes to the management station of which it is not yet aware. This reduces unnecessary SNMP traffic on the network.

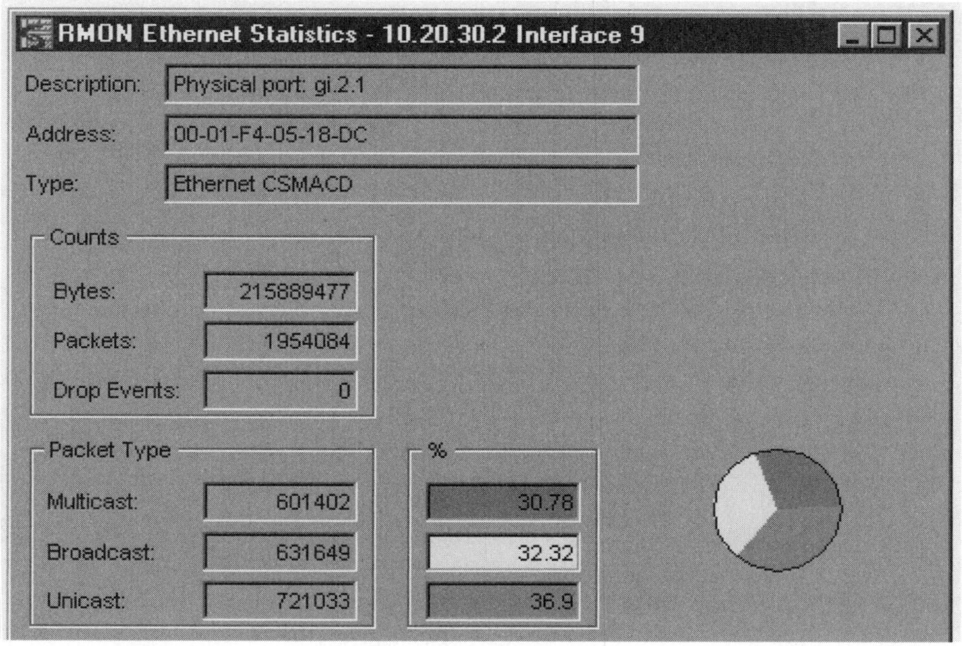

Figure 16.6 The Ethernet Statistics window accessed from Enterasys Networks' NetSight Element Manager. This window would be used to view a detailed statistical breakdown of traffic on the monitored Ethernet network segment. The data provided applies only to the interface or network segment.

Host Top N Group. The Host Top N Group extends the host table by providing sorted host statistics, such as the top 10 nodes sending packets or an ordered list of all nodes according to the errors sent over the last 24 hours. Both the data selected and the duration of the study are defined by the user at the network management station, and the number of studies is limited only by the resources of the monitoring device.

When a set of statistics is selected for study, only the selected statistics are maintained in the Host Top N Group counters; other statistics over the same time intervals are not available for later study. This processing—performed remotely in the RMON MIB agent—reduces SNMP traffic on the network and the processing load on the management station, which would otherwise need to use SNMP to retrieve the entire host table for local processing.

Alarms Group. The Alarms Group provides a general mechanism for setting thresholds and sampling intervals to generate events on any counter or integer maintained by the agent, such as segment statistics, node traffic statistics defined in the host table, or any user-defined packet match

TABLE 16.1   Ethernet Statistics

Statistic	Definition
etherStatsDropEvents	The total number of events in which packets were dropped by the probe due to lack of resources. This is not necessarily the number of packets dropped, just the number of times this condition has been detected.
etherStatsOctets	The total number of octets of data (including those in bad packets) received on the network (excluding framing bits but including FCS octets).
etherStatsPkts	The total number of packets (including error packets) received.
etherStatsBroadcastPkts	The total number of good packets received that were directed to the broadcast address.
etherStatsMulticastPkts	The total number of good packets received that were directed to a multicast address. This number does not include packets directed to the broadcast address.
etherStatsCRCAlignErrors	The total number of packets received that had a length (excluding framing bits, but including FCS octets) of between 64 and 1518 octets, inclusive, but were not an integral number of octets in length or had a bad Frame Check Sequence (FCS).
etherStatsUndersizePkts	The total number of packets received that were shorter than 64 octets (excluding framing bits, but including FCS octets) and were otherwise well formed.
etherStatsOversizePkts	The total number of packets received that were longer than 1518 octets (excluding framing bits, but including FCS octets) and were otherwise well formed.
etherStatsFragments	The total number of packets received that were not an integral number of octets in length or that had a bad FCS, and were shorter than 64 octets (excluding framing bits but including FCS octets).
etherStatsJabbers	The total number of packets received that were longer than 1518 octets (excluding framing bits, but including FCS octets), and were not an integral number of octets in length or had a bad FCS.
etherStatsCollisions	The best estimate of the total number of collisions on this Ethernet segment.
etherStatsPkts64Octets	The total number of packets (including error packets) received that were 64 octets in length (excluding framing bits but including FCS octets).
etherStatsPkts65to127Octets	The total number of packets (including error packets) received that were between 65 and 127 octets in length inclusive (excluding framing bits but including FCS octets).
etherStatsPkts128to255Octets	The total number of packets (including error packets) received that were between 128 and 255 octets in length inclusive (excluding framing bits but including FCS octets).
etherStatsPkts256to511Octets	The total number of packets (including error packets) received that were between 256 and 511 octets in length inclusive (excluding framing bits but including FCS octets).
etherStatsPkts512to1023Octets	The total number of packets (including error packets) received that were between 512 and 1023 octets in length inclusive (excluding framing bits but including FCS octets).
etherStatsPkts1024to1518Octets	The total number of packets (including error packets) received that were between 1024 and 1518 octets in length inclusive (excluding framing bits but including FCS octets).

counter defined in the Filters Group. Both rising and falling thresholds can be set, each of which can indicate network faults. Thresholds can be established on both the absolute value of a statistic or its delta value, so the manager is notified of rapid spikes or drops in a monitored value.

Filters Group. The Filters Group provides a generic filtering engine that implements all packet capture functions and events. The packet capture buffer is filled with only those packets that match the user-specified filtering criteria. Filtering conditions can be combined using the boolean parameters AND or NOT. Multiple filters are combined with the boolean OR parameter.

Users can capture packets that are valid or invalid, or that are one of the five error packet types (discussed previously). With the proper protocol decoding capability at the management station, this filtering essentially provides distributed protocol analysis to supplement the use of dispatched technicians with portable protocol analyzers.

The monitor also maintains counters of each packet match for statistical analysis. Either an individual packet match, or a multiple number of packet matches through the Alarms Group, can trigger an event to the log or the network management system using an SNMP trap. Although these counters are not available to the History Group for trend analysis, a management station may request these counters through regular polling of the monitor so that trend analysis can be performed.

Packet Capture Group. The type of packets collected depends on the Filters Group. The Packet Capture Group allows the user to create multiple capture buffers and to control whether the trace buffers will wrap (overwrite) when full or stop capturing. The user may expand or contract the size of the buffer to fit immediate needs for packet capturing, rather than permanently commit memory that will not always be needed.

The network manager can specify a packet match as a start trigger for a trace and depend on the monitor to collect the results without further user involvement. The RMON MIB includes configurable capture slice sizes to store either the first few bytes of a packet (where the protocol header is located) or to store the entire packet (which may reveal application errors). The default slice setting specified by the RMON MIB is the first 100 bytes.

Notifications (Events) Group. In a distributed management environment, traps can be delivered by the RMON MIB agent to multiple management stations that share a single community name destination specified for the trap. In addition to the three traps already mentioned—rising threshold and falling threshold (see Alarms Group) and packet match (see Packet

Capture Group)—there are seven other traps: coldStart, warmStart, linkDown, linkUp, authenticationFailure, egpNeighborLoss, and enterpriseSpecific.

- *ColdStart.* This trap indicates that the sending protocol entity is reinitializing itself such that the agent's configuration or the protocol entity implementation may be altered.

- *WarmStart.* This trap indicates that the sending protocol entity is reinitializing itself such that neither the agent configuration nor the protocol entity implementation is altered.

- *LinkDown.* This trap indicates that the sending protocol entity recognizes a failure in one of the communication links represented in the agent's configuration.

- *LinkUp.* This trap indicates that the sending protocol entity recognizes that one of the communication links represented in the agent's configuration has come up.

- *AuthenticationFailure.* This trap indicates that the sending protocol entity is the addressee of a protocol message that is not properly authenticated. While implementations of the SNMP must be capable of generating this trap, they must also be capable of suppressing the emission of such traps via an implementation-specific mechanism.

- *EgpNeighborLoss.* This trap indicates that an EGP neighbor for whom the sending protocol entity was an EGP peer has been marked down and the peer relationship is no longer valid.

- *EnterpriseSpecific.* This trap indicates that the sending protocol entity recognizes that some enterprise-specific event has occurred.

The Notifications (Events) Group allows users to specify the number of events that can be sent to the monitor log. From the log, any specified event can be sent to the management station. Events can originate from a crossed threshold on any integer or counter or from any packet match count. The log includes the time of day for each event and a description of the event written by the vendor of the monitor. The log overwrites when full, so events may be lost if not uploaded to the management station periodically. The rate at which the log fills depends on the resources the monitor dedicates to the log and the number of notifications the user sends to the log.

Traffic Matrix Group. The RMON MIB includes a traffic matrix at the MAC layer. A traffic matrix shows the amount of traffic and number of errors between pairs of nodes—one source and one destination address per pair. For each pair, the RMON MIB maintains counters for the number of packets, number of octets, and error packets between the nodes. This allows network operators to sort this information by source or destination address.

Full compliance with the Ethernet RMON MIB specification requires that the vendor provide support for every object within a selected group. Since each group is optional, when selecting RMON MIB agents, Information Technology (IT) managers should determine the features they require and verify that those features are included in actual products.

Applying remote monitoring and statistics-gathering capabilities to the Ethernet environment offers a number of benefits. The availability of critical networks is maximized, since remote capabilities allow for more timely problem resolution. With the capability to resolve problems remotely, operations staff can avoid costly travel to troubleshoot problems on-site. With the capability to analyze data collected at specific intervals over a long period of time, intermittent problems can be tracked down that would normally go undetected and unresolved.

16.5.4   Token ring extensions

Initially, RMON defined media-specific objects for Ethernet only. Later, media-specific objects for token ring, as well as other networks, became available.

Token ring MAC-layer statistics. This extension tracks statistics, diagnostics, and event notification associated with MAC traffic on the local ring. Statistics include the number of beacon, purge, and 803.5 MAC management packets and events; MAC packets; MAC octets; and ring soft error totals. A complete list of token ring MAC-layer statistics and their definitions is provided in Table 16.2.

Token ring promiscuous statistics. This extension collects utilization statistics of user data (non-MAC) traffic on the local ring. Statistics include the number of data packets and octets, broadcast and multicast packets, and data frame size distribution. Table 16.3 lists the token ring promiscuous statistics and their definitions.

Token ring MAC-layer history. This extension offers historical views of MAC-layer statistics based on user-defined sample intervals, which can be set from one second to one hour to allow short-term or long-term historical analysis.

TABLE 16.2   Token Ring MAC-Layer Statistics

Statistic	Definition
tokenRingMLStatsDropEvents	The total number of events in which packets were dropped by the probe due to lack of resources. This number is not necessarily the number of packets dropped, just the number of times this condition has been detected.
tokenRingMLStatsMacOctets	The total number of octets of data in MAC packets (excluding those that were not good frames) received on the network (excluding framing bits but including FCS octets).
tokenRingMLStatsMacPkts	The total number of MAC packets (excluding packets that were not good frames) received.
tokenRingMLStatsRingPurgeEvents	The total number of times that the ring enters the ring purge state from normal ring state. The ring purge state that comes in response to the claim token or beacon state is not counted.
tokenRingMLStatsRingPurgePkts	The total number of ring purge MAC packets detected by probe.
tokenRingMLStatsBeaconEvents	The total number of times that the ring enters a beaconing state (beaconFrameStreamingState, beaconBitStreamingState, beaconSetRecoveryModeState, or beaconRingSignalLossState) from a nonbeaconing state. A change of the source address of the beacon packet does not constitute a new beacon event.
tokenRingMLStatsBeaconTime	The total amount of time that the ring has been in the beaconing state.
tokenRingMLStatsBeaconPkts	The total number of beacon MAC packets detected by the probe.
tokenRingMLStatsClaimTokenEvents	The total number of times that the ring enters the claim token state from normal ring state or ring purge state. The claim token state that comes in response to a beacon state is not counted.
tokenRingMLStatsClaimTokenPkts	The total number of claim token MAC packets detected by the probe.
tokenRingMLStatsNAUNChanges	The total number of Nearest Active Upstream Neighbor (NAUN) changes detected by the probe.
tokenRingMLStatsLineErrors	The total number of line errors reported in error reporting packets detected by the probe.
tokenRingMLStatsInternalErrors	The total number of adapter internal errors reported in error reporting packets detected by the probe.
tokenRingMLStatsBurstErrors	The total number of burst errors reported in error reporting packets detected by the probe.
tokenRingMLStatsACErrors	The total number of AC (Address Copied) errors reported in error reporting packets detected by the probe.
tokenRingMLStatsAbortErrors	The total number of abort delimiters reported in error reporting packets detected by the probe.
tokenRingMLStatsLostFrameErrors	The total number of lost frame errors reported in error reporting packets detected by the probe.
tokenRingMLStatsCongestionErrors	The total number of receive congestion errors reported in error reporting packets detected by the probe.

TABLE 16.2   Token Ring MAC-Layer Statistics (*Continued*)

Statistic	Definition
tokenRingMLStatsFrameCopiedErrors	The total number of frame copied errors reported in error reporting packets detected by the probe.
tokenRingMLStatsFrequencyErrors	The total number of frequency errors reported in error reporting packets detected by the probe.
tokenRingMLStatsTokenErrors	The total number of token errors reported in error reporting packets detected by the probe.
tokenRingMLStatsSoftErrorReports	The total number of soft error report frames detected by the probe.
tokenRingMLStatsRingPollEvents	The total number of ring poll events detected by the probe (i.e., the number of ring polls initiated by the active monitor that were detected).

Token ring promiscuous history. This extension offers historical views of promiscuous statistics based on user-defined sample intervals, which can be set from one second to one hour to allow short-term or long-term historical analysis.

Ring station control table. This extension lists status information for each ring being monitored. Statistics include ring state, active monitor, hard error beacon fault domain, and number of active stations.

Ring station table. This extension provides diagnostics and status information for each station on the ring. The types of information collected include station MAC address, status, and isolating and nonisolating soft error diagnostics.

Source routing statistics. The extension for source routing statistics is used for monitoring the efficiency of source-routing processes by keeping track of the number of data packets routed into, out of, and through each ring segment. Traffic distribution by hop count provides an indication of how much bandwidth is being consumed by traffic-routing functions.

Ring station configuration control. The extension for station configuration control provides a description of the network's physical configuration. A media fault is reported as a "fault domain," an area that isolates the problem to two adjacent nodes and the wiring between them. The network administrator can discover the exact location of the problem—the fault domain—by referring to the network map. Faults that result from changes to the physical ring, including each time a station inserts or removes itself

TABLE 16.3   Token Ring Promiscuous Statistics

Statistic	Definition
tokenRingPStatsDropEvents	The total number of events in which packets were dropped by the probe due to lack of resources. This number is not necessarily the number of packets dropped, just the number of times this condition has been detected.
tokenRingPStatsDataOctets	The total number of octets of data in good frames received on the network (excluding framing bits but including FCS octets) in non-MAC packets.
tokenRingPStatsDataPkts	The total number of non-MAC packets in good frames received.
tokenRingPStatsDataBroadcastPkts	The total number of good non-MAC frames received that were directed to a Logical Link Control (LLC) broadcast address.
tokenRingPStatsDataMulticastPkts	The total number of good non-MAC frames received that were directed to a local or global multicast or functional address. This number does not include packets directed to the broadcast address.
tokenRingPStatsDataPkts18to63Octets	The total number of good non-MAC frames received that were between 18 and 63 octets in length inclusive, excluding framing bits but including FCS octets.
tokenRingPStatsDataPkts64to127Octets	The total number of good non-MAC frames received that were between 64 and 127 octets in length inclusive, excluding framing bits but including FCS octets.
tokenRingPStatsDataPkts128to255Octets	The total number of good non-MAC frames received that were between 128 and 255 octets in length inclusive, excluding framing bits but including FCS octets.
tokenRingPStatsDataPkts256to511Octets	The total number of good non-MAC frames received that were between 256 and 511 octets in length inclusive, excluding framing bits but including FCS octets.
tokenRingPStatsDataPkts512to1023Octets	The total number of good non-MAC frames received that were between 512 and 1023 octets in length inclusive, excluding framing bits but including FCS octets.
tokenRingPStatsDataPkts1024to2047Octets	The total number of good non-MAC frames received that were between 1024 and 2047 octets in length inclusive, excluding framing bits but including FCS octets.
tokenRingPStatsDataPkts2048to4095Octets	The total number of good non-MAC frames received that were between 2048 and 4095 octets in length inclusive, excluding framing bits but including FCS octets.

TABLE 16.3   Token Ring Promiscuous Statistics (*Continued*)

Statistic	Definition
tokenRingPStatsDataPkts4096to8191Octets	The total number of good non-MAC frames received that were between 4096 and 8191 octets in length inclusive, excluding framing bits but including FCS octets.
tokenRingPStatsDataPkts8192to18000Octets	The total number of good non-MAC frames received that were between 8192 and 18000 octets in length inclusive, excluding framing bits but including FCS octets.
tokenRingPStatsDataPktsGreater-Than18000Octets	The total number of good non-MAC frames received that were greater than 18000 octets in length inclusive, excluding framing bits but including FCS octets.

from the network, are discovered by comparing the start of symptoms with the timing of physical changes.

The RMON MIB not only keeps track of the status of each station but it also reports the condition of each ring being monitored by an RMON agent. On large token ring networks with several rings, the health of each ring segment and the number of active and inactive stations on each ring can be monitored simultaneously. Network administrators will be alerted to the location of the fault domain when any ring goes into a beaconing (fault) condition. As a network manager, you can also be alerted to any changes in backbone ring configuration, which could indicate loss of connectivity to an interconnect device such as a bridge or to a shared resource such as a server.

Ring station configuration.   The ring station group collects token ring specific errors. Statistics are kept on all significant MAC-level events to assist in fault isolation, including ring purges, beacons, claim tokens, and such error conditions as burst errors, lost frames, congestion errors, frame copied errors, and soft errors.

Ring station order.   Each station can be placed on the network map in a specified order relative to the other stations on the ring. This extension provides a list of stations attached to the ring in logical ring order. It lists only stations that comply with the 802.5 active monitoring ring poll or IBM station advertisement conventions.

16.6   RMON II

The RMON MIB is basically a MAC-layer standard. Its visibility does not extend beyond the router port, meaning that it cannot see beyond individual LAN segments. As such, it does not provide visibility into conversations across the network or connectivity between the various network segments. Given the trends toward remote access and distributed work groups, which generate a lot of intersegment traffic, visibility across the enterprise is an important capability for you, as a network manager, to have available.

Complementary to the RMON MIB is the RMON II MIB, which extends the capability of the original RMON MIB to include protocols above the MAC level. Because network-layer protocols such as IP are included, a probe can monitor traffic through routers attached to the local subnetwork. RMON II data would be used to identify such things as traffic patterns and slow applications. Specifically, an RMON II probe can monitor the sources of traffic arriving by a router from another network and the destination of traffic leaving by a router to another network.

Analysis tools that support the network layer can sort traffic by protocol, rather than just report on aggregate traffic. This means that as a network manager, you can determine, for example, the percent of IP versus Internet Packet Exchange (IPX) traffic traversing the network. In addition, these higher-level monitoring tools can map end-to-end traffic, giving network managers the ability to trace communications between two hosts—or nodes—even if the two are located on different LAN segments. RMON II functions that allow this level of visibility include:

- *Protocol directory table.* Provides a list of all the different protocols that an RMON II probe can interpret.

- *Protocol distribution table.* Permits tracking of the number of bytes and packets on any given segment that have been sent from each of the protocols supported. This information is useful for displaying traffic types by percentage in graphical form.

- *Address mapping.* Permits identification of traffic-generating nodes, or hosts, by Ethernet or token ring address in addition to MAC address. It also discovers switch or hub ports to which the hosts are attached. This is helpful in node discovery and network topology applications for pinpointing the specific paths of network traffic.

- *Network-layer host table.* Permits tracking of bytes, packets, and errors by host according to individual network-layer protocol.

- *Network-layer matrix table.* Permits tracking, by network-layer address, of the number of packets sent between pairs of hosts.

- *Application-layer host table.* Permits tracking of bytes, packets, and errors by host and according to application.

- *Application-layer matrix table.* Permits tracking of conversations between pairs of hosts by application.

- *History group.* Permits filtering and storing of statistics according to user-defined parameters and time intervals.

- *Configuration group.* Defines standard configuration parameters for probes that include such parameters as network address, serial line information, and SNMP trap destination information.

RMON II is focused more on helping network managers understand traffic flow for the purpose of capacity planning rather than for the purpose of physical troubleshooting. The capability to identify traffic levels and statistics by application has the potential to greatly reduce the time it takes to troubleshoot certain problems. Without tools that can pinpoint which software application is responsible for gobbling up a disproportionate share of the available bandwidth, as a network manager, you can only guess. Often it is easier just to upgrade a server or buy more bandwidth, which inflates operating costs and shrinks budgets.

16.7 Conclusion

Today's network management systems have demonstrated their value in permitting technicians to control individual segments or the entire network remotely. In automating various capabilities, network management systems can speed up the process of diagnosing and resolving problems with equipment and lines. The capabilities of network management systems permit maximum network availability and reliability, thus enhancing the management of geographically dispersed operations, while minimizing revenue losses from missed business opportunities that may occur as a result of network downtime. Some network management systems offer customizable Web-based reporting, which provides managers with convenient, on-demand insight into network performance from virtually any location. This capability can also be used by service providers to offer their customers personalized insight into their outsourced managed environments.

By itself, SNMP only offers basic capabilities and cannot be used to perform the higher-level management functions. Currently, SNMP is only capable of tracking activity on the network and taking corrective action when problems arise. Applying more extensive remote monitoring and statistics-gathering capabilities to various network environments via the

RMON MIB offers a number of benefits. The availability of critical networks is maximized, since remote capabilities allow for more timely problem resolution. With the capability to resolve problems remotely, operations staff can avoid costly travel to troubleshoot problems on-site. With the capability to analyze data collected at specific intervals over a long period of time, intermittent problems can be tracked down that would normally go undetected and unresolved. And with RMON II, these capabilities are enhanced and extended across the enterprise for network environments operating at Layer 3 and higher.

17

Network Design and Optimization

17.1 Introduction

A typical corporate network consists of different kinds of transmission facilities, equipment, Local Area Network (LAN) technologies, and protocols—all cobbled together to meet the differing needs of work groups, departments, branch offices, divisions, subsidiaries, and, increasingly, strategic partners, suppliers, and customers. Building such networks presents special design challenges that require comprehensive design tools.

Fortunately, a variety of automated design tools have become available in recent years. With built-in intelligence, these tools take an active part in the design process, from building a computerized model of the network, validating its design and gauging its performance, to quantifying equipment requirements and exploring reliability and security issues before the purchase and installation of any network component. Even faulty equipment configurations, design flaws, and standards violations are identified in the design process.

17.2 Data Acquisition

The design process usually starts by opening a blank drawing window from within the design tool into which various vendor-specific devices—workstations, servers, hubs, routers—can be dragged from a product library and dropped into place (see Fig. 17.1). The devices are further defined by type of components, software, and protocols as appropriate. By drawing lines, the devices are linked to form a network, with each link assigned physical and logical attributes. Rapid prototyping is aided by the ability to copy objects—devices, LAN segments, network nodes, and subnets—from one drawing to the next, editing as necessary, until the entire network is built.

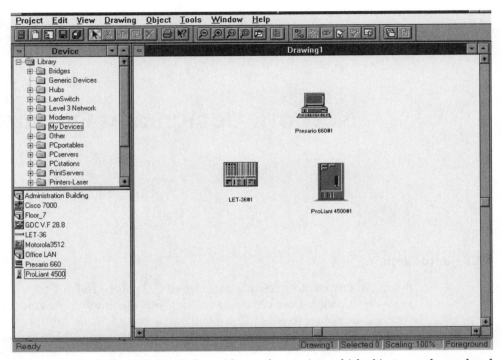

Figure 17.1   Typically, a design tool provides workspace into which objects are dragged and dropped from device libraries to start the network design process from scratch. (*Source*: NetFormx.)

Along the way, various simulations can be run to test virtually any aspect of the design.

The autodiscovery capabilities found in such management platforms as Hewlett-Packard's OpenView—which automatically detects various network elements and represents them with icons on a topology map—are often useful in accumulating the raw data for network design. Some stand-alone design tools allow designers to import this data from network management systems, which eases the task of initial data compilation. Although these network management systems offer some useful design capabilities, they are not as feature rich as high-end stand-alone tools, which are also able to incorporate a broader range of network technologies and equipment makes and models.

Designing a large, complex network requires a multifaceted tool—ideally, one that is graphical, object-oriented, and interactive. It should support the entire network life cycle, starting with the definition of end-user requirements and conceptual design, to the very detailed vendor-specific configuration of network devices, the protocols they use, and the various links between them. At each phase in the design process, the tool should be able to test different design alternatives in terms of cost, performance, and validity. When the design checks out, the tool generates network diagrams and a bill of materials—all this before a single equipment vendor or carrier sales representative is contacted or Request for Proposal (RFP) is written.

With the right tools, modules, and device libraries, every conceivable type of network can be designed, including legacy networks such as Systems Network Architecture (SNA) and DECnet; voice networks including Integrated Services Digital Network (ISDN), T1, X.25, and Asynchronous Transfer Mode (ATM); and Transmission Control Protocol/Internet Protocol (TCP/IP) nets. Some tools even take into account the use of satellite, microwave, and other wireless technologies.

The designer can take a top-down or bottom-up approach to building the network. In the former, the designer starts by sketching out the overall network; subsequent drawings add increasing levels of detail until every aspect of the network is eventually fleshed out. The bottom-up approach might start with a LAN on a specific floor of a specific building; subsequent drawings are linked to create the overall network structure.

As the drawing window is populated, devices can be further defined by type of component such as chassis, interface cards, and daughter boards. Even the operating system can be specified. Attributes can be added to each device taken from the library—to specify a device's protocol functionality, for example. Once the devices have been configured, a simulation profile is assigned to each device, which specifies its traffic characteristics for purposes of simulating the network's load and capacity.

With each device's configuration defined, lines are drawn between them to form the network. With some design tools, the links can be validated against common protocols and network functions. This prevents NetWare clients from being connected to other clients instead of servers, for example. Such on-line analysis can also alert the designer to undefined links, unconnected devices, insufficient available ports in a device, and incorrect addresses in IP networks. Some tools are even able to report violations of network integrity and proper network design practices.

17.3   Network Simulation

Once the initial network design is completed, it can be tested for proper operation by running a simulation that describes how the actual network devices behave under various real-world conditions. The simulator generates network events over time, based on the type of device and traffic pattern recorded in the simulation profile. This enables the designer to test the network's capacity under various what-if scenarios and fine-tune the network for optimal cost and performance. Simulators can be purchased as stand-alone programs, or may be part of the design tool itself.

Some tools are more adept at designing Wide Area Networks (WANs), particularly those that are based on Time Division Multiplexers (TDMs). With a TDM component taken from the device library, for example, a designer can build an entire T1 network within specified parameters and constraints. The designer can strive for the lowest transmission cost that supports all traffic, for instance, or strive for line redundancy between all TDM nodes. By mixing and matching different operating characteristics of various TDM components, overall design objectives can be addressed, simulated, and fine-tuned. Some tools come with a tariff database to price transmission links and determine the most economical network design.

Such tools may also address clocking in the network design. Clocks are used in TDM networks to regulate the flow of data transmitted between nodes. All clocks on the network must therefore be synchronized to ensure the uninterrupted flow of data from one node to another. The design tool automatically generates a network topology synchronization scheme, taking into account any user-defined criteria, to ensure that there are no embedded clock loops.

17.4 Bill of Materials

Once the design is validated, the network design tool generates a bill of materials that includes order codes, prices, and discounts. This report can be exported to any Microsoft Windows application, such as Word or Excel, for inclusion in the proposal for top management review or an RFP issued to vendors and carriers who will build the network. Through the tool's capability to render multiple device views, network planners can choose either a standard schematic or an actual as-built rendering of the cards and the slot assignments of the various devices. Some tools also generate Web-enabled output, which allows far-flung colleagues to discuss and annotate the proposal over the Internet—even allowing each person to drill down and extract appropriate information from the network device library.

17.5 Network Drawing Tools

An alternative to expensive network design tools is network drawing tools, which now offer capabilities that facilitate the design and documentation of large networks. Drawing tools started out providing a quick and convenient way to illustrate and share ideas about how a network should be laid out. As these inexpensive drawing tools became more sophisticated, they took on many of the characteristics of high-end network design tools.

Network managers and planners faced with building or expanding corporate networks can start their work using network drawing tools to manipulate detailed and often large quantities of information about local and enterprise networks. Drawing tools are now capable of depicting complex infrastructures such as frame relay and ATM networks, including the paths and addresses of virtual circuits and the lines and equipment at each end. While the automatic discovery capabilities of high-end network management systems can help in this regard, they are not very useful for documenting the equipment at the level of detail that is now required by network planners.

A variety of drawing tools have become available that can aid the network design process. Such tools provide the five major features considered critical to network planners:

- An easy-to-use drawing engine for general graphics

- An extensive library of predrawn images representing vendor-specific equipment

- A drill-down capability, which allows multiple drawings to be linked to show various views of the network

- A database capability to assign descriptive data to the device images

- A high degree of embedded intelligence that makes images easy to create and update

17.6   Drawing Techniques

Most network drawing tools are Windows based and employ the drag-and-drop technique to move images of network equipment from a device library to a blank workspace. Many also allow network designs to be published on the corporate intranet or the public Internet, allowing any authorized user to view them with a Web browser and make changes. Some drawing tools can automatically discover devices on an existing network to ease the task of drawing and documenting the network.

17.6.1   Device library

A device library holds images of such things as modems, telephones, hubs, Private Branch Exchanges (PBXs), and Channel Service Units/Data Service Units (CSUs/DSUs) from different manufacturers. Representations of LANs and WANs, databases, buildings and rooms, satellite dishes, microwave towers, and a variety of line connectors are included. There are also shapes that represent such generic accessories as power supplies, Personal Computers (PCs), towers, monitors, keyboards, and switches. There are even shapes for racks, shelves, patch panels, and cable runs (see Fig. 17.2).

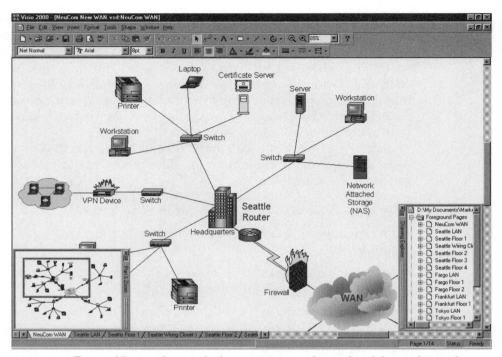

Figure 17.2   From a library of network shapes, items are dragged and dropped into place as needed to design a new node or build a whole network. [*Source*: Visio Corp., a division of Microsoft Corp. (As of January 2000, Visio Corp. became the Visio Division operating within Microsoft's Business Productivity Group.)]

Typically, an annual subscription provides unlimited access to the hundreds of new network devices, adapters, and accessories added to the device library. Depending on the drawing tool vendor, new objects may even be downloadable from the company's Web site.

While many drawing tools offer thousands of exact-replica hardware device images from hundreds of network equipment manufacturers, some tools have embedded intelligence into the shapes, which enables components such as network cards to snap into equipment racks and remain in place even when the rack is moved.

In addition, each shape can be annotated with product-specific attributes, including vendor, product name, part number, and description (see Fig. 17.3). This permits users to generate detailed inventory reports for network asset management.

The shapes are even programmable so they can behave like the objects they represent. This reduces the need for manual adjustments while drawing, and ensures the accuracy of the final diagram. For example, the shape

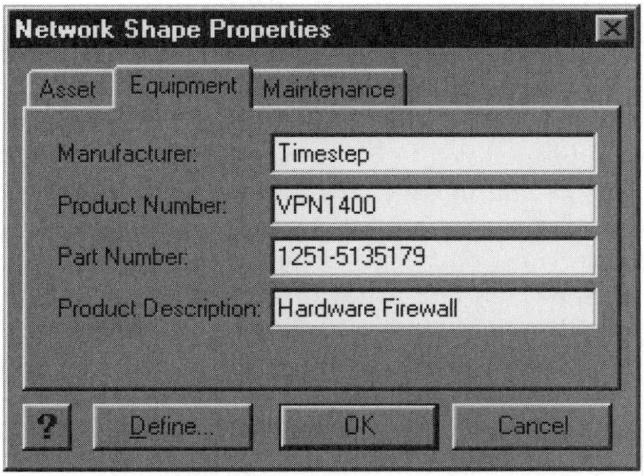

Figure 17.3 Details about network equipment can be stored using custom property fields. Device-specific data for each network shape keeps track of asset, equipment, and manufacturer records that can be accessed from within network diagrams. (*Source:* Visio Corp., a division of Microsoft Corp.)

representing an equipment rack from a specific vendor can be programmed to know its dimensions. When the user populates the drawing with multiple instances of this shape, it could issue an alert if there is a discrepancy between the space available on the floor plan and the space requirements of the equipment racks.

Each shape can also be embedded with detailed information. For example, the user can associate a spreadsheet with any network element—to provide cost information on a new switch node or LAN segment, for example—along with a bar chart to perhaps illustrate the cost data by system component. The spreadsheet data can be manipulated until costs fit within budgetary parameters. The changes will be reflected in the bar chart the next time it is opened.

17.6.2 Template usage

To start a network diagram, typically the user opens the template for the manufacturer whose equipment will be placed in the diagram. This causes a drawing page to appear that contains rules and a grid. The drawing page itself can be sized to show the entire network or just a portion of it.

Various other systems and components can be added to the diagram using the drag-and-drop technique. The user has the option of having (or not having) the shapes snap into place within the drawing space so they will be precisely positioned on grid lines. Once placed in the drawing space, the shapes can be moved, resized, flipped, rotated, and glued together. Expansion modules, for example, can be dragged onto the chassis so that the modules' endpoints glue to the connection points on the chassis expansion slots. This allows the chassis and modules to be moved anywhere in the diagram as a single unit. Via the cut-and-paste method, the user can add as many copies of the component as desired to quickly populate the network drawing.

To show the connections between various systems and components, the user can choose shapes that represent different types of networks, including LANs, X.25, satellite, microwave, and radio. Alternatively, the user can choose to connect the shapes with simple lines that can have square or curved corners.

Each network equipment shape has properties associated with it. Custom properties can be assigned to shapes for use in tracking equipment and generating reports, such as inventories. Text can be added to any network system or component, including a Lotus Notes field, specifying font, size, color, style, spacing, indent, and alignment. Text blocks can be moved and resized. Some tools even include a spell checker and a search-and-replace tool. The user can add words that are not in the standard dictionary that comes with the program. The user can specify a search of the entire drawing, a particular page, or selected text only.

AutoCAD files and clip art can be added to network drawings. The common file formats usually supported for importing graphics from other applications, including Encapsulated PostScript (.EPS), Joint Photographic Experts Group (.JPG), Tag Image File Format (.TIF), and ZSoft PC PaintBrush Bitmap (.PCX).

The various shapes used in a network drawing can be kept organized using layers. A *layer* is a named category of shapes. For example, the user can assign walls, wiring, and equipment racks to different layers in a space plan. This allows the user to:

- Show, hide, or lock shapes on specific layers so they can be edited without affecting shapes on other layers

- Select and print shapes based on their layer assignments

- Temporarily change the display color of all shapes on a layer to make them easier to identify

- Assign a shape to more than one layer, as well as assign the member shapes of a group to different layers

The user can also group shapes into customizable stencils. If the same equipment is used at each node in a network, for example, the user can create a stencil containing all the devices. All of the graphics and text associated with each device will be preserved in the newly created stencil. This saves time in drawing large-scale networks, especially those that are based on equipment from a variety of manufacturers.

At any step in the design process, the user can share the results with other network planners by sending copies via email. The diagram is converted to an image file, which is displayed as an icon in the message box, and sent as an attachment. When opened by the recipient, the attachment with all embedded information is displayed. The document can then be edited by creating a separate layer for review comments, each of which is done in a different color. The use of separate layers and colors protects the original drawing and makes comments easier to view and understand.

Some network drawing tools provide a utility that converts network designs and device details into a series of hyperlinked Hypertext Markup Language (HTML) documents that can be accessed over the Web. These documents show device configurations, port usage, and even device photographs. Users can activate the links to navigate from device to device to trace connectivity and review device configurations (see Fig. 17.4). In addition to supporting fault identification, the hyperlinked documents aid in planning design changes.

There are several ways that the network diagrams can be protected against inadvertent changes, especially if they are shared via email or posted on the Web:

- The shapes can be locked to prevent them from being modified in specific ways.

- The attributes of a drawing file (styles, for example) can be protected against modification.

- The file can be saved as read-only, so it cannot be modified in any way.

- The shapes on specific layers can be protected against modification.

Users can password-protect their work to prevent attributes of a drawing file from being changed. For example, a background containing standard shapes or settings can be password-protected. Users can also set a password for a drawing's styles, shapes, backgrounds, or masters. A password-protected item can be edited only if the correct password is entered.

17.7   Embedded Intelligence

Some products are so intelligent that they can no longer be considered merely drawing tools. Visio 2000 Enterprise Edition, for example, supports switched WANs through its AutoDiscovery feature. This technology includes

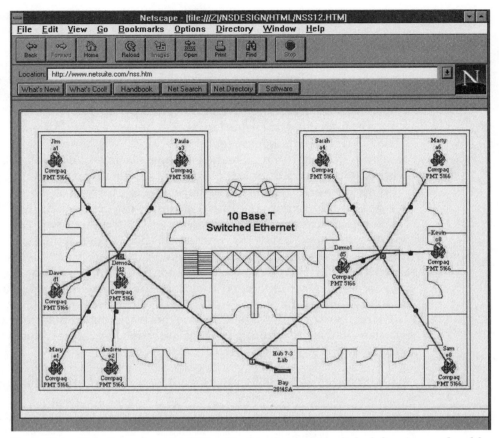

Figure 17.4 This floor plan of a 10BaseT network is a hyperlinked drawing rendered by Netscape Navigator.

support for Layer 2 (data link) and Layer 3 (network) environments. AutoLayout technology makes it simple for users to automatically generate network diagrams of the discovered devices—including detailed mappings.

In addition to allowing Information Technology (IT) specialists to create conceptual, logical, and physical views of their information systems, Visio 2000 Enterprise Edition owners can purchase Visio's add-on solution for monitoring network performance. Working with Enterprise Edition's AutoDiscovery, a feature called Real-Time Statistics documents the behavior of a network environment, capturing real-time data from any SNMP-manageable device on LANs and WANs. Because you, as a network manager, can monitor the network's performance, you have the information you need to redistribute network traffic and prevent overloads. Real-Time Statistics then turns this performance data into graphs that can be printed or exported for analysis.

Enterprise Edition enables developers to visualize and quickly start software development projects. They can visualize the design architecture of existing systems by reverse-engineering source code from Microsoft Visual Studio. They can also decrease development time by generating fully customizable code skeletons for Visual Basic, C++, and Java from Unified Modeling Language (UML)* class diagrams.

17.8 Bandwidth Optimization

Designing a network and fully documenting it is not a one-shot deal. Networks have a tendency to change over time as organizations add employees, applications, equipment, lines, and services. Network managers must continually work to optimize the network so that it meets the needs of individual users, work groups, departments, and divisions.

Complicating matters is that the continued growth of business-critical Internet applications, corporate intranets, storage services, Internet Protocol (IP) telephony, conferencing, and B2B e-commerce has prompted the need for innovation to improve traffic flow. Increasingly, organizations of all types and sizes are turning to management products that allocate bandwidth according to various application performance criteria in an effort to reduce congestion, ensure delivery of priority messages, and support real-time multimedia traffic over IP networks.

The use of bandwidth management products can improve the performance of IP networks through such means as traffic shaping, queuing, load balancing, and caching. For companies that do not want to be bothered with managing their own information flow, preferring instead to focus on core business issues, there is the alternative of outsourcing this function to content delivery networks.

Corporate operations and profits increasingly rely on network performance. But with new applications and more users constantly being added to IP networks, the result is a mix of critical and noncritical traffic sharing the same resource. Routing delays and lost packets, due to transient network congestion, impose long round-trip times that severely limit the usability of IP networks. Peak data rates and bursts in network traffic are hard to predict, making it difficult to guarantee the performance of mission-critical applications. The greater the traffic mix, the greater need for controlling resources so that all applications get the bandwidth and response time they deserve.

Consequently, Internet Service Providers (ISPs), IP carriers, and corporations are increasingly turning to bandwidth managers to help them uti-

---

*Pioneered by Rational Software Corp. and officially adopted as a standard by the Object Management Group (OMG), the Unified Modeling Language (UML) is an industry-standard language for specifying, visualizing, constructing, and documenting the elements of software systems. UML simplifies the complex process of software design, making a "blueprint" for construction.

lize network resources more efficiently. These products employ a variety of traffic control techniques—including TCP rate control, queuing, and policy definition—for ensuring that essential traffic makes it to its destination, even during periods of network congestion. The need for these capabilities depends on the following factors:

- The variety of traffic types that are run over an intranet or IP WAN service
- Whether certain types of traffic are valued more highly than others
- The delay characteristics of the various types of traffic, which may require applications to be prioritized
- The number of users who can be expected to require network bandwidth at the same time

If an organization is running multiple traffic types that are of equal value, some of these capabilities may not be needed. But if an organization finds itself running more multimedia applications on its intranet—such as streaming audio or video, IP telephony, and collaborative computing—then adding such capabilities will improve network performance more economically than upgrading the network with more bandwidth. Of course, the number of users accessing the network at the same time requires consideration as well, and this may dictate the amount of bandwidth that must be available to meet peak demand.

17.8.1   Approaches

Bandwidth management products come in both hardware and software versions, or a combination of both. Hardware products must be installed at each corporate location on the network, while software products can be loaded to existing network routers and/or switches.

Hardware-based bandwidth managers tend to offer the best performance, since they rely on Application Specific Integrated Circuits (ASICs) and dedicated memory to handle bandwidth management. They can be expensive and hard to upgrade, however. Software systems provide considerable flexibility and ease in upgrading, but lag in performance compared to dedicated hardware-based systems.

There are a number of specific bandwidth management situations that need to be addressed. Setting a hard upper bandwidth limit for a given application is the simplest, but bandwidth managers also need to enforce multiple priority levels. Prioritization must also be dynamic, so that bandwidth is available when priority applications compete for access. Bandwidth managers must also work with TCP traffic, which is self-tuning for traffic conditions, resulting in fluctuation. Finally, they need to scale from a few sessions to several hundred and ensure that all applications are served and none are "starved," even if they have a low priority.

It is important that any bandwidth management product be able to handle traffic in all possible situations that may be encountered, that it does nothing to compromise security, and that it works with existing network management products. Most bandwidth managers also offer remote management and reporting features, which can be integrated with other management systems.

To minimize the administrative burden, the bandwidth management system should integrate with a directory service; specifically, one that supports the Lightweight Directory Access Protocol (LDAP). This makes it easier to allocate bandwidth to all members of a work group, department, or business unit because changes can be implemented globally based on directory entries.

17.8.2   Traffic shaping

The general principle of traffic shaping is to alter the traffic characteristics of the stream of cells and move high-priority traffic through network bottlenecks more efficiently. This can be accomplished by controlling outgoing flows based on a First-In, First-Out (FIFO) queuing mechanism such as "leaky bucket" or token, which receives varying incoming flows. In particular, the leaky bucket uses a fixed transmit rate mechanism to smooth out the traffic, while the token type gets a fixed number of tokens to control outgoing flows with burst capability.

Another way of controlling outgoing flows is TCP rate control, a technique that paces or smooths the IP flow by detecting a remote user's access speed, factoring in network latency, and correlating this data with other traffic flow information. It is designed to evenly distribute packet transmissions by controlling TCP acknowledgments to the sender, causing the sender to throttle back and avoid packet loss when there is insufficient bandwidth.

Traffic shaping also refers to certain Quality of Service (QoS) capabilities, most notably limiting the rate at which traffic generated from a given application can flow. This prevents a single application from becoming a bandwidth hog during periods of congestion, while moving applications such as email to the front of the queue. For instance, the QoS interfaces of Windows 2000 allow applications to request a certain level of bandwidth from the network, giving more important applications the service they need to run efficiently.

17.8.3   Load balancing

Another way to improve handling of IP traffic on both ISPs and corporate intranets is to make use of load-balancing systems. On the WAN, load balancing is usually carried out by a router equipped with Border Gateway Protocol, version 4 (BGP4). There are two types of load balancing: per packet and per session. In packet load balancing, packets are released onto two separate lines to balance the load. In session load balancing, the router makes a decision on the best route for the session. Cisco Systems, for example, supports both types of load balancing across its lines of routers.

Load balancing can also occur between servers. While bandwidth management tools enable allocation of portions of available bandwidth to different users, load balancers operate on the server side, routing traffic to the best server available to handle a job. In a load-balanced network, incoming traffic is distributed among replicated servers, thus permitting server clusters to share the processing load, provide fail-back capability, and speed response time for users.

With Allot Communications' hardware-based NetEnforcer, for example, advanced load-balancing policies can be defined that reflect the capabilities of individual servers on the network. NetEnforcer allows policies to be defined that redirect traffic based on the capabilities of the servers. For instance, all video can be redirected to the video server, all Web traffic to the Web server, and all employees in marketing to the marketing server. NetEnforcer continuously adjusts both the flow and prioritization of applications through the network and the distribution of those applications to servers.

Traffic can be balanced between available servers using algorithms such as:

- *Round robin.* Each server is treated with equal priority.

- *Weighted round.* Each server is given an individual weight or priority based on its ability to deliver specific applications.

- *Maintenance rerouting.* Traffic is rerouted to another server when an originally targeted server becomes unavailable.

The selection of the right load-balancing approach results in efficient utilization of bandwidth and other resources, and improves traffic flow throughout the organization.

17.8.4   Caching

A *cache* is temporary storage of frequently accessed information. Caching has long been used in computer systems to increase performance. A cache can be found in nearly every computer today, from mainframes to PCs. More recently, caching is used to improve the performance of corporate intranets. Many vendors of bandwidth management products offer network caching as well. Instead of users accessing the same information over the WAN, it is stored locally on a server. This arrangement gives users the information they need quickly, while freeing the WAN of unnecessary traffic, which improves its performance for all users.

Caching is frequently applied to the Web, especially for e-commerce applications. When users visit the same Web site, the browser first looks to see if a copy of the requested page is already in the computer's hard disk cache. If it is, the load time is virtually instantaneous; if not, the request goes out over the Internet. At the backbone level, lack of sufficient bandwidth is a global problem. Internet telephony, videoconferencing, and multimedia applications are consuming even greater amounts of bandwidth.

Network caching offers an effective and economical way to offload some of the massive bandwidth demand. This allows ISPs and corporations with their own intranets to maintain an active cache of the most often visited Web sites so that when these pages are requested again, the download occurs from the locally maintained cache server instead of the request being routed to the actual server. The result is a faster download speed.

Caches can reside at various points in the network. For ISPs and backbone providers, caches can be deployed in practically every Point of Presence (PoP). For enterprises, caches can be deployed on servers throughout campus networks and in remote and branch offices. Within enterprise networks, caches are on the way to becoming as ubiquitous as IP routers. Just about every large company now depends on Web caches to keep its intranet running smoothly. There are two types of caching: passive and active.

Passive caching.   In passive caching, the cache waits until a user requests the object again, then sends a refresh request to the server. If the object has not changed, the cached object is served to the requesting user. If the object has changed, the cache retrieves the new object and serves it to the requesting user. However, this approach forces the end user to wait for the refresh request, which can take as long as the object retrieval itself. It also consumes bandwidth for unnecessary refresh requests.

Active caching.   With active caching, the cache performs the refresh request before the next user request—if the object is likely to be requested again and the object is likely to have changed on the server. This automatic and selective approach keeps the cache up to date so the next request can be served immediately. Network traffic does not increase because an object in cache is refreshed only if it has a high probability of being requested again, and only if there is a statistically high probability that it has changed on the source server.

Active caches can achieve hit ratios of up to 75 percent, meaning that a greater percentage of user requests is served locally, instead of from remote resources. If the requested data is in the cache and is up to date, the cache can serve it to the user immediately upon request. If not, the user must wait while the cache retrieves the requested data from the network. Passive caches, on the other hand, typically achieve hit rates of only 30 percent. This means that users are forced to go to the network 2.5 times more often using passive caches to get the information they need.

17.8.5   Client-aware acceleration

One of the leading vendors of bandwidth management solutions is Packeteer, which offers caching technology that is client aware. The technology optimizes and accelerates both static and dynamic content. As such, it can be used as a stand-alone product or it can augment conventional caching products that only optimize static content. AppCelera ICX can dynamically determine a user's remote access speed, device type, and browser brand and version. It uses the information to serve content appropriate for that Web application session. This removes the need for businesses to author and maintain multiple versions of each site resource.

For dynamic content, knowledge of the actual connection speed enables the accelerator to calculate the net benefit of performing real-time optimizations. In understanding the rendering habits of popular browsers and leveraging their built-in capabilities, performance can be improved without requiring any special client-side software.

The result is a reduction in the amount of time it takes to load a Web page by 35 to 50 percent, and an increase in the speed with which the page is displayed. Additionally, client-aware acceleration reduces the amount of bandwidth required to service each transaction, giving businesses and service providers the potential to service more users with the existing bandwidth, or to increase the richness of the content knowing that it will be automatically optimized for users connecting at slower speeds.

As client devices increase in diversity, this type of client-aware acceleration will become increasingly important to the performance of business-critical Web applications, including those that reside on corporate intranets accessed by telecommuters and mobile employees.

17.8.6   Queuing

Queuing techniques may be used separately or with TCP/IP rate control. Queuing types include priority, weighted, and class-based. Priority queuing sets queues for high and low priority, and empties high-priority queues first. Cisco's Weighted Fair Queuing (WFQ), for example, assigns traffic to priority queues and also apportions a bandwidth share. Its Class Based Queuing (CBQ) guarantees a transmission rate to a queue, and other queues can borrow from unused bandwidth.

WFQ classifies traffic into "conversations" and applies priority (or weights) to identified traffic to determine how much bandwidth each conversation is allowed relative to other conversations. Conversations are broken into two categories: those requiring large amounts of bandwidth and those requiring a smaller amount of bandwidth. The goal is to always have bandwidth available for the small bandwidth conversations and allow the large bandwidth conversations to split the rest proportionally to their weights.

Without adding excessive bandwidth, WFQ provides consistent response time to heavy and light network users alike. It is a flow-based queuing algorithm that schedules interactive traffic to the front of the queue to reduce response time, and fairly shares the remaining bandwidth between other high bandwidth flows.

Low-volume traffic streams—which comprise the majority of traffic—receive preferential service, transmitting their entire offered loads in a timely fashion. High-volume traffic streams share the remaining capacity proportionally between them. WFQ works with both of Cisco's primary QoS signaling techniques, IP Precedence and RSVP (Resource Reservation Protocol).

17.8.7   Outsourcing content delivery

Another way of improving information flow over IP networks is to use a Content Delivery Network (CDN) such as Akamai, Inktomi, and Speedera Networks. Simply, this "edge network" solution acts as a distribution mechanism for Web content, which is replicated on cache servers at many PoPs on different backbone providers' networks so that content can be delivered directly from these servers without needing to traverse the frequently congested Internet backbone. Since users no longer have to go to origin sites across the Internet backbone to access specific content, content is delivered faster and more reliably, greatly improving the user's Web experience.

CDNs employ various technologies to improve the performance of Web sites, reduce hardware and bandwidth costs, and boost reliability by mirroring a Web site's content on distributed servers. With infrastructure technologies like caching, CDNs push replicated content closer to the network edge to minimize delay. Global load balancing ensures that users are transparently routed to the nearest content source.

Speedera's network, for example, is a distributed, robust system that leverages a worldwide set of probes and global traffic managers to make real-time decisions to intelligently route users' content requests to the best server and best location. Traffic routing decisions are made in real time, based on preset policies and real-time data. Key criteria for traffic routing decisions include network health (such as packet loss and latency), server health (such as availability and load), application health (such as the ability of a streaming server or Web server to deliver valid content), user geographical location, origin site persistence requirements, and other metrics.

Typically, large enterprise Web site owners subscribe to the CDN and determine the content it will serve. This can be done by selectively reassigning Uniform Resource Location (URLs) to embedded objects. That way, dynamic or localized content—such as banner ads, Java applets, and graphics, which represent 70 percent of a typical Web page—can be served up locally by the company's own Web site, avoiding the CDN, while static and easily distributed content can be retrieved from the nearest CDN server. Although such services are currently used mostly for Internet applications, there is great potential for the technology in intranets and extranets as well.

Content delivery networks can also address the security needs of organizations that subscribe to this type of service. The CDN does this by supporting the Secure Sockets Layer (SSL) protocol for encrypting information during delivery. Web properties are also protected against hack attacks by packet filtering that occurs at the CDN's edge servers. At the same time, Web servers are shielded from flash crowds of any kind, whether caused by legitimate or malicious traffic. In a malicious context, a flash is a burst of traffic from one or more sources. The aim is to overload the Web server with bogus requests so that it cannot service legitimate requests. This makes the server unavailable. The CDN, however, can mitigate such attacks by using its network of servers to dissipate such actions.

17.9 Optimization Issues

Because of the way network and transmission protocols work, IP networks are not a particularly good choice for transporting real-time traffic, particularly voice. With Voice over Internet Protocol (VoIP), delay is a serious impediment to a conversation. As speech is digitized and turned into discrete packets for transmission, there is processing delay. And since the packets may travel over different routes to their destination, there is more delay at the receiving end while the packets are put back in order. Packets that do not arrive in time to be resequenced must be dropped rather than retransmitted, a situation that can occur when too many users try to access the available bandwidth at the same time. The resulting packet loss results in clipped speech, another impediment to a conversation.

Even when the packets are not lost or dropped, they can travel at uneven rates through multiple networks to get to their destination. This situation is particularly disruptive to streaming applications, such as the audio and video clips, which require a constant flow rate to ensure smooth playback. Variability in flow rate is especially noticeable on Web sites that contain multimedia content. Text usually arrives first to the user, followed by simple graphics and then by complex graphics, such as a Macromedia Flash movie or an animated GIF. The Web site may be using Java applets or ActiveX controls as well. The uneven flow of the various content types often tests the patience of users who must wait for all the content to load before the Web site can be fully appreciated. But by then, most users have gone elsewhere to find what they want—the average user can only stand to wait a maximum of eight seconds for a Web page to load.

Dealing effectively with these and other performance-related problems calls for the use of IP bandwidth management tools of the kind previously discussed, which give network managers the means to prioritize business-critical traffic over less important traffic to give all applications the quality of service they deserve. The ability to prioritize traffic improves the performance of vital applications, increases transactions per second, optimizes bandwidth usage, and improves user satisfaction. The ability to use bandwidth efficiently can also minimize the need for more bandwidth and contain telecommunications costs. Bandwidth management tools ensure that users and applications share this resource appropriately.

17.10   New Directions

Bandwidth managers have emerged as a quick fix for Internet/intranet congestion problems, providing a way to introduce policy-based routing at a relatively low cost and with minimal impact on existing infrastructure. The future of these devices and software systems, however, is toward direct incorporation of these capabilities in routers and switches, and bandwidth management integration with directory services. These trends are already evident and can be expected to accelerate in the future.

Meanwhile, some equipment vendors, notably Cisco, are implementing the Internet Engineering Task Force (IETF) QoS standards, Differentiated Services (DiffServ), and Multiprotocol Label Switching (MPLS). DiffServ uses the Type of Service (TOS) field to carry packet service requirement information. MPLS specifies how Layer 3 traffic can be mapped to Layer 2 transports, such as ATM and frame relay, by adding a routing label to each IP packet. It also provides additional capabilities such as traffic engineering to boost IP routing efficiency. DiffServ will most likely be used on the enterprise where a LAN meets a WAN or a service provider network, since it specifies QoS at Layer 3 and runs without modification on any IP Layer 2 infrastructure. MPLS will be used first at the core of carrier networks because it requires investment in sophisticated label-switching routers.

17.10.1   Differentiated services

Differentiated Services (DiffServ) is an IETF standard for assigning a class of service to the different types of network traffic. This mechanism might seem to dispense with the need for some of the key capabilities offered by bandwidth management systems, since it entails marking the headers of each packet to indicate relative priority while traversing the WAN. If successful, all WAN equipment will interpret the setting in the same manner, allowing the WAN to prioritize more important traffic.

DiffServ uses the existing TOS field in the IP packet header, but allows for more possible values. However, DiffServ does not discover or classify traffic or handle LAN-to-WAN congestion, as many bandwidth management solutions do. At the discretion of the network administrator, bandwidth management systems can read, set, ignore, and sanitize DiffServ settings.

Of note to network managers is the fact that implementing the DiffServ standard may entail a hidden cost. Since DiffServ enforces most policies at the edge of the network, where older routers are most likely to reside, the routers may not be equipped to take on the increased processing burden and still handle peak traffic loads. Therefore, it may be necessary for you to take stock of your edge equipment to ascertain the extent to which upgrades of the router operating systems and memory are required to support DiffServ.

17.10.2   Multiprotocol Label Switching

With the explosive growth of the IP networks in recent years, there is growing dissatisfaction with its performance, especially as more applications are added. Among the new techniques for improving performance is Multiprotocol Label Switching (MPLS), which delivers Quality-of-Service (QoS) and security capabilities over IP networks, including Virtual Private Networks (VPNs).

MPLS attaches tags, or labels, to IP packets as they leave the edge router and enter the MPLS-based network. The labels eliminate the need for intermediate router nodes to look deeply into each packet's IP header to make forwarding and class-of-service handling decisions. The result is that packet streams can pass through an MPLS-based WAN infrastructure very quickly, and time-sensitive traffic can get the priority treatment it requires.

The same labels that distinguish IP packet streams for appropriate class-of-service handling also provide secure isolation of these packets from other traffic over the same physical links. Since MPLS labeling hides the real IP address and other aspects of the packet stream, it provides data protection at least as secure as other Layer 2 technologies, including frame relay and ATM.

To enhance the performance of IP networks, the various routes are assigned labels. Each node maintains a table of label-to-route bindings. At the node, a Label Switch Router (LSR) tracks incoming and outgoing labels for all routes it can reach, and it swaps an incoming label with an outgoing label as it forwards packet information (see Fig. 17.5). Since MPLS routers do not need to read as far into a packet as a traditional router does and perform a complex route lookup based on destination IP address, packets are forwarded much faster, which improves the performance of the entire IP network.

Although MPLS routers forward packets on a hop-by-hop basis, just like traditional routers, they operate more efficiently. As a packet arrives on an MPLS node, its label is compared to the Label Information Base (LIB), which contains a table that is used to add a label to a packet, while determining the outgoing interface to which the data will be sent. After consulting the LIB, the MPLS node forwards the packet toward its destination over a Label Switched Path (LSB). The LIB can simplify forwarding and increase scalability by tying many incoming labels to the same outgoing label, achieving even greater levels of efficiency in routing. The LSBs can be used to provide QoS guarantees, define and enforce Service Level Agreements (SLAs), and establish private user groups for VPNs.

MPLS provides a flexible scheme in that the labels could be used to manually define routes for load sharing or to establish a secure path. A multilevel system of labels can be used to indicate route information within a routing domain (interior routing) and across domains (exterior routing). This decoupling of interior and exterior routing means MPLS routers in the middle of a routing domain would need to track less routing information. That, in turn, helps the technology scale to handle large IP networks.

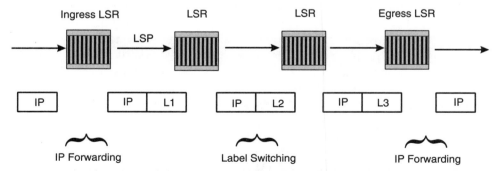

Figure 17.5 A label-switched route is defined by fixed-length tags appended to the data packets. At each hop, the LSR strips off the existing label and applies a new label, which tells the next hop how to forward the packet. These labels enable the data packets to be forwarded through the network without the intermediate routers having to perform a complex route lookup based on destination IP address.

MPLS could provide a similar benefit to corporations that have large ATM-based backbones with routers as edge devices. Normally, as such networks grow and more routers are added, each router may need additional memory to keep up with the increasing size of the routing tables. MPLS alleviates this problem by having the ATM switches use the same routing protocols as routers. In this way, the routers on the edge of the backbone and the ATM-based label switches in the core would maintain summarized routing information and only need to know how to get to their nearest neighbor—not to all peers on the network.

Finally, MPLS can be applied not only to the IP networks but also to any other network-layer protocol as well. This is because tag switching is independent of the routing protocols employed. While the Internet runs on IP, a lot of campus backbone traffic is transported on protocols such as Internet Packet Exchange (IPX), making a pure IP solution inadequate for many organizations.

17.11 Selection Criteria

Today's networks are more complex by orders of magnitude than networks envisioned only a few years ago. New Internet services, new technologies, new trends toward VPNs and voice-data convergence, plus the sheer number of new equipment offerings, have made reliance on traditional manual solutions to network engineering problems simply unworkable.

With the plummeting price of bandwidth, it might be tempting for network managers to simply add more of it to improve application response time over these networks. But under this approach, certain applications would simply have more bandwidth to hog, leaving mission-critical and real-time applications gasping for more. Without careful bandwidth management, routine Hypertext Transfer Protocol (HTTP) traffic, for example, can make it impossible to implement VoIP with any degree of efficacy. Getting ahead of the performance curve requires a more practical solution, which entails the use of bandwidth management tools that can enable the network to effectively support many more users and applications than it could otherwise.

Organizations seeking bandwidth management solutions should give preference to those that can be easily integrated into their current management platform or added to existing routers and switches. Depending on the traffic volume between the corporate intranet and the public Internet, a bandwidth management product that is integrated into a firewall might be justified, as is bandwidth management products that include security features.

If a bandwidth management product is warranted, preference should be given to vendors that can both clearly demonstrate cost-effectiveness and also can integrate with the products that are already used on the corporate intranet today. Ideally, the product will integrate seamlessly into the existing network without requiring extensive changes in topologies or hardware. Deviating from this recommendation could result in higher costs in the long run if time and resources must be allocated to a prolonged integration effort.

Along with bandwidth management, as a network manager, you should give consideration to implementing active caching. Caching not only limits the amount of traffic that traverses the IP net, resulting in improved performance for all users, it provides dramatic improvements in response time, since frequently accessed information is stored locally and updated automatically without imposing an undue burden on the network.

17.12 Conclusion

Intelligent design tools with built-in error-detection, simulation, and analysis capabilities, and plug-in modules for ancillary functionality are now available. They do not require managers and planners to be intimately familiar with every aspect of their networks. The essential information can be retrieved on a moment's notice—often with point-and-click ease—analyzed, queried, manipulated, and reanalyzed if necessary, with the results displayed in easy-to-understand graphical form or exported to other applications for further manipulation and study.

Unlike traditional CAD programs, today's drawing tools are specifically designed for network and IT planners. They can improve communications and productivity with their easy-to-use and easy-to-learn graphics capabilities that offer seamless integration with other applications on the Windows desktop. Their graphical representations of complex projects also enable more people to understand and participate in the planning process. Despite their origins as simple drawing tools, this new generation of tools provides a high degree of intelligence, programmability, and Web awareness that makes them well suited for the demanding needs of planners.

Bandwidth managers identify and manipulate traffic classes by looking at the TOS bit in the IP header, the IP address, the TCP or User Datagram Protocol (UDP) port number, the Domain Name System (DNS), the application, or the URL. Traffic shapers, using the TOS bit, can identify and categorize traffic without adding extra overhead to the IP header. This eliminates the need to change routers and switches.

In prioritizing traffic, bandwidth managers make use of a variety of different strategies, including queuing, changing the size of the TCP/IP window (TCP rate control), or a mixture of both. TCP/IP rate control adjusts the intervals at which TCP information on reducing or increasing the window size is sent, so that the window size is always under control. Some bandwidth managers can handle protocols other than IP and some offer advanced traffic identification.

Large enterprises will usually want to take responsibility for bandwidth management. But with so many devices to manage, administrators can easily get bogged down performing manual configurations to fully optimize the enterprise network. The use of policy-based bandwidth management solutions makes this task less tedious and error prone. Smaller firms that are more resource-constrained might be better off subscribing to the managed IP services of an Integrated Communications Provider (ICP) that can handle the growing number of diverse business applications that are being extended to distributed locations.

Managing Technology Transitions

18.1 Introduction

Most organizations recognize the need to transition to advanced technologies and understand the relationship between successful technology implementation and competitive advantage. The transition process, however, involves a fair amount of risk, mainly because new technologies are not easy to integrate with existing legacy environments. For example,

- How can a network manager ensure that existing systems, Personal Computers (PCs), peripherals, interfaces, and development tools will work well together once deployed in a client-server network?

- How can network management and help desk personnel be trained to deal with a new Wide Area Network (WAN) architecture designed to integrate different data types in support of multimedia applications?

- How can an organization transform a rigid predefined Information Technology (IT) architecture that imposes limitations on business processes to a flexible one that supports true reengineering and ongoing fine-tuning of business processes?

Addressing such challenges requires a view of information systems and networks that includes not only technology solutions but also solutions to the equally critical people and process issues that are an inseparable part of any technology transition. To mitigate transition risk, a plan must be devised and a methodology followed that leads organizations through a process to identify business problems and requirements, establish clear goals, evaluate technology options, and choose and implement the most appropriate solution.

Today's fast-paced, global business marketplace requires that companies respond to changing customer needs and react quickly to dynamic market conditions. However, IT organizations that depend on proprietary architectures to provide them with real-time information about their businesses are often hard-pressed to meet these challenges. Not only does vital data frequently seem to be locked up on mainframe and proprietary midrange architectures, but the delays in processing often render information meaningless when it finally does arrive. Transitioning from legacy systems and mature networks to open, distributed systems and higher-speed, standards-based networks can free up information bottlenecks, enabling companies to be more responsive to changing business needs. Companies that can improve the timeliness of information retrieval and distribution are better capable of supporting new lines of business and making more informed decisions about distribution, pricing, and new product development.

For many organizations, planning technology transitions is a continuous process that closely parallels the drive to stay competitive in global markets. Assessing business requirements, evaluating emerging information technologies, and incorporating them into the existing IT infrastructure has become an endless, complex exercise aimed at pushing the price and/or performance curve to stay one step ahead of the competition. Whether the business is growing or restructuring, the pressures to contain costs—to do more with less—remain considerable whether the economy is booming or faltering. For companies that do not have the expertise or the resources to spare for planning technology transitions, this function can be outsourced. There are firms that specialize in analyzing the existing and proposed technical environment and identifying the steps the organization can take to make a successful transition. The assessment includes a summary of estimated costs and savings, financing options, a technology road map, and supporting documentation.

18.2 The Transition Plan

Once an organization has made the decision to adopt a new technology, the main challenge is to make the transition as smooth as possible, with the least disruption to existing business operations, and within budget. Organizations are more successful when they define the future environment in overall technology terms, not in terms of individual products. This target environment should be business-driven and based on current and future business needs.

Developing a transition plan helps start the move toward a new target environment. In most cases, there will be a number of iterations of the plan implemented, and the route may be circuitous. Different products will be used to populate the respective aspects of the architecture over time. Business needs will change, and the target architecture will have to evolve.

Having a transition plan is important to ensure that the environment designed, and subsequently implemented, supports and enhances the business, and is not merely a pursuit of the latest technology. It is also important that the design consider the unique needs of the people who interface with the environment—application users as well as application developers and network managers.

The transition plan defines what deliverables are to be produced during the development process as well as the tasks to be performed and who is to perform them. Although the plan does not define how the deliverables are to be created, it should identify specific techniques that are to be employed in a task. It is left to those managing the development process to identify the specific tools to be used.

The transition plan should not be viewed as an optional process. It provides a uniform life-cycle approach to managing systems development, yet provides sufficient flexibility for IT staff to choose the techniques and tools best suited for a specific objective. The use of the plan also ensures the use of uniform terminology and procedures among all staff engaged in the development process.

A good plan can accommodate the needs of all types of systems development efforts, including those which use purchased application software packages and turnkey solutions that might be available from Application Service Providers (ASPs), as well as those information systems developed entirely in-house or customized with outside assistance. The essentials of the life cycle should remain consistent in their use throughout the enterprise, regardless of project size or system solution.

The life cycle contains optional or conditional deliverables to allow flexibility in the system development process. This optional feature enables the system developer to produce only those deliverables that are appropriate to the complexity of the project.

This flexibility ensures the plan's use for both very short projects such as enhancements to existing systems as well as for major development efforts.

It will usually be the responsibility of project leaders to tailor the deliverables and tasks to the specific information system project. The project leaders may consolidate tasks in order to focus on producing the minimum set of deliverables for the particular type of project. The goal here is to eliminate inappropriate tasks and deliverables, thereby increasing productivity and containing costs.

For example, the selected system solution may include the purchase of an application software package. In this case, responsibility for the deliverable should fall on the vendor instead of the IT department. Management of the development process requires assurance that the plan's recommended deliverables are present in the vendor-supplied product. But it is not usually cost-effective for the IT department to be responsible for those deliverables.

The transition plan will have several implementation stages, each of which consists of one or more tasks that describe what is to be performed in the pursuit of the stage's objectives. The tasks are broken down into discrete steps. Execution of the tasks creates deliverables that consist of components. The components can be described simply by listing their elements.

For example, the scope of the transition plan might be to overhaul the organization's information systems. The objective of this stage is to identify those business areas (broad, major information system projects) that are significant to the strategic information needs of the enterprise. Stage inputs consist of the organization's business functions, information requirements, and perceived need for increased technology support.

Stage outputs address one or more defined business areas—each with a defined set of business activities and data. For example, a stage output can be budgetary approval for a slate of one or more information systems projects. Another stage output might be the formulation of a request for the creation of an information system to produce needed business information. Table 18.1 puts these concepts into perspective.

18.3 Transition Methodology

The transition plan should articulate a methodology for implementation that takes into consideration the current base of users. In fact, most new systems and networks will be run by the same employees who are quite comfortable with what they are currently using. They have invested considerable time, energy, and expense in learning how to use the current technology effectively, and may feel insecure about abandoning it for something less familiar.

Instituting change can also be perceived as very threatening to those who are most affected by it. The adoption of new technology may very well involve a redistribution of responsibilities and, consequently, of power and influence. Therefore, the "soft" issues of feelings and perceptions must be addressed first to ensure success with the "hard" issues that address the technical aspects of implementation.

It is also necessary for the organization or company to inventory its current IT environment. It should include profiles of processes and procedures of current systems, networks, applications, operations, and the overall organization. This type of detailed information will help to identify areas of cost savings and operational efficiency.

All of these issues can be effectively addressed by implementing a transition methodology that starts with a participative approach to planning

TABLE 18.1   Stages of the Technology Transition Plan

Stage scope	Overhaul organization's strategic information systems
Stage input	Organization's business functions, information requirements, and perceived need for increased technology support
Stage tasks	Plan the migration project Create broad enterprise information model Approve broad enterprise information model Refine enterprise information model Approve specific enterprise information model Inventory existing information systems Develop information systems architecture Develop action plan Approve action plan Identify information resource management principles Characterize current technology environment Characterize target technology environment Define technology migration plan Manage the introduction of new technology
Stage outputs	Migration project plan Inventory of enterprise directions and concerns Inventory of enterprise business locations and organizational units Inventory of business functions Inventory of subject areas Association of business functions with subject areas, enterprise directions, concerns, business locations, and organizational units Association of subject areas with enterprise directions, concerns, business locations, and organizational units Inventory of business processes Association of business processes with enterprise business locations, organizational units, entity types, enterprise directions, and concerns Association of entity types with enterprise directions, concerns, business locations, and organization units Inventory of data collection mechanisms Assessment of data collection mechanisms Association of existing information systems to enterprise activities, data, business locations, and organization units Information systems architecture Data architecture Prioritized business areas Approved strategic enterprise information plan Budgetary approval Information resource management principles Characterization of current technology environment Characterization of target technology environment Technology migration plan Inventory of emerging technologies

and ends with the development of a time line against which the project's progress is measured.

18.3.1   Participative planning

The best way to defuse emotional and political time bombs that may jeopardize the success of implementing a new technology is to include all affected employees in the planning process. The planning process should be participative and start with the articulation of the organizational goals that the move to the new technology is intended to achieve, outlining anticipated costs and benefits. This stage of the planning process is also intended to address the most critical concern of the participants: "How will I be affected?" Once the organizational goals are known, these become the new parameters within which the participants can influence their futures.

Department managers, too, may feel threatened. They might view the change in terms of having to give up something: resources, in the form of budget and staff; power and prestige; and control of various operations. These are very real concerns in this era of corporate downsizing. Perhaps as important, they see themselves as having to sacrifice an operating philosophy that they have invested considerable time and effort to construct and maintain throughout much of their tenure. To suddenly put all this aside for something entirely new may be greeted with little or no enthusiasm, or worse—lack of cooperation, which can spoil the best transition plans.

This participative approach not only facilitates cooperation but it also has the effect of spreading ownership of the solution among all participants. Instead of a solution dictated by top management, which often engenders resistance through emotional responses and political maneuvering, the participative approach provides the most affected people with a stake in the outcome of the project. With success comes the rewards associated with a stable work environment and shared vision of the future; with failure comes the liabilities associated with a tumultuous work environment and uncertainty about the future. Although participative planning takes more time, its effects are often more immediate and long lasting than imposed solutions, which are frequently resisted and short-lived.

Another benefit of the participative approach to planning is that it gives all parties concerned a chance to buy into the new system or network and to recommend improvements that can benefit the entire organization. Consider including in the planning process Information System (IS) and department managers as well as representatives from the various business units.

In some cases, as with technologies that transcend organizational boundaries (i.e., electronic procurement, inventory, distribution, and payment systems), suppliers, customers, and strategic partners should even be included in the planning process—at their own corporate locations, if possible. This is particularly important for planning and implementing extranets—TCP/IP networks that span multiple companies. It is also necessary for planning and implementing systems sharing arrangements between organizations as part of an emergency restoration plan.

18.3.2   Education

A critical aspect of preparing personnel for change is managing the fear that change invokes. Fear of change has long been noted as a reason for delays in all phases of a new project. Compounding the fear of change is that many IT professionals have witnessed a short-lived resolve for certain decisions and remain hesitant to embrace new technologies for fear of wasting their efforts. In other cases, introductions of new technologies occur so rapidly that many IT personnel find it difficult to envision and comprehend practical applications. The hesitation to embrace new technologies becomes counterproductive and is often the result of obsolete management policies where change is perceived as an error-correction process. In such cases, if there is no clear problem to fix, change is discouraged, usually with the stale refrain, "If it ain't broke, don't fix it."

Organizations must develop a strategy for dealing with the technology transition period, during which the organization will be transformed from one with a set of well-developed, yet increasingly outdated skills, into an organization with a set of newly acquired skills, ready to redefine its future. A key element of any strategy for dealing with the technology transition period is training and education. This can be achieved in either of two ways.

One is to work with an experienced consulting team to create a transition solution that provides internal IT staff with the skills needed for success. This process includes assessing the impact of the proposed solution on the organization, performing a training needs analysis, and developing an appropriate education approach. Follow-up analysis can be performed at a later date to determine the effectiveness of training and education, and to identify new needs.

A second way to manage the transition period is to blend the skills of internal IT staff with the focused expertise of a third-party integration team through selective outsourcing. This coordinated, comprehensive approach ensures that the organization's internal staff—from executives to front-line professionals—all develop the attitudes and skills necessary to use and support the new technology.

Effective training not only provides a shared vision of how new systems and networks will strengthen the organization, it also allows faster development of the skills needed for the transition. Training also reduces resistance to change by removing doubts about the new technology. It builds confidence among staff, so they can make a valuable contribution in the new environment. Investing in training also goes a long way toward demonstrating the company's commitment to success.

An education program might include instruction on how to perform traditional data center functions in a client-server environment and explain the similarities and differences between mainframe and UNIX systems. This type of training can be provided by in-house specialists or a consulting firm. An outside consulting firm would be the better choice when there is a need to perform a thorough training needs analysis and periodic follow-up assessments to determine training effectiveness and the need for further training. An important early step involves preparing both business managers and IT staff for change by building a common awareness and understanding of open systems and client-servers and their impact on business strategy and goals. The objective is to align business and IT. This step raises the learning levels of employees, creates support for the technology vision, facilitates organizational change, and increases effective follow-up after implementation.

18.3.3 Develop solution design

The solution design of the transition plan requires a broad understanding of leading-edge technology. The components will include hardware and software, networks, applications, and administration tools. The process involves matching technologies to both the previously identified business and information goals to the needs of the proposed solution approach. The target architecture will address standards, technologies, products, and processes.

A critical step in this process is to determine which platform is best for the organization's applications and networking requirements, and which operating system standards will achieve maximum interoperability and portability. If applications will be replaced or reengineered for open, client-server computing, for example, it will be necessary to develop an application data model. This will be useful in formulating target application systems and the physical system design.

18.3.4 Evaluating alternatives

Objectively evaluating alternative IT strategies is an important element to achieving business goals. Alternatives should be evaluated and compared to determine the appropriate model that best meets the company's needs. These evaluations will help to formulate solution alternatives for implementing a transition. Transition recommendations ultimately may include rehosting, replacing, redesigning, or even outsourcing the existing IT environment.

With the completion of these evaluations, the next step is to develop a conceptual transition model. This model should be based on recommendations for alternative solution approaches. It should include a description of the target architecture, diagrams of the transition approach, and a preliminary business case. With all the alternatives laid out, a risk assessment can be performed.

18.3.5   Risk assessment

Although change is constant and provides an opportunity to add greater value to the business, change also brings risk. Therefore, it is important to identify any risk elements of the transition plan—the potential risks in the transition of the staff and organization. Once the probability and impact of these risks are evaluated, a contingency plan can then be developed to offset each identified risk.

While limiting transition costs is an effective risk-management method, a number of others may be applied:

- Create a transition plan that can adapt over time.

- Take an evolutionary instead of revolutionary approach to change.

- Work with suppliers, consultants, and service organizations experienced in implementing technology transitions for a broad customer base to leverage their knowledge and experience.

An evolutionary approach is an effective method for managing risk. When moving from a mainframe to a client-server alternative, for example, the evolutionary approach often translates into a plan detailing the coexistence of mainframe and client-server technologies during a transition period. In some cases, the mainframe will continue to operate as a database server well into the transition period and beyond.

Another risk management strategy is to select a number of mainframe applications that could be simply rehosted or transferred to open systems platforms. Remaining essentially unchanged, the application would operate on a more cost-effective platform and become more widely accessible. The conversion of mainframe programs written in COBOL, for example, to an open systems platform is relatively easy: Numerous firms specialize in performing such conversions. In addition to minimizing risk, rehosting applications can provide cost benefits, which can then be extended to include networks.

For example, the majority of organizations considering a move to open networking have made a significant investment in the Systems Network Architecture (SNA) infrastructure. A transition needs to be made in carefully planned stages to minimize the risk and cost of upgrading the network and allowing the organization to continue leveraging its existing investments as long as possible. Organizations should therefore consider transitioning their networks in stages.

In stage one, as new work groups or subnets are added to the network, they should be based on TCP/IP, the most commonly used set of protocols for internetworking. These work groups can communicate with the established SNA network using SNA gateway products running on open systems servers.

In stage two, implemented over time, the legacy systems on the established SNA network will be replaced with open systems alternatives. The new systems can use TCP/IP for communication, and emulate SNA devices when dealing with remaining legacy systems. IBM 3270 terminals can remain on the network with the use of emulation and interface software, either indefinitely or until they are phased out.

In stage three, at the point when a significant number of nodes on the network are open systems communicating via TCP/IP, the SNA network backbone itself can be transitioned to a TCP/IP. This move will likely improve communications among the various systems, reducing the need for emulation software. Any remaining SNA-dependent systems can be migrated to TCP/IP by installing TCP/IP interfaces on the legacy systems.

After these transition stages have been achieved, what remains will be an open, flexible network positioned for cost savings and future technologies. For example, if the organization already uses frame relay for Local Area Network (LAN) interconnectivity between its corporate locations, SNA can be added at an incremental cost for bandwidth and upgrade to the router software at each location, if required.

18.3.6   Project time line

The final step of the transition methodology is to create a time line to identify milestones for various phases of the project. Activities should be prioritized according to overall requirements, application strategy, and solution availability. It also is useful to include an estimate of the time required for each component. In creating this time line, it is often helpful to work with external consultants and third parties as well as internal staff. Once the time line is developed and each party signs off on it, the time line is distributed and becomes the baseline against which progress is measured.

18.4   Role of Outsourcing

Flexible and adaptable information systems (and networks) are critical assets in today's increasingly competitive business environment. Because of this, many companies are migrating from legacy systems to new technologies in order to reduce cycle times, lower costs, and increase access to meaningful data.

For example, combining client-server technology with the reengineering of business processes can dramatically improve customer service and empower user departments. In addition, replacing network and/or hierarchical data structures with a relational model provides users with the flexibility to quickly query corporate information and produce ad hoc reports.

However, migrating to the best available technology infrastructure has consistently been a difficult challenge for IT managers. Some of the immediate problems often encountered include:

- Insufficient experience in the new technology

- Limited personnel resources

- Complex legacy systems with little documentation

- Lack of a migration methodology

To minimize risk, enhance productivity, and deploy personnel resources more effectively, many companies are outsourcing migration projects to experienced third-party services firms. During a major technology transition, such as the move to open systems client-server, outsourcing supplements the internal skill set with specific technical expertise, tools, and processes that will better manage the transition process. The right outsourcing firm can provide a full range of customized information systems support services for migrating from one hardware or software technical infrastructure to another, including migration planning, code conversion, restructuring and resizing, software reengineering, and forward engineering. In addition, the outsourcing firm can train technical and user staffs in the new environment.

Outsourcing firms use specific tools to assist companies with their technology migrations. Some have developed migration technologies and methods specifically geared to reengineer the core functionality of legacy applications to standards-based computing platforms. Such reengineering tools streamline and modularize all elements of a legacy application before regenerating the system into open, standards-based deployment formats. They automate all critical reengineering activities, including:

- *Database normalization.* This tool automatically normalizes all nonrelational COBOL substructures (e.g., OCCURS and REDEFINES). This tool may also perform data type conversions, construction of all primary and foreign key specifications, and generation of a Relational Database Management System (RDBMS)–specific data definition language.

- *Code restructuring.* This process automatically detects and eliminates common forms of dead code and restructures logic flow paths, significantly reducing code complexity and increasing maintainability. The tool also restructures all Input/Output (I/O) paths to replace nonrelational data access methods with optimized Structured Query Language (SQL) equivalents.

- *Logic partitioning.* Once existing logic flow paths are restructured and optimized, the reengineering tool provides an environment for partitioning logic elements based on a multitiered deployment strategy. Server-side components can be generated as RDBMS-stored procedures and in popular object model formats such as Microsoft's COM, OMG's CORBA, and Sun's Enterprise JavaBeans. Client-side components can be generated to support a variety of visual frameworks, interface standards (e.g., ActiveX) and application environments (e.g., Web browsers). The optimal deployment strategy is driven by specific business needs, performance parameters, and technology preferences of the organization or business unit.

In the past, many organizations have been hesitant to outsource anything because it meant relinquishing control to an outside vendor, possibly becoming locked into a long-term contract with hidden costs, and having no way to ensure quality performance. Moreover, companies were also afraid of exposing sensitive information to outsiders. Selective outsourcing, however, provides the benefits of traditional outsourcing while minimizing the risks.

Selective outsourcing involves a short-term, scalable, low-risk relationship in which IT management selects exactly which activities will be performed by one or more third-party firms. Selective outsourcing provides maximum flexibility and allows companies to complement internal resources with external expertise while maintaining control of IT responsibilities and strategies.

In some cases, the vendor supplying outsourcing services can be used as an interim solution until internal resources are hired and trained. In other cases, the organization may not want to invest internal resources in skills that will only be used for a transitional period, or for activities that are not considered strategic to new technology architectures.

Selective outsourcing allows a company to focus its internal staff on core competencies, or strategic business and IT issues, and to avoid permanent staffing costs in more transitional areas. Selective outsourcing also improves performance by providing needed technical skills and products, including access to technical experts, industry-leading systems management and network management technologies, and methods proven in other organizations in similar industries. In addition, such arrangements help to contain costs by avoiding expenses normally associated with bringing IT organizations up the learning curve.

18.5   Role of Process Reorganization

The success of transitioning to a new technology often hinges on the extent to which various business processes can be understood. There are several ways to go about this. The most obvious way is to learn how people do their jobs and by soliciting their input on how things can be improved. If there are procedural problems, they should be hashed out among all participants early on so that the new systems and networks can be used to maximum advantage. It makes no sense to adopt new technologies to improve workflows and ease performance bottlenecks, if the organization's outmoded practices and procedures are not changed first.

Departmental processes should be mapped out and completely understood well ahead of purchasing, installing, and implementing any new technology. Among the problems that are often revealed are processes that result in duplications of effort, employees working at cross-purposes, and situations that produce unnecessary reporting and filing requirements. It is sometimes possible to eliminate the need for some technology upgrades merely by overhauling outdated procedures. The environments best suited for technology transitions are those where:

- The business function is widely understood and the need for improvement in terms of quality and timeliness is clear.

- The workflows lend themselves to automation and distribution via networks.

- The time spent in paper handling can be dramatically reduced.

- A significant positive return on investment is likely.

- Early implementation mistakes will not jeopardize mission-critical functions and place the company at risk.

If the goal of the new technology is merely to overlay it on the existing organization, the benefits of the new technology are going to be minimal There are several levels of technology implementation that merit consideration, categorized by increasing levels of process reorganization:

- *Pilot projects.* These are short-term projects set up for limited use at a single location. The value of these projects is that they can reveal problems, provide a learning experience, and highlight the need for more resources without adversely affecting daily business operations. When the problems are fixed, the new technology can be phased in throughout the rest of the organization.

- *Internal integration.* In helping to move information between processes, this method may yield significant benefits. However, if existing processes are not made more efficient first, long-term gains will be limited and return on investment prolonged.

- *Process redesign.* This method restructures an entire organization or discrete departmental process to take advantage of the new technology. Although this method can produce noteworthy improvements in efficiency, it can also be difficult and time-consuming to execute.

- *Network engineering.* This method extends the reorganization process to locations outside the company's main location. It can produce enterprisewide benefits over the long term, but it is also more difficult to implement and manage.

To properly implement a technology at the right level of process, reorganization requires that companies strive to understand their business processes, which is not as easy as it sounds because it involves a commitment of staff, effort, and time. Complicating matters is the fact that many corporate managers may have a stake in preserving the status quo. To overcome these obstacles, it may be more effective to have an outside firm or consultant evaluate various business processes and workflows. That way, a more objective and accurate assessment can be rendered. This, in turn, helps ensure that the investment in new technology is targeted wisely.

Alternatively, an evaluation of business processes and workflows can be performed internally, either with third-party assistance or follow-up review to validate (or invalidate) the conclusions. Considering the high cost of today's systems and networks, these extended measures may be well worth the extra time and effort.

18.6   Topology Considerations

Most new technologies are brought in to solve specific business problems. This usually means that adoption of a technology will begin in a work group or department that has this particular problem. If the solution is successful and it can be applied elsewhere, the technology tends to spread to other departments. However, when the technology is ready to be deployed elsewhere, its impact on existing networks and applications must be assessed.

18.6.1   Impact on existing networks

It is important to assess how the deployment of new technologies, particularly client-server systems, will affect existing networks, both LANs and WANs. This is because files typically will be distributed among various servers, which increases the traffic on the network. The problem is exacerbated by image, video, and multimedia applications, which are becoming more popular. Careful planning is necessary to avoid traffic bottlenecks. Accordingly, the cost and method of adding extra bandwidth, or using existing bandwidth more efficiently, must be considered when making the transition to a new technology.

Among the methods for improving the performance of LANs with the transition to new technologies are the following:

- Take advantage of ubiquitous twisted-pair wiring to add more bandwidth between clients and servers and between servers and hubs. In the case of Ethernet, this means taking advantage of such things as 100BASE-T and 1000BASE-T or even Asynchronous Transfer Mode (ATM). In the case of token ring, look for opportunities to implement newer High-speed Token Ring (HTR) switching at 100 Mbps. And, if possible, use more reliable and secure optical fiber on the backbone to connect local LANs within a building or campus environment.

- Segment existing LANs into communities of interest and/or separate heavy network users from those who require only occasional access to the network. Segmenting overgrown LANs into smaller subnets through the use of bridges and/or routers can conserve bandwidth, improve overall network performance for everyone, and facilitate management and troubleshooting. Users can even be grouped together in ad hoc fashion (i.e., virtual LAN) for a particular application, such as a videoconference, so that only users on that segment get the video stream. This not only conserves bandwidth but it also allows applications on other segments to run unimpeded.

- Consider using LAN switches for Ethernet and token ring networks. They are more flexible than bridges and routers for segmenting LANs in that the connections can be shared or dedicated. If the connection is dedicated, it can be configured to provide the attached workstation with full duplex capability (send and receive at the same time), which effectively doubles the native speed of the LAN on that segment. In addition, power workstations and servers can be directly connected to the switch's high-speed backbone to provide maximum performance for high-bandwidth applications.

- Implement caching programs at the servers and clients to reduce network traffic. By temporarily storing frequently used files in cache, the processing load on clients and servers is also reduced. Caching is especially important with regard to Internet access and corporate intranets because frequent requests for the same information can bog down the network. Implementing a proxy server not only stores frequently used information and improves overall network performance, but offers security features as well.

- Make use of real-time data compression and/or decompression to efficiently use the available bandwidth. Image and graphics files lend themselves to the highest compression ratios.

- Manage the applications so the needs of all users are served. This can be done with middleware that determines the best way to deliver applications based on such factors as the current network load and the capability of the server to continue meeting the routine needs of clients. Prioritization schemes can also be implemented to ensure that multimedia applications are served before routine file transfers, or that multimedia traffic is delivered in packets that are spaced far enough apart to let other traffic jump into the open spaces. This allows different data types to share the same network without performance degradation.

There are a number of methods available for improving the flow of traffic with the transition to new technologies.

- Take advantage of high-capacity services such as frame relay or ATM, which run over T-carrier or Optical Carrier (OC) links for higher transmission performance, reliability, and cost savings.

- For data traffic between buildings within a metropolitan area, try Ethernet services instead of traditional T-carrier private lines. A point-to-point Ethernet link at 50 Mbps, for example, can cost much less than a T3 link at 45 Mbps over a comparable distance. Gigabit Ethernet is also available as a service in many cities.

- Consider Integrated Services Digital Network's (ISDN's) high-capacity channels (384 kbps and 1.536 Mbps) for dial-up bandwidth. This option can relieve congestion during peak traffic periods. The user pays only for the amount of bandwidth used and for the period it is used, just like an ordinary phone call.

- Use compression-equipped bridges or routers to minimize traffic on the WAN. The 4-to-1 compression applied to a 384 kbps fractional T1 line, for example, yields the throughput of a T1 line, but at much less cost than a T1 line.

- Use routers that have the capability to automatically reroute traffic around failed links, balance the traffic load across multiple paths, and prioritize traffic to service all applications based on Quality of Service (QoS) parameters.

- Use bandwidth reservation and conservation protocols such as resource Reservation Protocol (RSVP) and Protocol Independent Multicast (PIM) to support multimedia applications on IP-based networks. RSVP-equipped routers set up a dedicated path for the duration of the session. PIM replicates multimedia data streams at the very last node on the network, instead of allowing packets to be broadcast to all nodes whether or not they have attached stations that want the session. This prevents packets from flooding the network, a majority of which will be discarded anyway.

18.6.2 Impact on existing applications

With many users interested in transitioning to technologies that support multimedia applications, network managers must determine the impact of these applications on existing ones. Their concern is that mixed media applications might flood a LAN with traffic, disrupting the performance of mission-critical data applications. After all, the addition of continuous stream video traffic could lead to chronic overload, traffic delays, protocol time-outs, and general unrest among the user population.

Setting up a parallel network just for video is not a very appealing solution, even for a company with deep pockets. There is good reason for concern. Running video applications over legacy LANs—Ethernet (10 Mbps) and token ring (4 or 16 Mbps)—is fraught with potential problems. These problems stem from the fact that video applications have characteristics that cause them to behave differently from other applications that share the LAN. These characteristics include file size, flow rate, and sensitivity to delay.

A single video file can be hundreds of megabytes long, whereas most ordinary data files are typically shorter by orders of magnitude. Whereas a traditional data application runs over the LAN as a burst of packets, a video application puts out a continuous stream of data. If the bandwidth is not available to support this continual flow of data, or other applications are prevented from gaining access, the entire network can grind to a halt. Video applications are also more sensitive to delay than ordinary data. A few seconds of delay can cause video quality to degrade substantially, whereas the same amount of delay would have no impact on the final outcome of a file transfer or email message.

Simply adding LAN switches to give each user more bandwidth is not an adequate remedy, especially when audio and video are involved. For example, it is possible to give each user a dedicated 10 Mbps Ethernet segment and even increase the available bandwidth by choosing an Ethernet switch that offers full-duplex (simultaneous two-way transmission) operation. Sending and receiving segments would be interconnected via a shared high-speed fiber backbone. This is like giving everyone a dedicated on-ramp to a multilane freeway. Users wait their turn to squeeze into an open space. As network load increases, the number of open spaces that other users can squeeze into becomes too few and far between. The result is intolerable congestion, causing delay that plays havoc with streaming applications.

However, these problems need not arise. In fact, video, audio, and routine data applications can be made to work very well together on the same network. Some innovative techniques are becoming available to ensure that audio-video traffic does not flood the network and adversely affect the performance of mission-critical data applications. By actively managing the audio-video traffic at the client workstations, servers, and switches, as well as at the routers on the WAN, different data types can be made to coexist, regardless of the type of facilities or services used to run the applications.

Bandwidth prioritization protocols can help prevent contention between streaming and bursting data services. In a training video application, for instance, the protocol activates whenever a workstation requests access to a video sequence stored at the media server. First, the protocol checks to make sure that the server itself is not overloaded and can safely handle another video stream. A number of factors go into this decision, including the data rate and encoding type of the desired video service, the quality of service requested by the user, and the server's capability to fulfill other ongoing responsibilities.

Second, the bandwidth reservation protocol checks the total video traffic present on the network segment directly connected to the server. By limiting the amount of video that can be carried over that segment, the bandwidth reservation protocol guarantees access to the remaining bandwidth for use by bursting data services.

If either the server or its attached network segment is in danger of becoming overloaded with the addition of video, the bandwidth reservation protocol denies access to that service and returns a busy indication to the user. This admission control mechanism results in the setup of only good connections that are capable of delivering the required service. Marginal connections that could possibly cause problems never have the opportunity of becoming established.

Once the connection is established, the user bears responsibility for the traffic handling capability of his or her workstation. If multiple tasks are running simultaneously and the user requests more audio-video streams than the workstation can handle at any given moment, it is the responsibility of the user to either shut down unneeded applications or limit the number of simultaneous audio-video streams requested from the media server. In other words, the bandwidth reservation protocol does nothing to save individual users from making mistakes at their own workstation; it only prevents the actions of one user from adversely affecting other users on the network.

Another optimization technique—traffic shaping—works to limit the video transfer rate to only what is really needed. While the bandwidth reservation protocol establishes the minimum data transfer rate needed for a video application, it does not address the maximum data transfer rate. This is where traffic shaping comes in. It acts to limit the flow of outgoing packets so that the maximum data transfer rate, measured over any interval, does not exceed the minimum data rate by more than 50 percent. In releasing packets in fine, controlled streams, the overutilization of network bandwidth is prevented.

In addition to solving the problem of network congestion, traffic shaping also helps guard against packet loss, both in the receiver and in the network. If too many video packets are sent at once, packet loss can occur at the receiving workstation from buffer overflow or an interrupt overload at the Central Processing Unit (CPU). On the network, packet loss can result from excessive collisions, or cell or packet overload on a switched network. All of these conditions can be prevented from occurring because traffic shaping regulates the short-term maximum data transfer rate by evenly spacing the individual packets.

Traffic shaping does not restrict network access to other users. Although it significantly improves overall network performance, it does so at the expense of somewhat slower response to the video application. In most situations, this is an acceptable compromise.

18.6.3   Performance measurement tools

The use of performance measurement tools is important for the success of any technology migration plan. They can be used to simulate traffic loads to test various technology migration scenarios, particularly their impact on existing applications. Of course, performance measurement tools can also be used to create a baseline for network performance so that deviations can be spotted and addressed in a timely manner. This means that performance measurement tools play a continuing role in keeping systems and networks optimized for peak performance.

There are some very compelling reasons for companies to have internal programs in place that monitor, test, and analyze their networks. The first reason is to catch problems before they become serious enough to affect the productivity of end users and cause significant downtime. Keeping the network available and reliable means that the company will get the most out of its investments in technology and infrastructure. Such a program can also be used for capacity planning and for justifying additional system or network resources.

Another important reason for having an in-house testing capability is to keep carriers honest with regard to Service Level Agreements (SLAs). These are formal arrangements between carriers and customers (or IT department and user community) that provide performance guarantees of network reliability and availability. For example, AT&T, MCI, and Sprint build guarantees against network delay into their frame relay offerings that vary from 50 to 65 milliseconds, depending on the number of Permanent Virtual Circuits (PVCs) and the applications running over them. SLAs are available for other services as well, from traditional T-carrier services to Internet-based Virtual Private Networks (VPNs), intranets, and extranets. However, the SLA is unenforceable unless the customer has an independent monitoring and reporting system that is capable of verifying carrier performance and providing conclusive evidence that the problem is the fault of the carrier.

An alternative to buying an expensive network monitoring and reporting system that is compatible with the one used by the carrier is to track performance through software agents deployed across the enterprise network. Response Networks, for example, offers a performance-tracking system called ResponseCenter, which includes software agents that act as virtual users by creating live transactions that measure the availability and performance of key applications and services. The real-time performance tests determine actual application response time so meaningful reports can be generated on the quality of application delivery to end users. The system provides valuable capabilities to you, as a network manager, including the ability to:

- Monitor performance at end-user locations to get an early warning before users call to complain.

- Validate potential brownouts before critical services become unavailable.

- Eliminate finger-pointing by helping triage problems in network, server, Web, or database components.

- Benchmark circuit utilization, server availability, or Web site performance for peak times of day and then monitor against it to validate vendor performance claims or service levels.

- Verify that end-user service levels meet performance goals—in real time and historically.

Web-based tools are another alternative to expensive test equipment. However, while convenient, these tools do not provide the sophisticated functions that companies need. For the most part, they provide status monitoring of systems and networks and are not capable of more sophisticated filtering, decoding, and analysis functions. Often, Web-based tools provide performance snapshots in 15-minute intervals or they provide static reports compiled daily. For the most part, Web-based tools complement rather than substitute for traditional test and measurement tools.

18.7   Support Issues

There are a variety of support issues that deserve attention when transitioning to a new technology. Among them are systems reliability, systems integration, and training.

18.7.1   Systems reliability

When making the transition to new technologies, thought must be given to keeping systems and networks functioning: specifically, how these systems and the networks they run on will be protected against failure. Sometimes reliability issues can be addressed within a system or network, but other times they must be addressed by the addition of products or services. The possible solutions include, but are not limited to, the following:

- Uninterruptible Power Supplies (UPS), battery arrays, and generators to protect against commercial power outages

- Redundant components and subsystems to ensure maximum system availability despite hardware failures

- Alternative routes to corporate locations for access lines to offer some protection against cable cuts

- Load balancing data traffic among different carriers to protect against a total loss of communication should one carrier experience a link or node failure

- Mirrored servers to ensure database availability should the primary server suffer an outage

- Off-site data storage to guard against the complete loss of data occurring from a local or regional disaster

- Implementing a fiber network with a self-healing capability that keeps data flowing despite cable cuts or node failures, or subscribing to a carrier service that provides this capability

An assessment must be made as to what protective measures are already available and to what extent they can be applied to the new systems or networks. If new protective measures are needed, ideally they must be considered in the planning process so they can be factored into the project's budget.

18.7.2   Systems integration

When a network is being designed from scratch and is composed of products from different vendors, the task becomes one of tying together diverse systems to provide users with transparent access. Rarely does in-house staff have the depth and breadth of expertise to accomplish this formidable task alone. Systems integrators can help and be justified on the grounds of:

- Specialized industry knowledge, awareness of technology trends, and experiences gained from a broad customer base

- Knowledge of specific protocols, interfaces, and cabling requirements

- Reputation for doing quality work on time and within budget

- Capability to transfer knowledge to existing corporate staff so they can take over responsibility for ongoing management, administration, and control

- The need for an outside party to act as a catalyst in implementing change and to validate or fine-tune in-house plans

A systems integrator brings objectivity to the task of tying together diverse products and systems to build a seamless, unified network. To accomplish this, different physical connections and incompatible protocols must be reconciled. The systems integrator uses its hardware and software expertise to perform the necessary customization.

Qualified systems integrators have in place a stable support infrastructure capable of handling a high degree of ambiguity and complexity, as well as any technical challenge that may get in the way of the integration effort. In addition to financial stability, this support infrastructure includes staff that represents a variety of technical and management disciplines, computerized project management tools, and strategic relationships with carriers, vendors, and specialized service firms.

There are a number of discrete services that are provided by integration firms, which can aid in a successful technology transition process, such as:

- *Design and development.* Activities such as network design, facilities engineering, equipment installation and customization, integration, acceptance testing, and network management

- *Consulting.* Business planning, systems and/or network architecture, technology assessment, feasibility studies, Request For Proposal (RFP) development, vendor evaluation and product selection, quality assurance, security auditing, disaster recovery planning, and project management

- *Systems implementation.* Procurement, documentation, configuration management, contract management, and program management

- *Facilities management.* Operations, technical support, hotline services, change management, and trouble ticket administration

- *Systems reengineering.* Maintenance, systems and network optimization, remote monitoring and diagnostics, and automated design tools

18.7.3   Training

Another service that may be provided by integration firms is training, which can run the gamut from hands-on user training, technical and management courses, executive briefings, and industry trends seminars. Of course, these can be provided by other types of firms as well, some of which can work with internal departments of an organization to customize content. A training firm can help with the transition from Lotus Notes to Microsoft Outlook, for example. By working closely with the organization's IT group, the training firm can instruct users on the differences between the two applications and how to move their work from the old platform to the new platform. The training firm can also address the back-office issues and the time frames for the transition itself.

Although employee training is often viewed as discretionary, especially when budgets are tight, a well-designed training program can pay for itself and shore up the bottom line. The value of such programs to technology transitions is in being able to teach employees to do things right the first time, thus minimizing downtime, equipment damage, loss of data, and sometimes personal injury while maximizing productivity and profits. A formal training program ensures consistency of content and provides a means to objectively measure employee understanding or performance as training progresses. Depending on the subject, the objective might even be certification in a technical area such as network installation and support, network engineering and design, communications and services, systems administration, database administration, and security management.

When in-house technical staff has responsibility for first-level support of information systems or network equipment, often vendors require that they go through a formal course of instruction to become certified in various installation, diagnostic, and repair procedures. Without such certifications, equipment warranties could be voided if unqualified personnel tamper with the systems. Many manufacturers put a seal on the edges of their equipment; if the seal is broken and the company cannot verify that a certified technician is on staff, the warranty is voided when the company sends the item back for repair. Rather than getting a free repair or replacement, the company could find itself being billed on a time and materials basis for the repair.

18.8   Management

With the advent of new, open architectures, organizations must ensure that their mission-critical applications and data are adequately secured and managed. When organizations begin the transition to distributed information systems, for example, they need to consider how they will manage these systems with the same level of reliability, productivity, and security traditionally demanded in commercial mainframe environments. The operational freedom that is derived from UNIX systems and alternatives such as Windows 2000 must not be attained at the price of security breaches, lost backups, unexplained outages, and delayed workloads.

To address these management issues, attention must be given to procuring an integrated, robust set of tools for managing an open systems IT environment. The solution should provide mainframe-like control in such key areas as security management; storage management, including archiving, backup, and file restore; help desk and problem management; job scheduling; console automation; print spooling; performance monitoring; report distribution; and resource accounting. The integration of the components creates a synergistic relationship, making the complete management system much more powerful than a collection of individual products. With this approach, companies can use their existing IT operations staff to deploy their new IT architecture, thus saving valuable learning time, minimizing risk, and reducing the overall cost of deployment.

18.9   Conclusion

Many of the architectures that were announced, deployed, and implemented during the early 1980s and early 1990s are no longer effective. The solutions of several years ago worked well for the needs of several years ago. Today's solutions must meet expanded business requirements and be capable of growing with the corporation as it seeks to respond to changing customer demands and markets.

Effective IT management today demands a flexible, open environment that can guarantee the availability of mission-critical applications, databases, systems, and networks. Organizations seeking to remain competitive are driving the transition from proprietary, centralized—often mainframe-based—architectures to an open, distributed, and scaleable client-server architecture. Yet that transition carries with it inherent complexities and risks. By focusing not just on technology but also on the IT processes and people skills, organizations can build flexible enterprisewide networks that provide strategic advantages in increasingly competitive global markets.

Protecting the Business

Network Security

19.1 Introduction

Protecting vital information from unauthorized access has always been a high-priority concern among most companies, especially as more business applications migrate to the Internet, which is a notoriously insecure environment. While access to distributed data networks improves productivity by making applications, processing power, and mass storage readily available to a large and growing user population, it also makes those resources more vulnerable to abuse and misuse. The situation is even more pronounced when dealing with such services as Electronic Data Interchange (EDI), Electronic Funds Transfer (EFT), email, and other data networking environments where unauthorized use, information theft, and malicious file tampering can result in immediate financial loss and, in the long term, damage to competitive position.

Because EDI and EFT applications, for example, involve financial transactions and monetary payments, security is a paramount issue. Security for EDI and other messaging systems is basically the same as for other automated information systems, involving such considerations as message integrity and source authentication as well as controlling access to workstations, servers, and data centers that handle various portions of mission-critical applications. With the increase of e-business and e-commerce applications on the Internet, attention to security is especially important because the underlying protocols on which the Internet is based were not designed with secure transactions in mind.

19.2   Risk Assessment

Risk assessment begins with the assumption that a potentially hostile environment exists in which intruders are passively or actively trying to breach network security. There are two types of intruders to be concerned about. *Passive intruders* may browse through sensitive data files, monitor private conversations between other network users, intercept email messages intended for other users, or read restricted information. *Active intruders,* on the other hand, destroy information integrity by modifying data, denying others access to network resources, and introducing false data or unauthenticated messages onto the network. This type of intruder may even seek to destroy programs and applications by introducing viruses or worms into the network.

19.2.1   Securing the work environment

The physical environment is relatively easy to protect, starting with the building itself. This is important because most thefts of hard and soft assets are committed by the company's own employees. Such precautions as locking office doors and wiring closets, restricting access to the data center, and having employees register when they enter sensitive areas can greatly reduce risk. Issuing badges to visitors, providing visitor escorts, and having a security guard station in the lobby can reduce risk even further. A guard not only discourages intruders from entering the building to do mischief but also discourages employees from walking out the door with company property or sensitive information. The possibility of random searches as employees leave the building further dissuades violations of corporate security policies.

Securing the work environment requires the cooperation of management and staff. For example, employees who work near unprotected workstations should take note of authorized operators and their usual work shifts. However, such simple measures as keyboard and disk drive locks are even more effective in deterring unauthorized access to unattended workstations. In addition, locking down workstations to desks can help protect against equipment theft. These are important security features, especially since some workstations provide management access to wiring hubs, Local Area Network (LAN) servers, switches, routers, and other network access points.

Companies should also encourage users to log off whenever they leave the workstation, since an unattended workstation invites data theft. In fact, an unattended logged-on computer represents so great a risk that a company should monitor workstations, determine the last user who logged on, then take disciplinary action for failure to log off. Alternatively, Information Technology (IT) administrators can configure all computers to blank out the display after a specified period of inactivity. This would force the user to log on with his or her user identification (ID) and password in order to resume work.

Sound security policy dictates that no employee should leave a notebook computer unattended after normal work hours. Being small and lightweight, notebooks are too easy to grab and smuggle out of the building. A security sweep of the building should include confiscation of unattended notebook computers by authorized personnel, forcing offending employees to pick up their equipment the next day.

Network administrators should also implement controls that prevent tampering with the wires or cables linking the workstations to the network. In addition to examining the obvious wires linking telephones and data processing equipment, administrators should examine conduits, wiring closets, and patch panels where telephone and data wires traverse other floors in the building. A basement may have a wire room where all of the wires in a building terminate. Administrators should keep unattended wire rooms and closets locked and monitor any installation work that must be performed.

19.2.2 Securing the network

Beyond taking precautions to protect the work environment, an organization should evaluate the accessibility of all shared network resources. When assessing vulnerability, network administrators must determine whether the access controls in place effectively prevent unauthorized users from accessing the network and legitimate users from accessing unauthorized resources.

Wire line networks. Part of any risk assessment should include inspecting communications links for vulnerabilities and performing upgrades if necessary. The degree of difficulty encountered when tapping a line often depends on the type of wiring used. For example, tapping a fiber-optic line is considerably more difficult than tapping other media. This is because optical fiber does not radiate signals that can be collected through surreptitious means. Optical-fiber cores are surrounded with a less dense material called *cladding,* which prevents light from being radiated away, so there is nothing to collect. Therefore, other means of tapping must be used, which entail physically breaking the core and fusing a connection to it. Such means are routinely used to add nodes to the fiber cable. But during the procedure, no light can be transmitted past that point, which would make unauthorized access easy to detect. If the intruder were to perform a sloppy job of tapping, the increased error rate would indicate that a breach has occurred. Even if a skilled and properly equipped intruder can make a precise connection, the resulting signal loss could show up in a routine measurement, raising the possibility of a security breach.

Shielded copper wiring can also prevent signals from being collected. The shielding prevents signals from radiating from the wire. This not only eliminates crosstalk between adjacent wires but also helps prevent signal collection by intruders. To monitor transmissions, the intruder would have to break the wire to insert surveillance equipment, but this would raise an alarm at the management station. The use of a Time Domain Reflectometer (TDR) can pinpoint the exact location of the break on both fiber and copper transmission media.

All electronic devices emit electromagnetic radiation. Ultra-sensitive snooping devices can detect the signals radiating from LAN cabling or attached devices. The U.S. government's TEMPEST standards define acceptable emission levels for secure applications. Workstations and other network devices, as well as cabling, can be shielded to reduce signal emissions to a virtually undetectable level. However, since facilities used by common carriers are not under user control, data encryption must be employed to protect the transmission of sensitive information (discussed in Sec. 19.5).

Wireless networks.   The main security difference between wireless and wire line networks is that wire line networks propagate signals over a much larger area. In general, this makes signals easier to receive by intruders, unless encryption is added to make deciphering the signals impossible. Even without encryption, however, some wireless LAN technologies are inherently more secure than others.

For example, since infrared does not penetrate walls or ceilings, it offers more protection against unauthorized reception. Although direct sequence spread spectrum signals can penetrate walls, the signal spreading and despreading algorithms used at each end make casual eavesdropping very difficult. A high level of security can be achieved using a variation of spread spectrum technology called *frequency hopping.*

Frequency hopping means that the transmitter jumps from one frequency to the next at a specific hopping rate in accordance with a pseudo-random code sequence. The order of frequencies selected by the transmitter is taken from a predetermined set as dictated by the code sequence. For example, the transmitter may have a hopping pattern of going from channel 3 to channel 12 to channel 6 to channel 11 to channel 5, and so on, as shown in Fig. 19.1. The receiver tracks these changes. Since only the intended receiver is aware of the transmitter's hopping pattern, only that receiver can make sense of the data being transmitted.

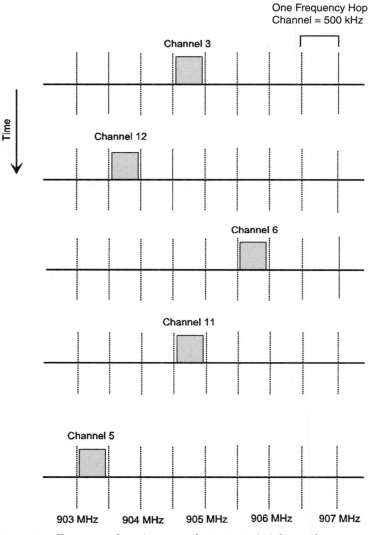

Figure 19.1 Frequency-hopping spread spectrum is inherently more secure than other wireless technologies.

An organization's other frequency-hopping transmitters will be using different hopping patterns that will be set for other, noninterfering frequencies. Should different transmitters coincidentally attempt to use the same frequency and the data of one or both become garbled at that point, retransmission of the affected data packets is required. Those data packets will be sent again on the next hopping frequency of each transmitter.

Most LAN protocols have an integral error detection capability. When the protocol's error-checking mechanism recognizes incoming packets that are bad or determines that there are missing packets, the receiving station requests a retransmission of only those packets. When the new packets arrive to rendezvous with those held in queue, the protocol's sequencing capability puts them in the correct order.

Of course, even frequency-hopping can be hacked. To completely safeguard sensitive information over wireless networks, encryption should be used. Some vendors of wireless networks offer integral signal scrambling coupled with a capability known as *dynamic path selection*, which is a variation of spread spectrum's frequency hopping. Under this scheme, the system continually changes transmission paths, making the interception of a complete transmission virtually impossible. Even if that were possible, any data fragments received would be unintelligible because the data is contained in a unique frame structure and is scrambled. Of course, the drawback to these methods is that they are proprietary, forcing IT organizations to depend on a single vendor for network upgrades and expansion.

While encryption can effectively protect sensitive data against external intruders, it will usually not be enough to stop threats originating from within the organization. Here is where access controls can help.

19.3   Access Controls

Often, the threat of intrusion originates from the public network. Access controls should prevent unauthorized local access to the network and control remote access through dial-up ports. The three minimum levels of user access usually assigned are public, private, and shared access. *Public access* allows all users to have read-only access to file information. *Private access* gives specific users read-and-write file access. *Shared access* allows all users to read and write to files.

Perhaps the most difficult application to secure is the shared database environment offered by LANs. When a company offers network access to one or more databases, it must restrict and control all user query operations. Each database should have a protective "key," or series of steps, known only to those individuals entitled to access the data. To ensure that intruders cannot duplicate the data from the system files, users should first have to sign on with passwords and then prove that they are entitled to the data requested.

19.3.1   Password security

The security features of a network operating system are crucial to preventing unauthorized access to network resources. Usually this entails the user having to enter a logon ID and password to access the system. Most passwords identify the user and associate the user with a specific workstation and perhaps a designated shift, work group, or department. Reliance on these mechanisms has drawbacks—chief among them is that users do not always maintain password confidentiality.

Passwords should have a minimum of six or seven characters—any fewer characters and the passwords can be too easily guessed by unauthorized users. Worth noting is that plain-text passwords are especially vulnerable on LANs, since each guess increases the chance of unauthorized entry by a factor of $1 \times n,$ where n equals the number of passwords on the LAN. To decrease the chances of a good guess, users should not be allowed to make up their own passwords. Random password generators can be used for this purpose. A user ID should be suspended after a certain number of passwords have been entered, further reducing the chance of a trial-and-error procedure accessing the operating system. Additionally, the network administrator should obtain a daily printout of the keys and/or passwords used to help track down any potential security breaches. Changing passwords frequently can help to enforce tight security. If users are allowed to choose their own passwords, the selected passwords should be filtered through a dictionary to eliminate guessable words from being used, to eliminate already used passwords, and to make sure the passwords meet minimum length requirements. And when passwords expire, users should be prevented from using any expired passwords as the new password.

There are two systems of password protection that companies can employ to maintain security: hierarchical or specific. Using *hierarchical passwords,* users can employ a defined password to gain access to a designated security level, as well as all lower levels. With *specific passwords,* on the other hand, users can access only the intended level and not the others above or below. Although specific-level passwords offer more security, they require that a senior, trusted employee have many passwords in order to work with the many databases and associated levels used throughout the day. Password levels, especially specific levels, also complicate the task of security administration.

Once an administrator implements a particular password security method, he or she should ensure that the connected workstations play an active role in supporting password usage. For example, as a user enters the characters of the password, the monitor screen should automatically blank out all key entries to minimize the risk of exposing the password to casual observers. Administrators can also install password routines that do not display any information on the screen or that sound an audible alarm while locking the keyboard after a specified number of failed entry attempts.

In addition, administrators should periodically survey the password master file, change or retire any infrequently used passwords, keep the updated file on disk and store it in a secure area, and reassess risks whenever a breach of security occurs, or is even suspected. In addition, when personnel leave the company, their user ID and password should be rendered inoperable. In the case of corporate layoffs, where hundreds or even thousands of people leave the company at once, some preplanning is necessary to ensure that corporate resources are protected.

19.3.2 Single point log-on

Distributed systems enable processing, data, and applications to be shared throughout the organization for the benefit of all users. However, the security features of a network should not impose limitations on the ability of users and network administrators to do their jobs. A capability called *single point log-on* or *sign-on* enhances network security by simplifying access, regardless of the user's location. Using the same user ID and password, single point log-on provides controlled access to applications and services residing on a local disk or file server, as well as to host-based applications, anywhere on the network. A desktop window provides the user with a set of icons that invoke access to enterprise applications and services. The single point log-on software does the work of controlling and managing all of the procedures that are required to access and execute the applications, regardless of their location on the network. In addition to user IDs and passwords, the software provides emulator selection, network navigation, and application subsystem selection. All of this takes place automatically and transparently, while ensuring security.

The single point log-on software permits the storage of remote log-on information and passwords in encrypted form, so that even the administrator does not know user passwords. The administrator can also establish customized password-aging policies of any length of time, setting up passwords to expire in a day—or to never expire. The administrator can control access to only the enterprise applications and services that are authorized for each user. When the administrator adds a service or application for a group, all users in that group have instant access to it. This ensures the administrator's ability to consistently provide the information that end users need while preventing inappropriate access to information that should remain secure.

19.4 Other Security Measures

Although passwords are the most frequently used access control method, their effectiveness depends on how carefully users protect them, and how rigorously network administrators enforce the procedures. For this reason, administrators should consider combining password security with another control measure, such as a keyboard lock, card reader, or even a biometric device. The choice will depend on the level of security desired, as well as the cost of implementation.

19.4.1 Key and card systems

Some of the simplest security systems require users to insert a key in a lock in order to turn on the computer. Some vendors also market external hard disks, keyboards, and modem switches that incorporate key locks. With lock-and-key systems, however, intruders can pick the lock or duplicate the keys. To help minimize this possibility, the network administrator should keep keys in a secure cabinet where the keys can be checked out and their use can be monitored.

Alternatively, magnetic card–reading systems can be used to control access to workstations. Card systems offer more flexibility than simple key systems in that they allow users to access data from any available workstation they are authorized to access simply by inserting a card into a reader attached to the workstation. Most banking Automatic Teller Machines (ATMs) use this type of security, combined with a system of user IDs and passwords. This card key system allows access-level definition for each user, rather than for each workstation. If users will be taking cards out of the building, the cards should not contain a company logo, street address, or anything that could identify the company. A simple printed statement to mail the card is sufficient. Administrators should reserve a box at a commercial mail drop for this purpose, listing it under a fictitious name to maintain complete security.

Although cards with magnetic stripes are the most popular, other types, such as bar-code, plastic, and proximity cards, are also available. Some companies favor plastic cards with embedded magnetic stripes containing authentication information, which can provide for employee entrance control as well as workstation control. Cards encoded with bar codes for optical readers can also be used, but this type can be relatively easy to duplicate. Proximity cards, which can be read by radio frequency at distances of a few inches to 10 feet, may be unsuitable for offices in which the workstations may be too close together. Administrators should not issue the same card for both workstation identification and access to other areas of the company. The workstation deserves a higher level of security and a closer control of card distribution than company access cards.

Network administrators can also consider so-called smart cards as possible security alternatives. These devices, which contain embedded microprocessors, can accommodate a range of security tasks, such as performing on-line encryption, recording a time-on/time-off log, and providing pass-word and biometric identification. Such devices offer a feasible security option for both local and remote access controls.

Some smart keys can provide very complex security solutions. For example, one of these products, used in conjunction with system software that generates a flashing pattern on a monitor, optically scans the pattern. The key device interprets the coded pattern and displays a remote access code on its Liquid Crystal Display (LCD). The user can then enter the code on a keyboard. This new generation of hand-held smart security system offers a wide variety of options, including cards (with synchronized host software) that internally generate new access codes every minute and devices that optically read user fingerprints to permit system entry. Most of these devices target organizations that require a high degree of information security.

19.4.2 Biometrics

The one drawback with both key and card systems is that access control can, willingly or unwillingly, transfer to someone other than the authorized user. Biometric devices, on the other hand, use an individual's unique physical attributes for identification, thereby providing a higher level of security. These devices can identify an individual based on characteristics that cannot be duplicated or forged such as a fingerprint, voice quality, or pattern of capillary blood vessels in the retina of the eye.

Fingerprint recognition, the most developed of these technologies, is now appearing in the corporate world. Some fingerprint scanners can be added to desktop computers via a Universal Serial Bus (USB) connection, and computer makers offer notebooks with built-in fingerprint scanners. To get started, the user enrolls one or more of his or her fingerprints in the unit's biometric database. During enrollment, the computer records the characteristic bends and endpoints of the fingerprint ridges. This process is repeated several times to verify the consistency of the results before the system registers the authorized user's profile.

To log in, the user's fingerprint is scanned and the system compares the new scan with the stored image. If the right number of fingerprint details match, access to the Personal Computer (PC) is granted. At this writing, however, fingerprint scanning is not 100 percent reliable. For some people, getting the scanner to accept their fingerprint is like struggling to feed a dollar bill into a vending machine over and over—sometimes it is accepted, other times it is not. With practice and patience, users can train themselves to present their finger more consistently, thus minimizing the number of access failures. As a backup, these systems also let the user enter a password to log in.

Another biometric security technique operates by sensing the user's habitual typing rhythms. To log on to the network using this type of security system, the user types a brief sentence that has previously been recorded in workstation memory, creating an identification pattern that is extremely difficult to forge. This method provides a cost-effective means of user identification because it does not require additional hardware and it is sufficiently transparent that users do not have to learn a new technique, yet secure enough to deter an intruder from invading the network by guessing a password.

19.4.3   Disk and drive controls

Local Area Network security is weakest at the desktop, not only because password confidentiality tends to be weak, permitting unauthorized access, but because removable disks make data vulnerable to theft. To avoid theft and unauthorized copying of removable disks, data cartridges, and portable hard drives, administrators should store them in a locked cabinet and store critical disks, such as backup copies of sensitive files, in the corporate safe or in another secure location within the data center. Users should create backup copies of sensitive files at weekly or daily intervals to provide a reliable source of archived data for restoration in the event of system disaster.

Creating backup copies of data also helps to prevent the spread of worms and viruses. In this way, if an infected disk does contaminate network resources, multiple backup copies dating back to a time before the virus infection occurred are available to restore the affected files, application programs, databases, and operating systems.

A removable hard disk is ideal for transferring large files between machines, for archiving and backup tasks, and for use as a secondary storage device. Some removable drives are entirely self-contained, while others use removable cartridges that contain only the disk itself. Removable cartridges are best for applications in which security and long-term portability are the most important considerations.

Disk-locking programs are also available to prevent program disks from operating properly if copied or used with an unauthorized host computer. Administrators can protect data disks and files with passwords or modify them so that they allow access to data only when used with a specific program disk.

Another method of securing workstations is to use diskless workstations. A diskless workstation relies entirely on the network server for boot-up, address assignment, applications, and data storage, thereby eliminating the need for local disks. These workstations offer several potential benefits from a security standpoint. For example, since diskless workstations do not contain disk drives, they eliminate the possibility of disk theft, unauthorized copying, or concealing information downloaded from the host computer. The absence of disk drives also reduces the risk of introducing a virus into the network through infected input disks.

In the final analysis, the type of disk storage facility required depends on the duties of each workstation user. Therefore, a variety of safeguards should be in place to accommodate the differing needs of users.

19.5   Data Encryption

Data encryption, a method of scrambling information to disguise its original meaning, provides an effective and practical means of protecting information transmitted over dispersed communications networks. Since intruders cannot read encrypted data, the information is not vulnerable to passive or active attack. When implemented along with error detection and correction, encryption offers a highly effective and inexpensive way to secure a communications link. For example, file encryption with decryption at the user workstation adds security to both the file server and the transmission medium. The decryption key can be a string of characters known only to the user. The data are secure from other workstations, illegal taps, and interception of spurious electromagnetic radiation.

Either the hardware or software can perform encryption, but hardware-based encryption provides more speed and security, since an intruder who is skilled in programming will usually not be able to interfere with the encryption hardware. Online encryption requires installation of encryption and decryption units at both ends of the communications link.

Cryptographic methods involve the use of an encryption key, which is a string of characters used in conjunction with an algorithm to scramble and unscramble messages. Cryptographic methods are designed so that even if intruders know the algorithm (a mathematical formula stored in electronic circuitry), they will not be able to decode the scrambled information unless they also have the specific encryption key. The more characters the key contains, the more difficulty an intruder will encounter when attempting to breach security. The strength of a cryptographic system lies in the quality and secrecy of the keys selected.

19.5.1   Symmetric versus asymmetric

Encryption systems are either symmetric or asymmetric. Symmetric crypto systems use the same key (the secret key) to encrypt and decrypt a message, while asymmetric crypto systems use one key (the public key) to encrypt a message and a different key (the private key) to decrypt it. Asymmetric crypto systems are also called *public key crypto systems*.

Symmetric crypto systems have a problem: How can the secret key be safely transported from the sender to the recipient? If the secret key could be sent securely, there would be no need for the symmetric crypto system in the first place because the same secure channel could be used to send messages. Bonded couriers are used as a solution to this problem.

Another, more efficient, and reliable solution is a public key crypto system. Such systems use a two-part key structure to eliminate the problem of sharing a single encryption-decryption key. This technology permits users to encode data files with a public key that is associated with a specific user. The public key can encrypt but not decrypt a file. A private key, associated with each set of public keys, enables users to decrypt files that have been encrypted using this technique. The key used to decrypt a file is associated with, and available to, only a single user; this minimizes the likelihood of a key being copied or discovered by unauthorized users. Furthermore, because the decryption key is valid for only one user, it cannot be used to decrypt files intended for a different user. There is no need to keep the public keys secret or to risk compromising private keys by transmitting them among various users.

19.5.2   Data Encryption Standard

One of the most thoroughly tested cryptographic algorithms available is the Data Encryption Standard (DES), developed by IBM and adopted in 1977 as a federal standard by the U.S. National Bureau of Standards, now known as the National Institute of Standards and Technology (NIST), a unit of the Department of Commerce. DES is one of the most important cryptographic algorithms because it and its successor, Triple DES, are the basis for hundreds of security products in widespread use today.

DES-based encryption software uses an algorithm that encodes 64-bit blocks of data and uses a 56-bit key; the length of the key imposes a difficult decoding barrier to would-be intruders because 72 quadrillion (72,000,000,000,000,000) keys are possible. The DES offers four different encryption modes. Direct mode is the easiest to implement but provides the least security because it allows independent coding of each of the blocks of a message. Independently coded blocks can develop coding patterns in lengthy transmissions; such patterns can make the encryption technique vulnerable to unauthorized access. In the other three modes, coding of each data block varies depending on the coding of one or more previous blocks, reducing the risk of revealing a pattern in encoding that could provide clues to the decryption key.

Despite its 20-year performance of foiling decodes, DES was finally broken in 1998 by the Electronic Frontier Foundation, which managed to break the algorithm in less than three days at a cost of less than $250,000.

The encryption chip that powered the so-called DES Cracker was capable of processing 88 billion keys per second. In addition, it was shown that for a cost of $1 million a dedicated hardware device can be built that can search all possible DES keys in about 3.5 hours.

Since 1998, DES has been replaced with Triple DES, which is based on the DES algorithm, making it very easy to modify existing software to accommodate the new standard. It also has the advantage of proven reliability and a longer key length that eliminates many of the shortcut attacks that can be used to reduce the amount of time it takes to break DES. Even this more powerful version of DES had a limited life span owing to the increasing processing power of today's computers, and Triple DES is being replaced by the Advanced Encryption Standard (AES).

In December 2001, NIST announced approval of the Advanced Encryption Standard, which specifies the Rijndael symmetric encryption algorithm developed by two Belgian cryptographers. Experts claim that the algorithm is small, fast, and very hard to crack, estimating that it would take 149 trillion years to crack a single 128-bit AES key using today's computers. Corporations using Triple DES technologies will have to wait until low-cost AES implementations become available before a migration to the new standard makes sense from a price perspective.

19.6   Virus Protection

Among the dangers of inadequate network security are the alteration and destruction of valuable records and whole databases by worms and viruses introduced onto the network through unauthorized programs brought into the workplace by unknowing users. Worms and viruses are usually differentiated according to their degree of transferability. A *virus,* for example, limits its damage to the LAN workstation through which it entered and is transported by external action, such as by disks and software downloads from bulletin boards. *Worms* are self-replicating and move throughout the network—node to node. Some viruses and worms are timed for activation far into the future, making it even more difficult to track down their source.

Viruses can have a multiplier effect on networks. If a stand-alone PC gets infected, there is little chance that the virus will spread to other machines, unless handing disks around causes further infection. If a mainframe gets infected, the damage can be more severe, but there is still only one system to disinfect. But when a virus invades a LAN, the damage can be far-reaching. Disks used to boot servers are a prime cause of LAN infections. Since workstations communicate with the network file server to obtain shared programs and data files, a virus can spread rapidly to every computer that accesses the server.

A virus presents a major threat to security because of the damage it can do to information, especially in distributed processing environments where any one user accesses an array of network resources on a regular basis. Once a virus program has been introduced into a system, it can cause such problems as the following:

- Longer than normal program load times
- Excessive disk accesses for simple tasks
- Unusual system error messages
- Disk activity for no apparent reason
- Reduced available Random Access Memory (RAM)
- Reduced available disk space
- Unexplained file or directory disappearances
- Changes in executable program size
- Changes in appearance of screen icons
- Screens that blank out or jitter
- Text that dribbles to the bottom of the screen

To protect against a catastrophic virus attack, network administrators should implement internal barriers between connecting systems. These barriers (e.g., different encryption codes for separate programs or network devices) will not completely insulate a network from a virus attack, but they will restrict damage only to that area of the network where the virus has entered. If a subsystem (for example, a LAN) is only a "pass through" from a data source to a major interior system, a virus can be detected and blocked from entering the interior system. The technique involves making hash totals at the input and output of the subsystem and matching them. If a virus has intervened, the total will not match and the data will be blocked from passage. Networks that can be accessed by dial-up lines should have a barrier, such as an encryption change, at the second entry port or interface of the access server.

Some antivirus data security software packages only identify changes being made to files, while others identify and remove viruses and repair the damage the viruses inflict. Boot sector viruses locate themselves on the first sector of a floppy disk, while file viruses invade files, particularly executable files. Sometimes File Allocation Tables (FATs), files, and directories can be recovered after a virus attack, but antivirus software can identify and eliminate viruses before the damage occurs. Some packages disinfect files, boot sectors, and memory without harming the infected portions of the system, while others are less sensitive. As yet, no product can guarantee complete protection against a virus attack, and new types of viruses are constantly being discovered.

Once viruses enter the network, they often begin by destroying crucial operating system files. Some virus protection programs can monitor the status of these files by periodically verifying byte count. Other programs can often isolate and detect an abnormal file condition, such as an excess byte count, but cannot resolve the problem. In these cases, companies must enlist the aid of skilled computer analysts and programmers to correct the problem. Since no product can completely secure the network, companies must strike a balance between data accessibility and the level of data protection needed to maintain security. Inevitably, cost will also become a consideration. Unfortunately, the cost of security measures can be difficult to justify, since the benefits of additional security (i.e., reduced exposure to security threats) cannot be predicted or directly measured until an attack actually occurs.

19.7   Firewalls

A *firewall* is a method of protecting one network from another untrusted network. The actual mechanism whereby this is accomplished varies widely, but in principle, the firewall can be thought of as a pair of mechanisms: one that blocks traffic and one that permits traffic. Some firewalls place a greater emphasis on blocking traffic, while others emphasize permitting traffic.

One way that firewalls protect networks is through packet filtering, which can be used to restrict access from or to certain machines or sites. It can also be used to limit access based on time of day or day of week, by the number of simultaneous sessions allowed, service host(s), destination host(s), or service type. This kind of firewall protection can be set up on various network routers, communications servers, or front-end processors.

Proxies are also used to provide secure out-bound communication to the internetwork from the internal network. The firewall software achieves this by appearing to act as the default router to the internal network. However, when packets hit the firewall, the software does not route the packets, but immediately starts a dynamic, transparent proxy. The proxy connects to a special intermediate host that actually connects to the desired service (see Fig. 19.2).

Proxies are often used instead of router-based traffic controls, to prevent traffic from passing directly between networks. Many proxies contain extra logging or support for user authentication. Since proxies must understand the application protocol being used, they can also implement protocol-specific security. For example, File Transfer Protocol (FTP) proxy might

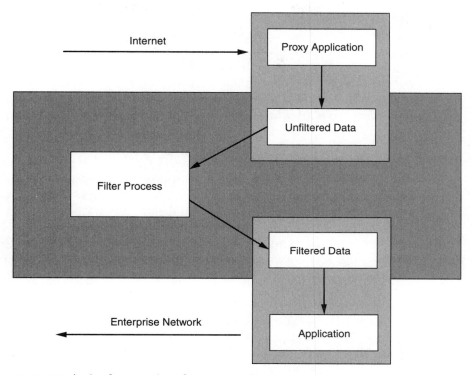

Figure 19.2 An implementation of a proxy application.

be configurable to permit incoming FTP traffic and block outgoing FTP traffic.

While firewall operation may look simple, it is quite complex in terms of initial configuration, fine-tuning, and ongoing administration. A firewall vendor might offer help with initial configuration and fine-tuning, but ongoing administration is the responsibility of the buyer. Modifying the attack detection parameters of a firewall to deal with new threats, for example, takes the knowledge and experience of a Certified Security Engineer (CSE), whereas changing a firewall rule set might require a Certified Security Administrator (CSA).

Since attacks can originate from anywhere at any time around the world, effective firewall operation requires 24 × 7 vigilance by expert staff. This means the IT organization must support three shifts of security personnel, which is expensive, or have someone on-call 24 hours a day, which can delay the response to threat situations. Either way, maintaining security staff is an expensive proposition. In fact, acquiring any level of management expertise is the biggest hidden cost of firewall ownership. The dearth of knowledgeable security personnel and the high salaries they command puts seasoned talent out of reach for many smaller companies. An effective and economical alternative, however, is a managed firewall solution that allows small and midsize firms to implement best-of-breed security solutions at a fixed monthly cost, and without the hassles of recruiting and retaining quality staff. These types of services could cost less than the equivalent salary of a half-time firewall administrator.

For effective security, a firewall solution will be required at each corporate location, including branch offices and the homes of telecommuters. Firewalls also protect the integrity of Virtual Private Networks (VPNs) and multicompany extranets. In addition to regulating traffic flow between public and private network environments, firewalls can be used to regulate traffic to and from internal company networks, such as the subnets of human resource, marketing, and legal departments.

A managed firewall service consists of hardware, software, consulting, monitoring, and management tools that continuously scan and analyze the vulnerability of an organization's Internet-connected systems. Firewall management usually must be ordered in conjunction with the provider's Internet services, including those provisioned over Asynchronous Transfer Mode (ATM), frame relay, Digital Subscriber Line (DSL), and integrated service offerings that combine voice and Internet services over the same access bandwidth. Assuming that the Internet service is already in place, a managed firewall solution can be up and running in 10 business days or less.

To configure a managed firewall appropriately, the service provider performs a comprehensive vulnerability analysis, starting with a port scan of the customer's network resources. A *port* is simply a place where information goes into and out of a device on the network, like a router or computer. Left unguarded, a port is a door through which a hacker can enter and gain access to other resources on the corporate network. After submitting the network to a battery of tests, sometimes using so-called hacker tools, the managed firewall service provider will present the customer with recommendations for fixing problems that have been identified. The recommendations will be codified in the form of rule sets that will be loaded into a firewall.

The firewall itself can be physically located at the company's location or the service provider's location. Regardless of location, the firewall will make decisions about what traffic to pass based on instructions contained in the rule set. Access control rules can be defined according to the source and destination of network traffic, specific applications, users or groups of users, and even time of day. If incoming traffic contains an executable file that has the signature of a known virus, for example, that traffic will not be allowed to pass beyond the firewall onto the corporate network where it can do harm when opened. The content security capabilities of the firewall can even spot suspicious Java applets and ActiveX controls, weed out undesirable Web content, and put limits on the size of files that are allowed onto the corporate network.

The service provider designs a firewall rule set in collaboration with the network manager, and there is usually a trial period to allow for minor changes to the rule set at no charge. During this period, as much as 14 days, real-world testing is performed and minor adjustments of the new firewall rule set are made, if necessary. Generally, customers may make unlimited changes during the trial period. After that, the service provider bills for further changes on a time and materials basis.

As new threats become known, the managed firewall service provider will take the appropriate course of action, which might entail adding a rule to the rule set or changing an existing port configuration on the firewall to thwart persistent access attempts. The changes are implemented remotely from the service provider's Network Security Operations Center (NSOC) over an encrypted Internet connection. If the customer's dedicated access connection is not available, perhaps due to an out-of-service transmission line or malfunctioning router, the service provider will use a dial-up connection to a modem attached to the firewall to upload the changes.

As an added precaution, the managed firewall service provider should take responsibility for maintaining backup copies of the customer's rule sets for locations, along with all the firewall passwords. A copy of the most recent router configuration might be kept as well, since this information is usually needed to reconfigure the firewall or router in case of a major system failure. It should take no more than four hours for the service provider to fully restore a firewall rule set and associated configuration files, assuming the dedicated connection is available or a dial-up modem link can be established.

The managed firewall service provider will usually be able to generate performance reports that can be accessed by the customer on a secure Web site using a browser that supports 128-bit key encryption. By entering a username and password, the customer can view high-level charts and graphs that summarize the quality of network and application resources. Comparative performance data on specific network resources and groups of resources should also be available.

Companies with branch offices, telecommuters, and mobile professionals should choose a service provider that can offer a range of security solutions, as well as connectivity services to suit the organization's various needs. A low-end firewall solution that protects corporate information stored on a telecommuter's PC, for example, might consist of firewall software loaded and configured in a DSL router. Since DSL connections are always on, hackers can come in through the Internet and mess with the personal data on a home computer as well as any corporate information it may hold. There is even the very real threat of hackers using the telecommuter's connection as a back door to launch an attack on the employee's company.

For companies with an installed base of Cisco routers, the provider may offer a combined firewall-router service that entails configuring the operating system's security features. For enterprise-level security, the choices might include Check Point Firewall 1. If separate devices provide firewall and router functionality, it might be preferable from a management standpoint to have both devices monitored by the same service provider.

Firewall implementation is not as simple as "set and forget." To maintain the highest degree of protection, security policy must be continually evaluated against the latest threats. For small and midsize companies, the value of a managed firewall service is in having an expert partner that will stay abreast of the latest developments and implement effective countermeasures to prevent unauthorized access to the organization's valuable network resources.

It is important to choose a service provider that will maintain contact with various network security watch groups, such as the Computer Emergency Response Team (CERT) Coordination Center and the National Infrastructure Protection Center (NIPC), to stay abreast of the latest security problems reported by the user community and the remedies proposed by the vendor community. This information helps the managed firewall service provider deal effectively with new types of security threats and become familiar with the associated attack profiles so the customer's firewall policies can be updated in an appropriate and timely manner.

19.8 Remote Access Security

With an increasingly decentralized and mobile workforce, organizations are coming to rely on LANs that provide remote access through a communications server that can be reached with a toll-free number or a series of local dial numbers. From time to time, telecommuters, traveling executives, salespeople, and remote offices all need access to the various resources that reside on headquarters' LANs and subnets. This calls for appropriate security measures to prevent unauthorized access to the corporate network.

19.8.1 Security measures

Depending on the size of the network and the sensitivity of the information that resides there, one or more of the following security methods can be employed:

- *Authentication.* This involves verifying the remote caller by user ID and password, thus controlling access to the server. Security is enhanced if ID and password are encrypted before going out over the communications link.

- *Access restrictions.* This involves assigning each remote user a specific location (i.e., directory or drive) that can be accessed on the server. Access to specific servers also can be controlled.

- *Time restrictions.* This involves assigning each remote user a specific amount of connection time, after which the connection is dropped.

- *Connection restrictions.* This involves limiting the number of consecutive connection attempts and/or the number of times connections can be established on an hourly or daily basis.

- *Protocol restrictions.* This involves limiting users to a specific protocol for remote access.

Among the most popular remote access security schemes is the Remote Access Dial-In User Service (RADIUS). Under this scheme, users are authenticated through a series of communications between the client and the server. When the client initiates a connection, the communications server puts the name and password into a data packet called the *authentication request,* which also includes information identifying the specific server sending the authentication request and the port that is being used for the connection. For added protection, the communications server, acting as a RADIUS client, encrypts the password before passing it on to the authentication server.

When an authentication request is received, the authentication server validates the request and decrypts the data packet to access the username and password information. If the username and password are correct, the authentication server sends back an authentication acknowledgment that includes information on the user's network system and service requirements. The acknowledgment can even contain filtering information to limit the user's access to specific network resources.

19.8.2   Callback security systems

Callback security systems, which are commonly used with password and/or ID security schemes, control remote dial-up access to hosts and LAN servers via modems. Typically, these systems use an interactive process between a sending and receiving modem. With callback security, the answering modem requests the caller's identification, disconnects the call, verifies the caller's identification against the user directory, and then calls back the authorized modem at the number matching the caller's identification.

Callback security ensures that data communication occurs only between authorized devices. The degree of security that callback modems provide is questionable due to the widespread use of telephone functions such as call forwarding. It is also possible for a knowledgeable intruder to stay on-line and intercept the return call from the answering modem. In addition, some data security centers may find callback security inappropriate for their networks, since it assumes that users always call from the same telephone number. Traveling employees and users who must connect to the LAN through a switchboard cannot use this technique.

Many popular modems can generate acoustic signals to indicate to the callback modem that a call was received and correctly linked. Administrators can also add callback modems to an established security system without disrupting existing procedures. Neither passwords nor callback modems provide complete security for remote communications, however. The difficulties inherent in remote authentication make a combined security method imperative. As a result, administrators often find using two identification methods (e.g., callback modems combined with data encryption) more effective than using a single method.

Other techniques are in use to enhance the security of modem access. Among them is the use of an ASCII password that is typically entered from a remote keyboard or automated log-on file. This method offers minimal security, since it is easy for computer hackers to crack the code using random number generators and launch attacks on system dictionaries. Moreover, it is limited to relatively slow asynchronous communications. The modems must establish the connection and enter the "pass data" mode, which typically takes 4 to 12 seconds on high-speed modems, before passwords can be verified.

A more reliable approach combines password security with the callback feature. With this technique the password is verified and the inbound call is disconnected. The security system then calls the remote user at a predetermined telephone number to enter the "pass data" mode. If a hacker breaks the password, the security system will call back a secure location, thereby denying the hacker access. This system is effective, but very slow, and is limited to a finite number of stored telephone numbers.

These approaches rely on establishing links to a central site device such as a front-end processor or communications controller. In all cases, security is implemented via American Standard Code for Information Interchange (ASCII) passwords or Dual Tone Multi Frequency (DTMF) tones after the modem handshake. A tenacious hacker will eventually break the security codes. To prevent this, security procedures can be implemented before the modem handshaking sequence, rather than after it. This effectively eliminates the access opportunity from potential intruders. In addition to saving time, this method uses a precision high-speed analog security sequence that is not even detectable by advanced line monitoring equipment.

With callback, the remote client's call is accepted, the line is disconnected, and the server calls back after checking that the phone number is valid. While this works well for branch offices, most callback products are not appropriate for mobile users whose locations vary on a daily basis. However, there are products on the market that accept roving callback numbers. This feature allows mobile users to call into a remote access server or host computer, type in their user ID and password, and then specify a number where the server or host should call them back. The callback number is then logged and may be used to help track down security breaches.

To safeguard very sensitive information, there are third-party authentication systems that can be added to the server. These systems require a user password and also a special credit card–size device that generates a new ID every 60 seconds, which must be matched by a similar ID number-generation process on the remote user's computer.

In addition to callback and encryption, security can be enforced via Internet Protocol (IP) filtering and log-on passwords for the system console and for Telnet and FTP server programs. Many remote-node products also enforce security at the link level using the Point-to-Point Protocol (PPP) with the Challenge Handshake Authentication Protocol (CHAP), and Password Authentication Protocol (PAP).

19.8.3 Link level protocols

When peers at each end of the serial link support the PPP suite, more sophisticated security features can be implemented. This is because PPP can integrally support the PAP or CHAP to enforce link security.

PPP is a versatile Wide Area Network (WAN) connection standard for tying dispersed branch offices to the central backbone via dial-up serial links. It is actually an enhanced version of the older Serial Line Internet Protocol (SLIP). SLIP is recommended for an IP-only environment, while PPP is recommended for non-IP or multiprotocol environments. Since PPP is protocol-insensitive, it can be used to access both AppleTalk and Transmission Control Protocol/Internet Protocol (TCP/IP) networks, for example.

PPP framing defines how data is encapsulated before transmission over the WAN. It supports multiple network-layer protocols, including TCP/IP and Internet Packet Exchange (IPX). PPP also offers remote protocol configuration, the capability to define the framing format over the wire, and password authentication.

PAP uses a two-way handshake for the peer to establish its identity. This handshake occurs only during initial link establishment. The peer repeatedly sends an ID-password pair to the authenticator until verification is acknowledged or the connection is terminated. However, network intruders send passwords over the circuit in text format, which offers no protection from playback.

CHAP periodically verifies the identity of the peer using a three-way handshake. This technique is employed throughout the life of the connection. With CHAP, the server sends a random token to the remote workstation. The token is encrypted with the user's password and sent back to the server. Then the server does a lookup to see if it recognizes the password. If the values match, the authentication is acknowledged; otherwise, the connection is terminated. Every time remote users dial in, they are given a different token. This provides protection against playback because the challenge value changes in every token.

Some vendors of remote-node products support both PAP and CHAP, while older, low-end products tend to support only PAP, which is the less robust of the two authentication protocols.

19.9   Policy-Based Security

With today's LAN administration tools, security goes far beyond mere password protection to include implementation of a policy-based approach characteristic of most mainframe systems. Under the policy-based approach to security, files are protected by their description in a relational database. This means that newly created files are automatically protected, not at the discretion of each creator, but consistent with the defined security needs of the organization.

Some products use a graphical calendar through which various assets can be made available to select users only during specific hours of specific days. For each asset or group of assets, a different permission type may be applied: Permit, Deny, and Log. *Permit* allows a user or user group to have access to a specified asset. *Deny* allows an exception to be made to a Permit, for example, not allowing writes to certain files. *Log* allows an asset to be accessed, but stipulates that such access will be logged.

Although the LAN administrator usually has access to a full suite of password controls and tracking features, today's advanced administration tools also provide the ability to determine whether or not a single login ID can have multiple terminal sessions on the same system. The LAN administrator also can specify an enforcement action to be taken when a user's login ID exceeds the system limit for violations, such as:

- *Cancel.* The access attempt is denied and the process that attempted the unauthorized access is canceled.

- *Log out.* The access attempt is denied and the process group and all child processes associated with it are canceled. If a logged-in user is associated with the attempt, he or she also will be logged out.

- *Suspend.* The access attempt is denied and the process group and all associated child processes are canceled. In addition, the login ID is suspended and the user will be locked out of the system until the lockout is explicitly lifted by the LAN administrator.

Through the console, the LAN manager can review real-time and historical violation activity online, along with other system activity.

19.10   Security Planning

With security issues becoming increasingly important among businesses of all types and sizes, it is a good idea to have a security plan in place. The plan's extensiveness is directly proportional to the network's complexity. At a minimum, the plan should address the following:

- Access levels and definitions

- Work groups and how they fit into the network's access rules

- Rules for access, such as password policy and disk volume access

- Physical installation and access guidelines, including approved add-in security protection

- Accountability guidelines

- Internet connections and access requirements

- Levels of administration and accountability

- Periodic auditing policy and software upgrade, auditing, and installation policies

The security plan should have the power of a corporate policy, signed and authorized by the highest level of management. Department heads should administer the plan. Every user should be required to read and sign off on the plan. Security plans should be reviewed and revised at least annually, perhaps more frequently if employee turnover is a problem.

The ideal network security plan must protect against theft, alteration, and interception of network information. It should also deter theft or vandalism of equipment, the duplication or resale of programs, and unauthorized use of computer time via wiretap or dial-in. Companies should first implement procedures to deter unauthorized users who may be tempted to breach the system. When properly implemented, such deterrents bolster confidence in the integrity of the organization, ensuring management, staff, and customers that the system is safe and dependable. When deterrence fails and an intruder attempts to breach the system, preventive techniques, including access controls, data encryption, interface barriers, and real-time network-intrusion detection guard against further damage.

Companies should choose a security method that offers password control, data encryption, alarm notifications of potential security breaches, and audit tools to track down the source of a breach. Companies should design a security plan that can expand to accommodate network growth. Because no security system can function effectively without the cooperation and support of end users, network support personnel, and upper-level management, companies should evaluate the effects that additional security measures will have on authorized users.

Network users usually resist improvements in network security, because security devices and controls can make a complex system even more difficult to operate and maintain. Trusted employees often resent having to use passwords to access the network. For these reasons, some security measures may cause delays and reduce productivity. Companies should expect resistance to added rules and regulations and counter this by providing security awareness training. As the network manager, with the aid of the human resources department, you must sell the long-term employee on the idea of additional security by implementing bulletin board campaigns and training sessions. At a minimum, security awareness training should include:

- A description of tangible security measures, such as locks, keys, card systems, and badges

- An explanation of password management, including password selection, access privileges, routine password changes, and the need to maintain the confidentiality of corporate information

- Definitions of sensitive data and procedures for keeping data confidential

- Procedures for reporting lost or stolen data, software, and hardware

- Mechanisms for encouraging and implementing employee suggestions for improving security

Companies should also develop security demonstrations, possibly in the form of a game in which one employee is assigned to penetrate the system while others try to prevent or detect the attack. Games demonstrate how well the system works and involve staff members with the security system. Whatever the result of the game, the company wins. If the attempt fails, the security system is performing effectively. If a security breach occurs, the company has identified a weakness and can take appropriate corrective measures.

Once an acceptable level of security awareness exists throughout the organization, the company should schedule periodic retraining sessions to maintain awareness. All aspects of a security system—the hardware, software, premises, facilities, and personnel—must work in unison and at a consistently high level to safeguard against threats to security.

19.11 Conclusion

Many companies incorrectly assume that they do not need to address security and related management issues because they do not have irreplaceable data and applications—after all they have backup systems in place that can instantly retrieve lost or corrupted files. Other companies rely on their low profile and relative anonymity to hackers as the means of avoiding security breaches. These attitudes ignore the fact that 80 percent of security breaches come from current and former employees, who may know the company's vulnerabilities and may be willing to exploit that knowledge to get even with their employer for some perceived wrong.

To protect valuable information, companies must establish a sound security policy before an intruder has an opportunity to violate the network and do serious damage. This means identifying security risks, implementing effective security measures, and educating users on the importance of security procedures. There is no way to bypass the human element—a commitment to secure computing and networking must be made at all levels of the organization if security provisions are to have the desired effect.

Ignoring security issues can result in information theft, malicious file tampering, and snooping by intruders inside and outside of the organization. In trying to deal with the consequences of leaked information, a company can experience immediate financial loss and, in the long term, damage to its competitive position. Depending on the nature of the security breach and the kind of information involved, such laxity can even expose the company to litigation, forced settlements, and government fines.

These consequences of lax security only scratch the surface. Not only are cyber-criminals constantly inventing new ways to steal information stored in electronic form, they are getting more nefarious in their use of it, making the theft of credit card information look like child's play in comparison. On the increase are cases of corporate blackmail, stock price manipulation, identity theft, and entering false records to make computers print out checks for bogus services.

Few companies would dispute the need for a secure enterprise network, particularly with the growing number of telecommuters and mobile professionals who need remote access to the corporate network, typically through Internet connections. But security is a full-time endeavor that requires a proactive approach, specialized technical staff, and appropriate tools. Organizations that lack these resources should consider outsourcing security to a qualified third-party firm.

Business Continuity Planning

20.1 Introduction

Business continuity is a proactive process that seeks to identify the key functions of an organization and the likely threats to those functions. From this information, plans and procedures can be developed to ensure that key functions can continue in the event of an emergency or disaster. For most companies, telecommunications systems and data centers are among the critical assets that must be addressed in the business continuity plan, since they are the platforms that support such necessary functions as applications usage, workflow, messaging, and other processes and services that are essential for participating in global commerce today. For this reason, network managers will play a pivotal role in developing the business continuity plan.

Associated with continuity planning is disaster recovery planning, which entails the development of an action plan to be followed in the event of a disaster or emergency that threatens to disrupt or destroy the continuity of normal business activities and which seeks to restore operational capabilities. Disaster recovery planning is not just meant to ensure survival in the face of a major event such as a fire, flood, or earthquake; it is equally important for ensuring smooth day-to-day operations in the face of telecommunications and information system failures. This type of planning is especially important in this age of "convergence," which has companies putting all their eggs—voice, data, and video—in one network basket. In such environments the downtime risks and resulting revenue losses can be quite substantial. To effectively deal with business-disrupting events, disaster recovery must be a major part of the continuity plan.

It has long been understood among companies that planning for disaster recovery is akin to buying life insurance—the expected rate of return is negative. This is often the reason companies put off the decision to invest in the systems, tools, and skills to develop and implement disaster recovery plans. Without this kind of insurance, however, the consequences can be quite severe. According to a 1995 study by the University of Texas, of companies that suffered a catastrophic data loss, 43 percent never reopened and 51 percent closed within two years. Only 6 percent survived.

Of course these figures have improved since then, largely because there is more awareness of the need for continuity planning, specifically, disaster recovery planning. To foster awareness, there are several industry consortia that address business continuity planning. In addition, numerous integrators, consultants, enterprise software vendors, and telecommunications carriers and other types of service providers provide business continuity planning services. There are even tools that facilitate in-house development of business continuity plans.

Companies that do not have a comprehensive business continuity plan leave the door wide open to disruptions in daily business operations, legal action from customers or strategic partners, and financial losses. If the adverse event is not properly addressed for a prolonged period of time, the enterprise may experience erosion of confidence among customers and stakeholders, which may result in diminished sales and eventually to loss of market share. The loss of business for Arthur Andersen, related to accounting irregularities associated with Enron and WorldCom, demonstrates what can happen when confidence in a company erodes to such low levels. Although Arthur Andersen's troubles did not result from the failure of its information systems or networks, its experience clearly demonstrates that once confidence in a company is severely shaken, it is very difficult to win back.

20.2   Planning Process

The first step in developing a business continuity plan is to identify and consider the potential impacts of each type of disaster or event. This is important because an organization cannot engage in proper planning for a disaster or event if no thought has been given beforehand to the likely impacts of various scenarios. Therefore, what has to be done first is generally referred to as *business impact analysis*.

Fundamentally, business impact analysis is a means of systematically assessing the potential consequences that could result from various events or incidents that have the potential of leaving systems or networks unavailable. Impacts resulting from other types of incidents, such as breach of confidentiality or loss of data integrity, are simultaneously explored, but need not be the focus when only considering business continuity planning or disaster recovery. The need to maintain the confidentiality of sensitive information, for example, can be addressed during orientation sessions for new employees and periodically reinforced with email messages. Data integrity is usually addressed within the framework of relational database management systems through such mechanisms as the rollback feature and real-time synchronization of data between multiple sites.

The business impact analysis is intended to help managers understand the degree of potential loss that could occur—not just the loss of systems and networks, but direct financial loss and the loss of customer and investor confidence, damage to reputation, exposure to litigation, and regulatory sanctions. The lack of a business continuity plan could be interpreted as carelessness, and leave the company less able to defend itself in court if the disaster or event caused damage to customers or partners.

Having determined the impacts, it is also important to consider the magnitude of the risks that could result in these impacts. This is a critical activity because it will determine which scenarios are most likely to occur and which should attract the most attention during the planning process. Various methodologies have emerged to simplify these tasks, which are available as templates that can be purchased on the Web and downloaded for immediate use. The templates are simply interactive Microsoft Word-based applications that guide network managers through the process of creating a business continuity plan.

The value of templates is that they provide a ready framework, which can be customized to suit the particular needs of virtually any business, thus getting the planning process off to a fast start. The templates provide guidelines that cover all functions required in the business continuity planning process. These include preparing detailed business risk assessments, developing strategic plans to mitigate potential crises, procedures to handle the disaster recovery and business recovery phases, separate phases for testing and training in simulated conditions, and instructions for keeping the plan up-to-date.

As noted, the first step in the continuity planning process is to assess the risks faced by the company and determine how exposed the company is to those risks, particularly those that threaten mission-critical processes. With increasing reliance on computer and telecommunications systems and services for mission-critical applications, these will be among the first assets to come under scrutiny.

The next step is to decide what measures can be put in place to minimize damage if a disaster does occur that disrupts computer and telecommunications systems and services. At this stage the elements of the continuity plan can be compiled into a document. Before the plan can be considered as complete, it must be tested to reveal any weaknesses. If weaknesses are revealed, the plan must be fine-tuned until it passes the test.

As more companies put important applications on their intranets and provide access to the public Internet for their employees, the need for continuity planning has never been more important. Protecting internal networks from unauthorized access and potential problems from viruses, and ensuring the confidentiality of electronic transactions are among the major challenges for businesses of all types and sizes. Therefore, continuity planning could reveal the need for authorization, encryption, firewalls, and Internet tunnels as solutions that need to be considered to thwart possible intrusions and attacks that can render these assets unusable.

Once the plan has tested successfully, it must not be allowed to gather dust. Plans should be reviewed periodically and changed as often as necessary to accommodate new corporate developments.

20.3 Staff Roles

Standards for continuity planning were prepared in 1997 jointly by the Business Continuity Institute and Disaster Recovery Institute International. The topics covered include project initiation and management, risk evaluation and control, business impact analysis, development of recovery strategies, emergency response, development and implementation of the plan, corporate awareness programs and training, and procedures to test, exercise, maintain, and update the plan. The following discussion adapts this model to the role of network managers in developing an appropriate business continuity plan to protect telecommunications systems and mission-critical information systems of the enterprise.

20.3.1 Overall responsibilities

Along with other corporate managers, senior telecom and Information Technology (IT) professionals are largely responsible for the design, development, and implementation of the business continuity plan and for ensuring that recovery of critical systems and networks occurs within the agreed-upon time parameters. As network manager, your role is to:

- Identify the components of the planning process
- Control the planning process and produce the plan
- Obtain executive-level approval and commitment to the plan
- Implement the plan
- Test the plan
- Maintain the plan to accommodate corporate changes

20.3.2   Project initiation and management

To establish the need for a continuity plan, including the disaster recovery component, your role as a network manager is to:

- Coordinate and organize the project
- Take the lead in defining objectives, policies, and critical success factors
- Present the project proposal to senior management and staff
- Develop the project plan and budget
- Define and recommend project structure and management
- Manage the planning process

20.3.3   Risk evaluation and control

Network managers must identify the events that can adversely affect the business processes of the organization, the potential damage such events can cause, and the tools and systems needed to prevent or minimize the effects of potential loss. The development of a cost-benefit analysis can justify investments to mitigate risks. Specifically, your role as network manager is to:

- Understand the function of risk reduction within the organization
- Identify potential risks to the organization
- Identify outside expertise required
- Identify vulnerabilities, threats, and exposures
- Identify risk reduction alternatives
- Identify credible information sources
- Interface with other corporate managers to determine acceptable risk levels
- Document and present findings

20.3.4   Business impact analysis

As a network manager, you must be able to predict the possible impacts that can result from disruptions to business operations, which can be done by formulating the most likely disaster scenarios that can affect the organization. Once critical functions are identified, recovery priorities can be established to minimize loss. Specifically, your role is to:

- Identify knowledgeable and credible functional area personnel. Identify and define criticality criteria.

- Present criteria to management for approval.

- Coordinate analysis.

- Identify interdependencies.

- Define recovery objectives and time frames, including recovery times, expected losses, and priorities.

- Identify information requirements.

- Identify resource requirements.

- Define management report format.

- Prepare and present business impact analysis.

20.3.5   Develop business recovery strategies

Network managers should be able to assist in determining and selecting alternative business recovery operating strategies that can be implemented within specific time objectives, while maintaining the organization's critical functions. In this area, your role is to:

- Understand available alternatives, their advantages and disadvantages, as well as the cost ranges of each

- Consolidate strategies, taking into consideration different business functional areas

- Identify off-site storage requirements and alternative telecommunications facilities

- Develop business unit consensus on recovery strategy

- Present strategies to management to obtain commitment

20.3.6   Emergency response

Attention must be given to developing and implementing procedures for responding to and stabilizing the situation following an incident or event, including establishing and managing an emergency facility to be used as a command center during the emergency. Depending on the organization, this facility might be the help desk or the network operations center. With regard to emergency response issues, your role as network manager is to:

- Identify potential types of emergencies (e.g., an attack from the Internet, failure of a carrier's frame relay network, loss of the email server or the entire data center, prolonged commercial power outage) and the responses that would be required

- Identify the existence of appropriate emergency response procedures

- Recommend the development of emergency procedures where none exist

- Integrate disaster recovery and business continuity procedures with emergency response procedures

- Identify the command and control requirements of managing an emergency

- Recommend the development of command and control procedures to define roles, authority, and communications processes for managing an emergency

- If appropriate, ensure that emergency response procedures are integrated with requirements of public authorities

20.3.7   Awareness and training programs

Network managers should participate with other corporate managers in preparing a program to create corporate awareness and enhance the skills required to develop, implement, maintain, and execute the continuity plan. Your role is to:

- Establish objectives and components of the training program with the human resources department or training group

- Identify functional training requirements

- Develop training methodology

- Acquire or develop training aids

- Identify external training opportunities

- Identify vehicles for corporate awareness, which can include periodic reminders sent via email

20.3.8   Maintenance and testing

Network managers should take the time to arrange and schedule tests of the continuity plan, and evaluate and document the results. Processes must be developed to maintain the currency of the plan and associated documentation in accordance with the organization's strategic direction. As a network manager, your role is to:

- Plan and coordinate testing of the plan
- Evaluate the test plan
- Perform the test
- Document and evaluate the results
- Report the results to corporate management
- Attend periodic strategic planning meetings
- Coordinate plan maintenance
- Fine-tune the plan as necessary and overhaul it in the event of a corporate acquisition or merger

20.3.9   Crisis coordination

Network managers can play a key supporting role in developing, coordinating, and evaluating plans to handle the media during crisis situations. Their technical knowledge can go a long way toward lending credibility to explanations of the problem and its solution. In turn, this can provide key customers, critical suppliers, strategic partners, and other stakeholders with assurance that the problem is understood and is being addressed in the most expeditious manner possible. An executive-level person, such as the Chief Information Officer (CIO) or Chief Technology Officer (CTO) should be the one who actually interfaces with the media. Senior network managers should play a supporting role and will have the responsibility to:

- Participate in the corporate public relations program for proactive crisis management
- Establish necessary crisis coordination with appropriate external groups, such as the Community Emergency Response Team/Coordination Center (CERT/CC), which acts as a clearinghouse for security matters
- Assist in establishing essential crisis communications with relevant stakeholder groups

20.3.10   Coordination with local authorities

Network managers should establish appropriate procedures and policies for coordinating service and power restoration activities with local authorities. This should be done in conjunction with the corporate public relations department. As network manager, your role is to:

- Coordinate emergency preparations, response, recovery, resumption, and restoration procedures with local authorities

- Establish liaison procedures for emergency and/or disaster scenarios

20.4 Security Risk Assessment

With network security of paramount concern among IT professionals, the continuity plan must include provision for scanning the network periodically to assess its vulnerabilities. Should hackers with malicious intent gain access to corporate resources by entering through open ports, for example, important servers could be disabled, operating systems could be tampered with, and databases wiped out. Because of the potential for catastrophic damage, network security has catapulted to the highest priority in the risk assessment process.

20.4.1 Layered approach

The inclusion of security risk assessment in business continuity planning requires that organizations adopt a layered approach that brings together technologies, expert support, and real-time intelligence and that addresses all the essential phases and processes of information risk management. This layered approach allows organizations, particularly large enterprises, to build and assure a continuously effective security posture.

Network managers must first measure risks to critical information assets. This can be done by:

- Identifying critical data, networks, applications, devices, and users

- Creating a detailed asset inventory

- Assessing and prioritizing risks

Next, the risks must be managed with a comprehensive security program. As a network manager, you can do this by:

- Creating a multidisciplined risk reduction plan

- Driving implementation with expert support

- Validating successful risk mitigation

- Defending networks with outsourced real-time security systems

Finally, risks must be continuously monitored so that the response to threats can be immediate. As a network manager, you can do this by:

- Tracking and analyzing emerging security risks
- Providing ongoing risk analysis and mitigation support
- Reporting compliance with essential security standards
- Monitoring the network on a 24×7 basis and responding to real-time threats

This layered approach to comprehensive security provides the essential foundation for companies that need to protect their information assets. It addresses security architecture and policies; the implementation of a preventive risk management program; and the management, maintenance, and real-time monitoring of critical security infrastructure.

20.4.2   Security benchmarks

A long-neglected element of Internet security has been the absence of useful, widely accepted, nonproprietary benchmarks for specifying in greater detail how systems should be configured and operated. This situation has been remedied by the Center for Internet Security (CIS), which has sponsored the development of measurable benchmarks based on recognized best practices. Until these benchmarks are widely used, however, IT managers may be forced to guess the answers to such important questions as the following:

- How can corporate information systems be made sufficiently reliable and secure, based on the cost of security measures versus the value of operating reliable systems?
- What method can be used to determine the minimum level of due care, based on best practice benchmarks, to reduce risk to an acceptable level?
- Who can I trust to provide information that will help protect my systems and networks?

Open security-enhancing benchmarks incorporate the knowledge of a wide range of recognized best-practice organizations and experts. As such, they are a more efficient and less costly alternative to each organization having to reinvent the wheel to determine how to implement effective security actions. According to the CIS, this approach helps compensate for the shortage of information security personnel in relation to the increasing demand for their expertise. By using the CIS benchmarks and scoring tools, organizations can expect to benefit in the following ways by:

- Reducing the likelihood of successful intrusions or attacks
- Verifying secure configuration of systems prior to network deployment

- Monitoring systems for ongoing conformity with CIS benchmark security configurations

- Generating reports for management to show how system security measures stack up against the benchmark

- Creating a framework of information security accountability and reporting that is shared by top managers, security professionals, system administrators, auditors, security consultants, and vendors

- Protecting against prosecution or regulatory sanctions by demonstrating compliance with an accepted prudent due care security standard

The CIS benchmarks spell out security configuration settings and actions that harden information systems against attack. They are unique because they represent the consensus among hundreds of security professionals worldwide. Two sets of benchmarks are available from the CIS at no cost. Level-I benchmarks are tuned for the prudent level of minimum due care, while Level-II benchmarks are tuned for prudent security beyond the minimum level.

Level-I benchmark settings and/or actions meet the following criteria:

- System administrators with any level of security knowledge and experience can understand and perform the specified actions.

- The action is unlikely to cause an interruption of service to the operating system or the applications running on it.

- The actions can be automatically monitored and the configuration verified by scoring tools that are available from the CIS or by CIS-certified scoring tools.

The CIS scoring tools provide a convenient way to evaluate systems and networks by comparing their security configurations against the CIS benchmarks. The scoring tools also create reports that provide guidance to system administrators on how to secure both new installations and systems currently in operation. The tool is also effective for monitoring systems to assure that security settings continuously conform to CIS benchmark configurations.

The Level-II security configurations vary depending on network architecture and server functionality. These benchmarks are intended for system administrators who have more granular security knowledge and can apply that knowledge to the operating systems and applications running in their particular environments. Among the systems for which benchmarks and scoring tools are available are those running HP-UX, Linux, Cisco IOS, Windows 2000, and Sun Solaris.

20.4.3   Security information resources

One of the most feared disasters among IT professionals is the loss of information systems from attacks by hackers who come into the corporate network from the Internet. One of the most effective ways to stay informed of potential threats of this kind is to make use of the services provided by the CERT/CC. This federally funded organization consists of network security experts who provide 24×7 technical assistance for responding to computer security incidents.

The CERT/CC charter is to work with the Internet community to facilitate its response to computer security events involving Internet hosts, to take proactive steps to raise the community's awareness of computer security issues, and to conduct research targeted at improving the security of existing systems. CERT/CC services also include product vulnerability assistance, technical documents, and seminars.

CERT/CC issues advisories—documents that provide information on how to obtain a patch or details of a workaround for a known computer security problem. It also works with vendors to produce a workaround or a patch for a problem, and does not publish vulnerability information until a workaround or a patch is available. A CERT advisory may also be a warning about impending or ongoing attacks.

Vendor-initiated bulletins contain verbatim text from vendors about a security problem relating to their products. These bulletins include enough information for readers to determine whether the problem affects them, along with specific steps that can be taken to avoid problems. The purpose behind these bulletins is to help the vendors' security information get wide distribution quickly.

The CERT Coordination Center provides seminars to help managers understand what needs to be done to ensure that their computer systems and networks are as securely managed as possible when operating within the Internet community. Attendees are provided with information that enables them to formulate realistic security policies, procedures, and programs specific to their operating environment.

20.5   Outsourcing Business Continuity

For organizations that do not have the internal resources to devote to business continuity planning, there is the option of outsourcing this process to a qualified software-centric enterprise solutions provider, telecommunications carrier, or a storage service provider. Network managers should be aware, however, that the primary purpose behind many of these outsourcing arrangements is to enable the providers to sell more of their products or services. For them, business continuity planning is simply a way to address this emerging market and generate new revenue. They begin the process of business continuity planning from the perspective of their own product or service portfolio.

20.5.1   Enterprise solutions providers

Software-centric enterprise solutions providers that address the business continuity market tend to showcase various combinations of their own products which, when put together by their experts, minimize the risk of business interruptions and protect the integrity of operations under threatening conditions. For example, the vendor may offer its enterprise management platform, security framework, and data storage-recovery systems, and tie them together with expert services. These product combinations are then marketed in such a way as to address business continuity best practices—from needs assessment and planning, to continuity command and control. They may also incorporate technology designed to provide a centralized common view of operational information, enabling organizations to proactively identify exposures and organizational threats and to create appropriate recovery plans.

Ostensibly, this leveraged approach enables IT organizations to gain significant additional value from their existing investments in the vendor's products. To reach as many customers as possible with the broadest range of business continuity expertise, enterprise solutions providers also work with their certified partners, including systems integrators and risk management specialists, to supplement their own technology implementation teams.

For existing customers, enterprise solutions providers offer a dynamic technology infrastructure, which is becoming essential for effective continuity planning, especially with infrastructure, applications, and processes rapidly evolving in today's highly fluid business environment. Preplanned emergency responses executed out of procedure books and other static representations of the enterprise have outgrown their usefulness. The leveraged approach adopted by enterprise solutions providers may offer a more effective alternative by creating an active representation of the enterprise that changes along with information systems, and encourages periodic review and fine-tuning.

20.5.2   Telecommunications carriers

For companies that need help approaching business continuity from the network perspective, telecommunications carriers—regional and national— offer packages of services designed to prevent catastrophic failure and ensure prompt recovery if service disruption does occur. Among the ways carriers achieve these goals is by offering off-site storage of customer configuration information, batteries and Uninterruptible Power Supplies (UPS) at the central office, diverse routing, redundant facilities, physical security and asset protection for collocation arrangements, secure system access and protection routines, and crisis management response.

While telecommunications carriers bring expertise in network disaster recovery services for critical voice, data, and video applications, they partner with other companies to provide alternative site recovery solutions for data centers, consulting services, and planning tools. The integration of carrier-provided services with those of its partners gives customers a more complete continuity solution that can be customized to meet unique business needs. In addition to alternative site recovery, such services may include voice and data redirect services and other services to protect critical elements of the customer's network, as well as continuity consulting services, continuity software planning tools, and mobile recovery solutions.

Most carriers offer a standard set of disaster recovery solutions that protect mission-critical applications, which constitute a key element of customers' business continuity plans. Dial backup for remote sites, for example, is a fairly standard offering among carriers, which protects remote locations against network failure. A reserved bandwidth option is available that protects the primary corporate site in both frame relay and private line networks. And then there is the option of ordering backup Permanent Virtual Circuits (PVCs) that allow companies to switch individual frame relay connections to an alternative data center if the primary data center goes down. Integrated Services Digital Network (ISDN) can also be used to back up frame relay PVCs, provided that the network manager had the foresight to deploy routers equipped with this capability. With Cisco routers, for example, if routing information is passed across the frame relay link, a floating static route can bring up the backup ISDN link if the frame relay link stops passing information.

20.5.3   Storage service providers

There is a growing movement to outsource storage and associated management requirements to a new class of supplier called the Storage Service Provider (SSP), which sells storage capacity as a pay-per-use utility. The SSP offers companies the means to store and access data without the expense of buying and maintaining hardware and software. The SSP also performs crucial functions such as easy expansion, data backup and restoration, security, disaster recovery, and capacity planning. The monthly cost of going with an SSP is ostensibly more attractive than establishing an in-house storage infrastructure, which can entail hundreds of thousands of dollars to establish a multisite installation, plus the ongoing cost of staff to manage the operation on a 24×7 basis. Storage-intensive enterprises should consider storage service providers in their business continuity plans.

For most organizations, data storage is not a core competency. Emerging Internet-related businesses are especially challenged, given their skyrocketing data storage requirements, combined with limited capital and human resources. Furthermore, network managers are under pressure to build a single infrastructure that allows their company to respond quickly to new competition, deal effectively with massive information growth, and generate new e-commerce revenue streams quickly. The infrastructure must be able to incorporate new storage technologies that allow interoperability with diverse platforms within the enterprise and among strategic partners. Not only is the procurement of hardware, bandwidth, and management tools costly, even when experienced IT professionals with storage and networking expertise are found, they are increasingly expensive and difficult to retain.

In response to these issues, SSPs emerged in 1999, creating a whole new market, which is being driven by three convergent trends. First, the level of expertise required for managing storage at the enterprise level has escalated beyond what many companies can afford. By outsourcing storage and management to an SSP, organizations are able to gain access to storage expertise at a more reasonable cost over the do-it-yourself approach.

Second, SSPs leverage Storage Area Network (SAN) architectures to establish enterprisewide networked storage. This widely accepted approach allows data to be delivered across the enterprise from a remote location that is managed by an SSP. Companies can then implement new technologies as needed, as opposed to limited upgrades because of distributed architectural and management complexity.

Finally, SSPs leverage connectivity capabilities resulting from Fibre Channel and new Infiniband protocols for multi-gigabit-per-second Input/Output (I/O) throughput. These architectures even allow gigabit-per-second transmission rates across metro area distances, allowing "local" storage performance levels from remotely administered facilities. With SSPs operating their own Wide Area Networks (WANs), storage facilities are linked up to enable seamless access to enterprise data from any corporate location.

SSPs address a variety of basic enterprise storage needs, starting with advising customers on enterprise storage strategies as part of business continuity planning. Most SSPs have a consulting capability or professional services unit that performs needs assessment and recommendations for on-campus storage needs as well as remote facility implementations.

In addition to providing storage space, SSPs provide customers with backup, disaster recovery, vaulting, migration, and hierarchical storage management solutions. For organizations with very high up-time requirements or multisite demand for data, SSPs provide a cost-effective alternative for heterogeneous data sharing, data replication, and multisite data consolidation and synchronization.

SSPs offer important capabilities that may be considered during business continuity planning. They offer services that provide organizations with managed second-site remote data replication, ensuring zero data loss and 24 × 7 data availability. They help customers determine and achieve the service levels necessary to meet their business requirements for continuity, flexibility, and data replication. The SSP also acts as a central source of support for maintenance and upgrades.

SSPs support both SAN and Network Attached Storage (NAS) environments via synchronous replication within a metro area network and asynchronous replication from any location via the WAN. In the event of disaster at the primary data center, for example, these arrangements provide remote mirror replication to a second site. Performance is guaranteed by a Service Level Agreement (SLA).

Some SSPs offer a Web-based application that gives subscribers greater control of data storage management services and storage resource analysis capabilities from a browser. The software enables organizations to closely monitor, in near real time, the end-to-end data storage environment, allowing them to focus on their applications, data, and business— not on hardware, interoperability, and ongoing storage management. The capabilities of these Web-based applications include:

- *Capacity reporting,* which allows you to see how much capacity is being used and how it is distributed among systems and organizational entities.

- *Data availability,* which allows you to monitor the availability of storage systems on the network and monitor the hosts that are running, the hosts that are down, and event logs. They can also monitor the performance of the SSP in meeting SLA commitments.

- *Performance metrics,* which allows you to analyze the performance of individual systems, see patterns of storage usage over time, and identify areas where performance bottlenecks can be eliminated by balancing busy and idle application schedules or server loads.

- *Backup reporting,* which gives you the ability to see backup job success and performance information.

- *Asset topology mapping,* which gives you a view of the autodiscovery topology tool, so that you can quickly troubleshoot infrastructure connectivity, assets, and the state of the assets at all times.

- *Charge back,* which gives you the ability to monitor and analyze usage of storage resources itemized by department, office location, work group, or individual for the purpose of billing for internal use of these resources.

- *Forecasting and provisioning,* which allows you to see trends in data growth and capacity, so that additional capacity can be provisioned as needed.

Some providers approach business continuity by providing direct links to customers' critical storage systems and servers. This helps prevent critical problems and facilitates repairs and restores systems or network devices if a problem should occur. Through the direct link, the provider's support engineers can analyze system data to deliver proactive services and to more effectively diagnose symptoms in the customer's IT environment. This capability helps companies minimize downtime, increase productivity, and yield a higher return on IT investments.

20.6 Training

All network managers and their staffs should be trained in the business recovery process, especially since the procedures are likely to be significantly different from those associated with normal operations. To be effective, business recovery training should be as meticulously planned and delivered as any other corporate training program. The program must have structure, be delivered by experienced instructors, include handout materials, and then be evaluated to determine if it has achieved its objectives and is relevant for the procedures involved. Training may be delivered using in-house resources or external resources, depending on available skills and resources.

With regard to addressing training in the business continuity plan, there should be a section that includes a clear statement of the objectives and scope of the training activities. This will enable the training to be delivered in a consistent and organized manner, so the results can be measured and the training fine-tuned, as appropriate, along with other elements of the business continuity plan. To reinforce formal training, a corporate awareness program should be developed to periodically remind staff of their roles in implementing procedures for the business recovery process. This can be done by sending interactive documents attached to email messages, which include check boxes throughout the text where recipients can indicate that they have read and understood their roles. A "submit" button at the end of the text is used to email the document back to the business continuity administrator. An application such as Microsoft Excel or Word can be used to build interactive features into documents.

Since it will be necessary to explain all new or revised processes to appropriate staff, the training section of the business continuity plan should specify the people who require training. For example, if a data center goes down for any length of time, it may be necessary to carry out some procedures manually and these procedures must be fully understood by the persons who are required to carry them out. For large organizations, it may be more convenient to carry out training in a classroom environment with appropriate training aids. For smaller organizations, training may be better handled in a workshop style in the data center and include visits to telephone closets, equipment rooms, or off-site collocation space at a service provider's facility.

Appropriate training materials need to be identified and developed. This can be a time-consuming task and unless priorities are given to critical training programs, it could delay the organization in reaching an adequate level of preparedness. If the company already has a training group, the infrastructure is in place for materials development and it is just a matter of scheduling time and resources for this task. If there is no training group within the organization, outside assistance may be necessary from a firm that specializes in developing training materials. This will require that a budget be created for development of PowerPoint handouts and a reference binder containing the business continuity plan. Once it has been decided who requires training and the training materials have been prepared, a training agenda should be drawn up and included in the business continuity plan.

After the training program is developed, it is necessary to inform employees of who is to attend, where the sessions will be held, and what is on the training agenda. The business continuity plan should include a draft communication to be sent, advising staff members about the training schedule. The communication should provide for feedback from staff members in case the training dates pose a scheduling conflict. To minimize delay in training delivery, employees should be given a choice of dates, times, and places where training will be held. The use of a calendar and scheduling tool, such as the one included in Microsoft Outlook, will expedite this process.

Whenever changes are made to the business continuity plan, they should be fully tested, and appropriate amendments should be made to the training materials. This will involve the use of a formalized change control procedure under your control, as the network manager.

20.7　Development Tools

For network managers who want to get started quickly with creating a formal business continuity plan, there are Web and Personal Computer (PC) tools available to assist in this process. Since the development process will usually involve inputs from many people, some from other departments at different locations, it is recommended that a network license be purchased to expedite development of the document over company networks or the Internet.

A good tool facilitates rapid production of comprehensive plans in a document-like format. Plan templates and styles, for example, allow the rapid design of report content and structures to be made on a plan-by-plan basis. The software should include a help system and extensive documentation to minimize the learning curve. It should allow direct editing of the individual plan sections, without requiring duplication of data input, and flexible data viewing to allow both "form" and "data" views to enhance usability. The tool should provide connectivity to external databases and include export and import functions with a large range of supported formats including Microsoft products such as Word, Excel, Access, and Project.

Finally, the business continuity-planning tool should come with built-in multilevel security access controls that allow user access to be restricted to individual forms or plans, giving them various read, write, and edit privileges. The tool should also be able to generate a detailed audit trail that tracks the contributions of all staff members participating in the development effort.

Some companies may find it advantageous to select a business continuity-planning tool that can be accessed over the Internet. Through the familiar browser interface, step-by-step questionnaires assist in building the plans. Planners also have full access to on-line help that guides them in plan development and assists them in educating recovery team members as well as distributing plan information and components throughout all company departments and locations. Central maintenance tables allow for periodic updating of the plans. Depending on the vendor, hosted and self-hosted versions of the business continuity-planning tool are available to meet company objectives and budgetary constraints.

20.8 Conclusion

A new awareness of enterprise vulnerability—brought into sharper focus by terrorist attacks, the increasing sophistication of viruses, and continuing onslaught of hacker activity around the world—is forcing corporate decision makers to embrace business continuity planning. Other trends are already emerging that further serve to reinforce the importance of business continuity planning in every industry sector. One trend is the desire among stakeholders for security, predictability, and risk reduction in companies they do business with. Ongoing company availability and resilience against most predictable threats are becoming important sales and client-retention tools, as well as prerequisites for investor interest. Failure to assess business risks and develop plans to return to normal operations as quickly as possible could erode customer and investor confidence.

As a consequence of this emerging high-level awareness of business continuity planning, network managers are finding themselves inundated with unanticipated, often nontechnical questions, the answers to which frequently reveal weaknesses in hardware and software recovery procedures that are attributable to lack of staffing, budget, or executive-level commitment. At the same time, line-of-business managers may find themselves hard-pressed to meet new planning demands to help get the company in synch with the rest of the industry, perhaps believing that their ability to conduct normal operations is being seriously hampered by what they view as yet another management craze. Sales managers, too, may also be affected. Their most difficult negotiations may suddenly shift from pricing to bonded SLAs designed to minimize customer risk.

If a company fails to take business continuity planning seriously enough, competitors can be counted on to pounce on the trend by turning it into an integral part of their value proposition. Therefore, companies should accept the costs of fully tested continuity plans and customer-oriented SLAs as the price of remaining viable within their industry. The trend among stakeholders—customers, partners, or investors—is to demand these measures as a standard method of transacting business, whether as suppliers or buyers.

In the development of a business continuity plan, take into account the organization's decision-making hierarchies and information distribution methods so that necessary corporate decisions can be made if the primary management team is unavailable or unreachable within a reasonable time frame. Wherever possible, use account teams to service customers to secure continuity of relationship in case of unanticipated or event-driven staff turnover. Failure to treat staff as potentially the weakest link in the recovery process may result in unnecessary delay in operations recovery.

Various options are available for developing the business continuity plan. A common approach is to hire specialist consultants, who are recognized experts in the field. Another is to brainstorm the plan internally via intensive meetings and workshops. A third approach employs prewritten checklists, questionnaires, forms, or templates packaged as toolkits. Such tools help network managers develop the document and examine the plan and support arrangements against best practices. In addition, some tools will perform a dependency analysis that assesses resource dependencies and time criticalities. Upon completion of the business continuity plan, it is advisable to perform an audit—not just initially, but at regular intervals. This helps ensure that the plan remains current, and that it stands up to rigorous examination under changed circumstances. The audit should also cover all of the plan's supporting continuity arrangements.

Index

ABOUT THE AUTHOR

Nathan J. Muller is Co-Founder and Senior Consultant of Ascent Solutions Group, a technical marketing firm in Sterling, Virginia, which implements custom programs for new and established technology companies that are designed to improve internal business processes, enhance corporate image, drive sales, and penetrate new markets. With 30 years of telecommunications industry experience, Mr. Muller has written extensively on many aspects of computers and communications, having published 24 books—including three encyclopedias—and more than 2000 articles in over 63 publications worldwide. He is a frequent speaker at industry trade shows, association meetings, and customer events. He can be reached via e-mail at nmuller@ascent-llc.com.